Springer Series in
CHEMICAL PHYSICS 94

Springer Series in

CHEMICAL PHYSICS

The purpose of this series is to provide comprehensive up-to-date monographs in both well established disciplines and emerging research areas within the broad fields of chemical physics and physical chemistry. The books deal with both fundamental science and applications, and may have either a theoretical or an experimental emphasis. They are aimed primarily at researchers and graduate students in chemical physics and related fields.

Please view available titles in *Springer Series in Chemical Physics* on series homepage http://www.springer.com/series/676

Kaoru Yamanouchi

Editor

Lectures on Ultrafast Intense Laser Science 1

With 305 Figures

Springer

Editor
Professor Kaoru Yamanouchi
The University of Tokyo
Department of Chemistry
7-3-1 Hongo, Bunkyo-ku, Tokyo 113-0033
E-Mail: kaoru@chem.s.u-tokyo.ac.jp

Springer Series in Chemical Physics ISSN 0172-6218
ISBN 978-3-642-26637-9 ISBN 978-3-540-95944-1 (eBook)
DOI 10.1007/978-3-540-95944-1
Springer Heidelberg Dordrecht London New York

Cover design: eStudio Calamar Steinen

Printed on acid-free paper

Springer is part of Springer Science+Business Media (www.springer.com)

Preface

When light is weak, atoms and molecules absorbing it may not change their original characters. In optical spectroscopy, an optical transition occurs from one eigenstate to a higher-lying eigenstate of atoms and molecules. From the spectrum, we can extract a variety of characteristic features of atoms and molecules. On the other hand, when light becomes much stronger, atoms and molecules interacting with such intense light fields behave in drastically different ways from those in weak light fields.

Ultrashort and intense laser technologies have developed rapidly, and recent investigations of atoms and molecules in such ultrashort intense laser fields have revealed the existence of characteristic phenomena such as above-threshold ionization, dressed-state formation, ultrafast structural deformation and ultrafast hydrogen migration in molecules, Coulomb explosion of molecules and clusters, and high-order harmonic and attosecond pulse generation. To better understand these phenomena, which carry rich information regarding light–matter interaction, we need to learn the fundamentals of nonlinear optical interactions, electronic structures in atoms and molecules, and advanced techniques in laser engineering.

The emerging research field of ultrafast intense laser science has an inherently interdisciplinary character, and the issues and phenomena being investigated may not be taught in standard physics, chemistry, or laser engineering courses. Therefore, in order to provide a standard basis for these newly emerging research fields, we held a school at which world-leading researchers in ultrafast intense laser science presented a series of lectures.

With the financial support of the JSPS (Japan Society for the Promotion of Science) Core-to-Core Program on Ultrafast Intense Laser Science, Japan Intense Light Field Science Society (JILS), Air Force Office of Scientific Research (AFOSR), and Asian Office of Aerospace Research and Development (AOARD), we held the COAST Autumn School on Ultrafast Intense Laser Science at the University of Tokyo, Japan, during November 24–26, 2006.

At this school, a total of 14 lectures were given by 12 invited lecturers, and a total of 32 young researchers from Canada, Germany, France, Italy, the United States, and Japan enjoyed the opportunity to learn about research topics at the forefront of ultrafast intense laser science.

On November 24, 2006, after the opening addresses by myself, Mr. Kazunori Higuchi (then Deputy Head, Research Cooperation Division II, International Program department, JSPS) and Mr. Nobuhiko Shimizu (then Head, International Affairs Division, The University of Tokyo), lectures were given by Prof. Farhad H.M. Faisal (Bielfeld University), Prof. Kazutaka G. Nakamura (Tokyo Institute of Technology), Prof. Jeff Squier (Colorado School of Mines), and Prof. Hidetoshi Nakano (NTT). On the next day, Prof. Howard Reiss (American University), Prof. Hirohiko Kono (Tohoku University), Prof. Pierre Agostini (Ohio State University), and Prof. Fumihiko Kannari (Keio University) gave lectures. On the third day, Prof. Gaetano Ferrante (Palermo University), Prof. Katsumi Midorikawa (RIKEN), Prof. See Leang Chin (Laval University), and Prof. Leonida Gizzi (CNR) gave lectures.

The lectures covered a wide range of subjects in ultrafast intense laser science such as fundamentals of atomic and molecular dynamics in intense laser fields, experimental and theoretical approaches to controlling molecular behavior and chemical reaction in intense laser fields, high-order harmonic generation, filamentation processes, quantum emission and X-ray emission from a solid target by the irradiation of intense laser light, and laser–matter interaction in the ultra-intense regime.

Since the lectures were so well prepared, and the attendees evaluated the lectures as very exciting and informative, we thought it appropriate to record the topics expounded on in the Autumn School in the form of a book for the benefit of researchers and graduate students in a variety of research fields.

In order to convey the excitement of the school, we arranged the chapters so that the readers could feel as if they were attending the lectures at the lecture room. Indeed, the chapters of this book are based on transcriptions of the actual lectures.

I greatly appreciate the effort and cooperation made by the invited lecturers to turn their transcribed manuscripts into the current form. I would also like to thank Ms. Chie Sakuta and Ms. Maki Oyamada for their help with the preparation of this book.

Finally, I would like to thank Dr. Claus Ascheron, Physics Editor at Springer, for kindly accepting this compilation of lectures as a volume in the Springer Series in Chemical Physics.

Tokyo, April 2010 *Kaoru Yamanouchi*

Contents

1 Introduction to Atomic Dynamics in Intense Light Fields 1
Farhad H.M. Faisal
1.1 Introduction 1
1.2 Historical Background 2
1.3 Virtual Absorption 5
1.4 Generalized Fermi Golden Rule 6
1.5 I^n Law 6
1.6 Above-Threshold Ionization 8
1.7 The Volkov State and KFR-Theory 11
1.8 High Harmonic Generation 16
1.9 Why Only Odd Harmonics? 16
1.10 Tests for the KFR-Model via the Floquet Theory and Experiments 17
1.11 Many-Electron Atomic Systems in Intense Light Fields 20
1.12 Intense-Field Processes in Many-Body Systems 22
1.13 Correlations: Static and Dynamic 22
1.14 Intense-Field Many-Body S-Matrix Theory 23
1.15 Nonsequential Double Ionization 26
1.16 The "CES" Diagram and "Mechanism" of Double Ionization 29
1.17 Comments on Sum-Momentum Distributions 34
1.18 Comments on Multiple Ionization 36
References 39

2 Foundations of Strong-Field Physics 41
Howard R. Reiss
2.1 Introduction 41
2.2 Special Features of Strong-Field Problems 42
2.3 General Quantum Transition Amplitude 45
2.3.1 Preliminaries 45
2.3.2 History of the S-Matrix 46
2.3.3 Derivation of the Transition Amplitude 47

2.4 Gauge Transformations 49
2.4.1 A Partial List of Gauge-Related Mistakes 54
2.4.2 Does a Laboratory Gauge Exist? 55
2.5 SFA (Strong-Field Approximation) 57
2.5.1 SFA Rates 58
2.5.2 SFA Spectra 60
2.5.3 SFA Momentum Distributions 63
2.6 Tunneling/Multiphoton Misconception 69
2.6.1 Tunneling and the Keldysh Parameter 71
2.7 Time Domains and Rescattering 73
2.8 Relativistic Effects 78
References 83

3 High Intensity Physics Scaled to Mid-Infrared Wavelengths 85
Pierre Agostini
3.1 Introduction 85
3.2 Mid-Infrared Sources at OSU 86
3.3 MIR Strong Field Ionization 87
3.3.1 Keldysh Parameter 87
3.3.2 Keldysh Scaling 90
3.3.3 Strong Field Ionization Photoelectron Energy Spectra 90
3.3.4 Wavelength Scaling of the Photoelectron Spectra 92
3.3.5 Wavelength Scaling of the Ionization Rate: TDSE vs. Tunneling Theory 94
3.3.6 Intensity Scaling of the Rescattering Plateau 94
3.3.7 Wavelength Scaling of the Rescattering Plateau 96
3.3.8 Ionization of Scaled Systems 96
3.3.9 The Low Energy Structure in the Photoelectron Energy Spectra 98
3.4 MIR High Harmonics and Attophysics 98
3.4.1 Scaling of the Harmonic Cutoff 100
3.4.2 Scaling of the Group Delay Dispersion 100
3.4.3 Scaling of the Harmonic Yield 105
3.5 Tomographic Reconstruction of Molecular Orbitals 106
References 108

4 How Do Molecules Behave in Intense Laser Fields? Theoretical Aspects 111
Hirohiko Kono
4.1 Introduction 111
4.2 Electronic and Vibrational Dynamics of H_2^+ in a Near-IR Field 112
4.3 Time-Dependent Adiabatic State Approach and Its Application to Large Amplitude Vibrational Motion of C_{60} Induced by Near-IR Fields 115

4.4 Bond Dissociation Dynamics of Ethanol: Branching Ratio of C–C and C–O Dissociation 130
References 133

5 Pulse Shaping of Femtosecond Laser Pulses and Its Application of Molecule Control 135
Fumihiko Kannari
5.1 Introduction 135
5.2 Femtosecond Laser Pulse Shaping with a $4f$ Pulse Shaper 136
5.3 Spatiotemporal Coupling at $4f$ Pulse Shapers 145
5.4 Replica Pulse Formation with a Pixelated SLM Pulse Shaper 149
5.5 Femtosecond Laser Pulse Shaping with an AOPDF 153
5.6 How to Generate the Desired Ultrashort Laser Pulse in an Actual Laser System: Case 1: We Know the Desired Pulse Shape 155
5.7 How to Generate the Desired Ultrashort Laser Pulse in an Actual Laser System: Case 2: We Do Not Know What the Desired Pulse Shape Is 165
5.8 Adaptive Pulse Shaping for Dissociative Ionization of Ethanol Molecules 166
5.9 Adaptive Pulse Shaping of Two-Photon Excited Fluorescence Efficiency 170
References 172

6 Nonlinear Interaction of Strong XUV Fields with Atoms and Molecules 175
Katsumi Midorikawa
6.1 Introduction 175
6.2 Generation of High-Power High-Order Harmonics 179
6.3 Spatial Properties of High-Order Harmonics 185
6.4 Characterization of Attosecond Pulses by PANTHER 187
6.5 Autocorrelation Measurement of Attosecond Pulses by Molecular Coulomb Explosion 195
6.6 Summary 200
References 200

7 Ultrafast X-Ray Absorption Spectroscopy Using Femtosecond Laser-Driven X-Rays 203
Hidetoshi Nakano
7.1 Introduction 203
7.2 Soft X-Ray Emission from Femtosecond Laser-Produced Plasma 207
7.3 Time-Resolved XAFS Measurement of Optically Excited Silicon 212

7.4 Spatiotemporally Resolved XAS 219
7.5 Summary 222
References 222

8 Quantum Emission and Its Application to Materials Dynamics 223
Kazutaka G. Nakamura
8.1 Introduction 223
8.2 Quantum Emission 224
8.3 Time-Resolved Imaging with Quantum Emission 226
8.4 Time-Resolved X-Ray Diffraction 231
8.5 Summary 237
References 239

9 Filamentation Nonlinear Optics 241
See Leang Chin
9.1 Introduction 241
9.2 Self-Focusing and Filamentation Physics 241
9.3 Theoretical Model and Simulation 247
9.4 Background or Energy Reservoir 249
9.5 Extraordinary Properties of Filaments 252
9.6 Long-Distance Propagation in Air 253
9.7 Clean Fluorescence 254
9.8 Self-Pulse Compression 256
9.9 Self-Spatial Filtering 257
9.10 Self-Group Phase Locking 258
9.11 Nonlinear Optics Inside the Filament Core 260
9.12 Four-Wave Mixing Inside the Filament Core 260
9.13 Detection of Chemical and Biological Agents in Air Based on Clean Fluorescence 264
9.13.1 Halocarbons 264
9.13.2 CH_4 264
9.13.3 Ethanol Vapor 265
9.13.4 CH_4 in air 266
9.13.5 Bio-agents: Egg White and Yeast Powders 266
9.13.6 Summary of Remote Sensing Feasibility Using Only One Laser 269
9.14 Super-Excited States of Molecules Inside a Filament 271
9.15 Looking Ahead and Conclusion 272
References 274

10 Diagnosing Intense and Ultra-intense Laser–Matter Interactions: Status and Future Requirements 277
Leonida A. Gizzi
10.1 Introduction on Ultra-Intense Laser–Matter Interactions 277
10.2 Optical Interferometry and Propagation Issues 287

10.3 Time-Resolved X-Ray Spectroscopy and Imaging 293
10.4 Fast Electron Production and Characterization 298
10.5 Summary and Future Instrumentation Requirements 306
References 306

Index 309

Contributors

Pierre Agostini The Ohio State University, Columbus, OH, USA

See Leang Chin Laval University, Quebec, Canada

Farhad H.M. Faisal Bielefeld University, Bielefeld, Germany

Leonida A. Gizzi Intense Laser Irradiation Laboratory, Istituto Nazionale di Ottica, Consiglio Nazionale Delle Ricerche, Area della Ricerca CNR, Pisa, Italy, and INFN, Pisa, Italy

Fumihiko Kannari Keio University, Yokohama, Japan, kannari@elec.keio.ac.jp

Hirohiko Kono Tohoku University, Sendai, Japan

Katsumi Midorikawa Laser Technology Laboratory, RIKEN Advanced Science Institute, Saitama, Japan

Kazutaka G. Nakamura Materials and Structures Laboratory, Tokyo Institute of Technology, Yokohama, Japan

Hidetoshi Nakano NTT Basic Research Laboratories, NTT Corporation, Kanagawa, Japan

Howard R. Reiss American University, Washington, DC, USA, and Max Born Institute, Berlin, Germany

Chapter 1
Introduction to Atomic Dynamics in Intense Light Fields

Farhad H.M. Faisal

Abstract A brief description is given of the early developments leading to the discovery of some of the basic phenomena that occur during the interaction of intense light fields with atomic systems – such as multiphoton ionization (MPI), above-threshold ionization (ATI), and high harmonic generation (HHG). Also outlined are the related theoretical concepts and the nonperturbative "KFR model" for interpreting the phenomena. Recent observation of the breakdown of a popular "tunneling model" (Keldysh $\gamma < 1$) is noted and the role of discrete photon effects that can account for the observed parallel momentum distribution is given. Next, the role of *e–e* correlation in intense-field processes in many-electron atomic systems is discussed. A systematic analysis of intense-field problems using the "intense-field many-body S-matrix theory" (IMST) is outlined. To this end, the basic motivation behind and the derivation of IMST is discussed, and its use is illustrated by application to the problem of nonsequential double ionization (NSDI) that provided a fully quantum analysis of the mechanism behind the NSDI process. Also discussed is how NSDI and its opposite, sequential double ionization (SDI), can both occur in intense fields depending, respectively, on near-infrared and ultraviolet wavelengths of the light field used. A mechanism for multiple ionization is suggested by the IMST diagrams, which can provide a good estimate of the known experimental observations of the ion states vs. intensity in noble gases. Finally, an example is given of the spatial distributions of the charge states that can be produced in the focal region (a "flying reaction vessel"!) of an intense Ti:sapphire laser.

1.1 Introduction

This presentation has two parts and is intended as an introduction to the subject of intense-field physics. I shall begin by briefly mentioning the historical development of the field of intense-field physics – or intense-field science in a broader

F.H.M. Faisal
Bielefeld University, Bielefeld, Germany

K. Yamanouchi (ed.), *Lectures on Ultrafast Intense Laser Science 1*, Springer Series in Chemical Physics 94, DOI 10.1007/978-3-540-95944-1_1,

sense – with an overview of the historical growth of the maximum intensities that have been achieved in laboratories worldwide, as well as a brief description of some early works.

In Part I, I shall consider the main ideas of the conventional perturbation theory and its failure when the field becomes very strong. Several approaches towards facing this problem will be mentioned. The mathematical aspects of the problem as well as its physical implications will be considered. We will then discuss a simple but fruitful nonperturbative model for the intense-field processes and point out several important nonlinear parameters that emerge naturally from it. The significance of the so-called "single active electron" (SAE) hypothesis in single-electron processes in intense fields will be discussed and illustrated.

In Part II, I shall consider the difficulties faced in many-body systems and the role of *e–e* correlation in intense-field processes. Certain "exact" methods of solving the problem and their difficulties will be mentioned. I shall then discuss the *intense-field many-body S-matrix theory* (IMST), which provides a systematic practical approach for tackling such problems. I shall illustrate the theory in some detail with respect to *nonsequential double ionization* (NSDI), and then discuss the related phenomena.

1.2 Historical Background

In 1905, Einstein's four famous papers were published, which included one on the photoelectric effect and the light quantum hypothesis. This paper marked the beginning of the modern era of quantum physics of light–matter interaction. Around 1931, Maria Goeppert-Mayer hypothesized a two-photon process in which not just one quantum but two light quanta are absorbed in "a single quantum act." This was a very fruitful idea since it allowed one to understand some of the earliest problems of light–matter interaction in "intense" fields, self-consistently.

In the late 1950s, Basov and Prokhorov, in the then Soviet Union, and Townes and Schalow, in the United States, proposed the principle of a "laser." The first laser was built in around 1960–1961 by Maiman, and ever since then the laser has been used widely, among other purposes as the main source of controlled intense light fields in a laboratory.

The first most relevant empirical finding in the development of intense-field science was the observation of breakdown of atomic gases in air when a laser was shone on it. The observed discharges with the optical wavelength light fields, implying ionization of the gases, were at first puzzling to the people and seemed difficult to understand in terms of the prevailing intuitions. This was because the laser photon energy was much lower than the ionization potentials of the gases, and the textbook orthodoxy would not permit one to violate the Einsteinian law of photoelectric effect, which said that photo-ionization occurs only if the energy of the *photon* becomes greater than or at least equal to the ionization potential.

It was argued – in accordance with Einstein's 1905 explanation of the photoelectric effect – that for ionization to take place and for the K.E. law of photoelectric effect to apply, the photon energy could not be lesser than the threshold energy for electron emission, i.e., the ionization potential of gases (or the work function for the metals). After all, the light–quantum hypothesis, or the concept of discrete energy (and momentum) of a "photon" has been one of the cornerstones of the quantum revolution itself and must not be abandoned.

The first well-controlled experiment of ionization of atoms by optical laser photons was performed by Voronov and Delone in 1965, using a ruby laser with photons of about 2 eV. It is of course much smaller than the ionization potentials of noble gas atoms Ar and Xe (that are greater than 10 eV) which served as targets. They reported not only the observation of ions but also a nonlinear dependence of the ion yields on the laser intensity. In the same year, Hall, Branscomb, and Robinson observed the photo-detachment effect, in which a negative ion emitted an electron by interaction with a beam of subthreshold laser photons.

Electron detachment is similar to the ionization process, except that a residual neutral atom is produced, instead of a residual positively charged ion, in this process. Thus, in the photo-detachment process, the electron that leaves the atom experiences only a short-range interaction potential with respect to the neutral atom. This is unlike the long-range Coulomb potential of the residual ion that the ejected electron experiences in the ionization process (Fig. 1.1).

The early available intensities of laser light were of the order of KW/cm^2 or so. In time, intensities roughly of the order of a megawatt, gigawatt, and several TW/cm^2 were produced in the laboratory and employed with ever-increasing vigor to investigate the laser–atom interaction processes in the new domains of light intensity and observed what soon become known as multiphoton processes, i.e., if one light quanta is not sufficient to overcome the threshold, then many photons may join to make the necessary energy and act together for the ionization process.

The higher is the field strength, and hence the higher is the interaction energy, the greater is the probability of such joint transition processes eventually making

A SHORT PRE-HISTORY

EINSTEIN (1905)
...Light Quantum & Photoeffect.
MARIA GÖPPERT MAYER (1931)
...Virtual Absorbtion.
BASOV & PROKHOROV + TOWNES & SCHAWLOW (1958)
...Theory of Laser.
MAIMAN (1960-1961)
...First Laser.
VORONOV & DELONE (1965)
...Multiphoton Ionization.
HALL, BRANSCOMB & ROBINSON (1965)
...Multiphoton Detachment.

Fig. 1.1 A synoptic prehistory of intense-field processes

it possible to observe the signal as a breakdown of gases by multiphoton ionization. When the laser–atom interaction energy became still higher and comparable to the photon energy itself (in the late 1970s), new phenomena such as the so-called "above-threshold ionization" (ATI) were observed.

In this case, electrons were found to be ejected from an atom not at a single energy but at a sequence of energies separated by the photon energy, as predicted theoretically earlier. In the late 1980s and early 1990s, lasers of $\mathrm{PW/cm^2}$ intensities became available.

In this domain the laser field strength and the Coulomb binding forces on an electron could be within an order of magnitude of each other. Thus, by the end of the 1980s, intense-field processes emerged as one of the most vigorous themes of research in atomic and molecular physics. Beyond this domain of intensity, the ionization of a valence electron essentially becomes saturated, and dense microplasmas are formed in the interaction volume. (The characteristic "atomic intensity" is approximately 35.1 $\mathrm{PW/cm^2}$.)

At still higher intensities (near and above $10^{19}\,\mathrm{W/cm^2}$, which are currently available) intensity-induced relativistic motion of free electrons becomes dominant, primarily in connection with collective plasma dynamics. Single-electron processes such as nonlinear Thomson–Compton scattering and generation of high Compton-harmonics can also occur. Furthermore, direct observation of processes related to the electrons in the inner shells of atoms becomes relevant, even at a lower intensity. For intensities above $10^{22}\,\mathrm{W/cm^2}$ – which are not yet available – one may expect to observe the effect of laser coupling with the nucleons of an atomic nucleus, opening the door to a new field of nuclear laser science. Ultimately, at about 100,000 times this intensity, the quantum (electromagnetic) vacuum would break down by copious production of electron–positron pairs (Fig. 1.2).

Growth of Laboratory Intensity:

$I = 10^3$ W/cm^2 .....(1961)
$I = 10^6$ W/cm^2 .....(1965)
$I = 10^9$ W/cm^2 .....(1975)
$I = 10^{13}$ W/cm^2 .....(1980)
$I = 10^{16}$ W/cm^2 .....(1987)
$I = 10^{18}$ W/cm^2 .....(1996)
$I = 10^{19}$ W/cm^2 .....(1998)
$I = 10^{20}$ W/cm^2 .....(2000)
$I = 10^{21}$ W/cm^2 .....(2003)
....

Atomic Intensity: 3.51×10^{16} W/cm^2
Relativistic Intensity: $U_p = I/4\omega^2$
(e.g. $I \geq 10^{19}$ W/cm^2)

Fig. 1.2 Historical growth of maximum available laser intensity shows every 5 years an increase of intensity by around two orders of magnitude

1.3 Virtual Absorption

How do photons of energy less than the ionization potential ionize the atom? Maria Goeppert-Mayer's idea of two-photon absorption "in one quantum act" effectively introduced the concept of a virtual absorption. Usually, for the absorption of a photon by an atom, one requires that the photon energy be equal to the difference in the energy levels between the ground state and a real excited state. If the photon energy does not match or "resonate" with a pair of stationary eigen-states of the atom, the absorption cannot occur.

However, the concept of a virtual absorption implies that even if the real states do not exist for the resonance condition to be satisfied, the electron can for a very brief period of time (as permitted by the uncertainty principle) absorb a photon temporarily and thus be excited to a "virtual state" – that is, a state having the energy equal to the photon energy above the ground state of the atom. If the field is strong enough to have a sufficient number of photons per unit volume, then the virtually excited electron may absorb another available photon even during the very short "lifetime" of the virtual state.

Thus, in principle, after one or more successive virtual absorptions, the electron can always reach the continuous stationary (long-lived or real) eigen-states of the atom above the ionization threshold. Therefore, if the light field is strong enough, i.e., the number density of photons is high enough, laser photons of any frequency and polarization can ionize the atom by successive intermediate virtual absorptions followed by the final transition to the real continuum states. The old law of Einsteinian photoelectric effect is thus to be generalized as follows: if the energy of n photons of the incident field is greater than the ionization potential $I_{\rm p}$ of atoms or molecules (or the work-function of solid materials), photoelectrons can be ejected with a finite kinetic energy:

$$\text{K.E.} = n \times (\text{photon energy}) - I_{\rm p}, > 0, \text{ for any integer } n.$$

The mathematical formulation of the probability of such "n-photon (or multiphoton) ionization" phenomenon is most simply formulated by generalizing the simultaneous two-photon process, first treated by Goeppert-Mayer using the second-order perturbation theory, to the nth order. The lifetimes of the virtual states are typified by the difference-energy between a virtual state and the nearest stationary state. They appear inversely as the energy denominators in the perturbation expressions for the multiphoton ionization (excitation) amplitudes (see, e.g., [1], p. 120).

This frequency difference (or the detuning) corresponds typically to the inverse of the duration of an electron in the virtual (transient or nonstationary) state. For example, if the energy detuning is of the order of 1 eV, the lifetime of the virtual state is of the order of 10^{-16} s. If it is possible to resolve in an experiment such a short time interval, the energy in such an experiment will not be conserved to within the uncertainty proportional to the inverse of the time interval. Thus, another way of understanding the concept of a virtual absorption is that the real states are excited energetically within the energy uncertainty permitted by the short lifetime of the virtual states (Fig 1.3).

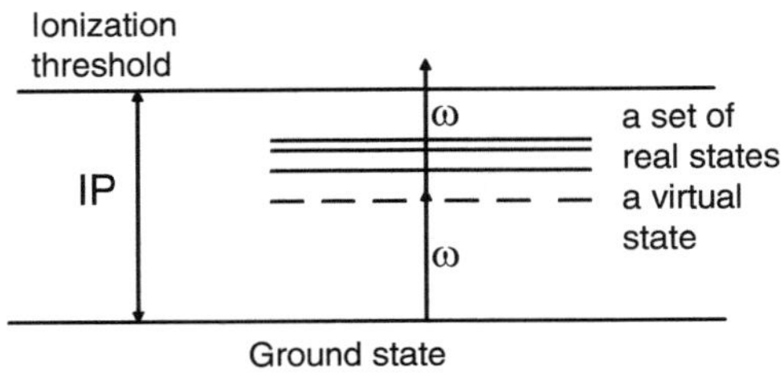

Fig. 1.3 A schematic diagram showing a "virtual state" and a virtual-absorption process

1.4 Generalized Fermi Golden Rule

The fundamental information for any process that the quantum theory can provide is the probability of occurrence of the process. As we all know, one of the most useful formulas for determining the quantum probability for any transition process from an initial state, $|i>$, to a final state, $|f>$, due to the action of an interaction of strength V, is given by the famous Fermi golden rule. This rule says that the probability per unit time, or the rate, Γ, of the process is given by (in a.u.): $\Gamma = 2\pi|<f|V|i>|^2 \cdot \rho$, where ρ is the density of the final states. The same formula also applies to the multiphoton transition probability, provided the first-order matrix element $<f|V|i>$ is replaced by the matrix element $<f|T(n)|i>$ where $T(n)$ is the nth order generalization of V, in which the interaction occurs n times; the operator T is often referred to as the T-matrix (T stands for "transition").

Since the light–matter coupling strength V is proportional to the electric field strength of the laser field, it is proportional to the nth power in the electric field, E, and thus, the probability of the n-photon ionization, which is proportional to the absolute square of the amplitude, is proportional to the nth power of the laser intensity I (since $I = E^2$ (in a.u.); 1 a.u. of I corresponds to $3.5 \times 10^{16}\,\mathrm{W/cm^2}$). This gives the famous I^n law for multiphoton ionization by the absorption of n photons, in the perturbative regime of intensity.

The generalized Fermi golden rule, with V replaced by T, holds even for highly nonperturbative situations, provided that the T-matrix is determined by appropriate nonperturbative methods (Fig. 1.4). The only condition for the existence of a steady "rate" of transition is that the interaction time should be long compared to the internal period associated with the transition. We note that for very short interaction times, the instantaneous transition probability, rather than the steady transition rate, is the appropriate parameter to characterize the process.

1.5 I^n Law

When plotted on a double logarithmic scale, the I^n law gives a straight line with slope given by the power n. From the slope, one can thus determine the number of photons absorbed for the process to occur, even when the threshold energy is unknown (Fig. 1.5).

Fig. 1.4 The Fermi golden rule and its generalization

Transition Rate Γ = Probability per Unit Time.

Fermi Golden Rule:

$$\Gamma^{(1)} = 2\pi \,|< \phi_f | V | \phi_i >|^2 \rho_f$$

(a.u. = atomic units: $e = m = \hbar = \alpha c = 1$)
Density of final states = $\rho_f = \kappa$ a.u.
Interaction $V = -\frac{1}{2}$ E · d

....

Generalized Fermi Golden Rule:

$$\Gamma^{(n)} = 2\pi |< \phi_f | T^{(n)} | \phi_i >|^2 \rho_f$$

An Example:
n-Photon LOPT Transition Matrix Element

$$< \phi_f | T^{(n)} | \phi_i > = \Sigma_{(all j)} < \phi_f | V \frac{|j_{n-1}><j_{n-1}|}{E_i - E_{n-1}} V .... V \frac{|j_1><j_1|}{E_i - E_1} V | \phi_i >.$$

....

I^n-LAW: $\Gamma^{(n)} = \sigma^{(n)} \, I^n$,
$n = 1, 2, 3, ...\infty$,
$\sigma^{(n)}$ = Generalized 'Cross-section'.

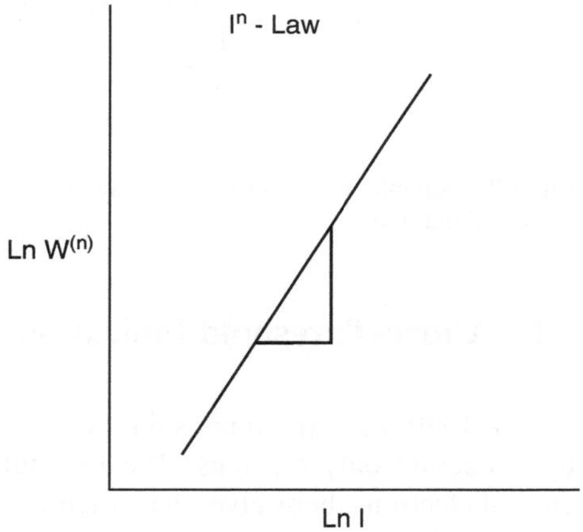

Fig. 1.5 A schematic diagram illustrating intensity dependence of a perturbative n-photon process

The rate divided by F^n – where $F = I/$(photon energy) is the flux of the incident laser photons – gives what is often called a *generalized cross section*, $\sigma(n)$, which is independent of intensity. It is found empirically that the atomic parameter $\sigma(n)$ follows roughly the power law 10^{-33n} (Fig. 1.6).

When the ionization data for Cs Xe, Kr, Ar, Ne, and He were plotted first by Mainfray et al., in semi-logarithmic scales, as a function of the multiphoton order of the process, N, the yields were found to lie on a straight line with a negative slope of -33 (Fig. 1.7).

Fig. 1.6 A power-law for the *generalized cross section*

"10^{-33N}-LAW"

$$\sigma^{(N)}(\omega) = 10^{-33N}[\mathrm{cm}^{2N}\,\mathrm{s}^{N-1}]$$

$\sigma^{(N)}(\omega)$ = 'generalized' cross section;

$W^{(N)} = \sigma^{(N)}(\omega)F^N$ = N-Photon Rate

$F = I/\hbar\omega$ = Incident Photon Flux,

N = Threshold Photon Number.

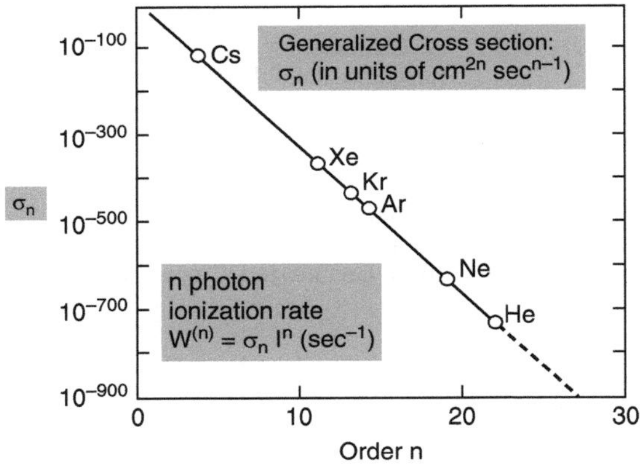

Fig. 1.7 Example of a "−33N" power law for generalized cross sections (suggested by Mainfray and collaborators)

1.6 Above-Threshold Ionization

Before 1979, the experiments on the multiphoton ionization processes were done by collecting only the ions. The first fully controlled experiment to detect the ejected electrons themselves was performed in that year by Agostini et al. [2], who discovered a second electron peak above the usual photoelectron peak nearer to the threshold. The latter peak corresponded to the lowest nonvanishing order (the minimum order) *multiphoton* ionization process.

The extra peak that appeared precisely at one photon energy above the first peak was dubbed *above-threshold ionization*, or ATI peak (by Sacley physicists Gontier and Trahin) who attempted a perturbation theoretical interpretation of the phenomenon. Soon after that a complete sequence of peaks in the atomic continuum were observed by the FOM-Amsterdam group. Subsequently, the Bell Labs group discovered the so-called "Freeman-resonances" associated with them. The latter is another intense-field phenomenon that arises from a laser intensity-induced intermediate resonance into which an initially virtual state moves into resonance with a real (Rydberg) state of the atom by dynamic Stark shift.

Since its first observation, the ATI process became one of the most ubiquitous intense-field phenomena to be investigated in the laboratory and analyzed theoretically. It has remained an attractive research topic for nearly three decades now. Originally it was thought to arise from the inverse-Bremsstrahung process in which an already free electron (e.g., at the first peak) absorbs an extra photon. (This is opposite of the process of emission of a photon by a free electron in the usual Bremsstrahlung process.) The early nonperturbative models, later to become well known as the strong-field "KFR model" [3–5], already predicted the emission of electrons from a bound state, with a sequence of kinetic energy separated by the photon energy, by direct absorption of an arbitrary number of photons from an intense laser field (Fig. 1.8).

This model implied that the energy of the electron was absorbed directly from the bound state, since the distribution of momentum of the bound electron in the field of the atomic nucleus could supply the momentum change required in the transition process (much more efficiently than a free electron), while only the energy conservation could be satisfied in the process by the (dipole) photons. The KFR theory also predicted that during the process the threshold of ionization would be shifted upward by what is now popularly known as the *ponderomotive energy* U_p.

This led to the new threshold law of intense-field ionization (ATI):

$$\text{K.E.} = [n \times (\text{photon energy}) - (I_p + U_p)] > 0, \quad \text{for all integers } n.$$

The predicted threshold-shift by U_p initially remained undetected, but eventually it was observed when the laser pulse durations became sufficiently short (e.g., picosecond), that unmasked the compensating effect of the ponderomotive acceleration in a long pulse (e.g., nanosecond), as clarified by Muller and others.

A characteristic feature of ATI is that the intensities of many successive peaks with increasing n are about the same order of magnitude (Fig. 1.9). This qualitative feature of ATI is indicative of its highly nonlinear nature, very unlike the I^n law. The latter is now known to be valid only for *weak-field* multiphoton processes.

Another remarkable phenomenon that was observed in the second half of the 1980s was *high harmonic generation* (HHG) (Fig. 1.10). A simple quantum picture of the process is the idea of "fusion" of many identical photons into a single photon of an enhanced energy. This might sound surprising at first since photons do not interact with one another and hence cannot "fuse" on their own. But, in an intense laser field, a bound electron can absorb many photons, and can be excited into energetic states with $n \times$ (the photon energy), and return to the same bound state by emitting a single photon of precisely the same excitation energy. The frequency of the emitted photon is then a harmonic of the laser frequency.

Note that in the HHG process, the active electron does not change its state and acts only as a "catalyst" of the process. The HHG process was soon observed in a number of laboratories in Saclay, Chicago, and Amsterdam, approximately about the same time.

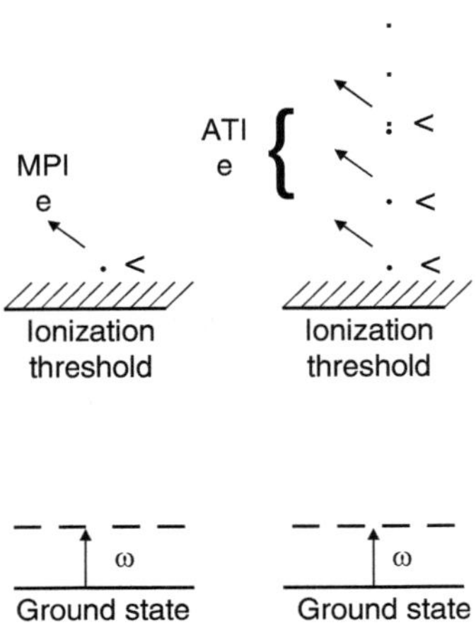

Fig. 1.8 Schematic diagram of multiphoton ionization (*left*) and above-threshold ionization (*right*)

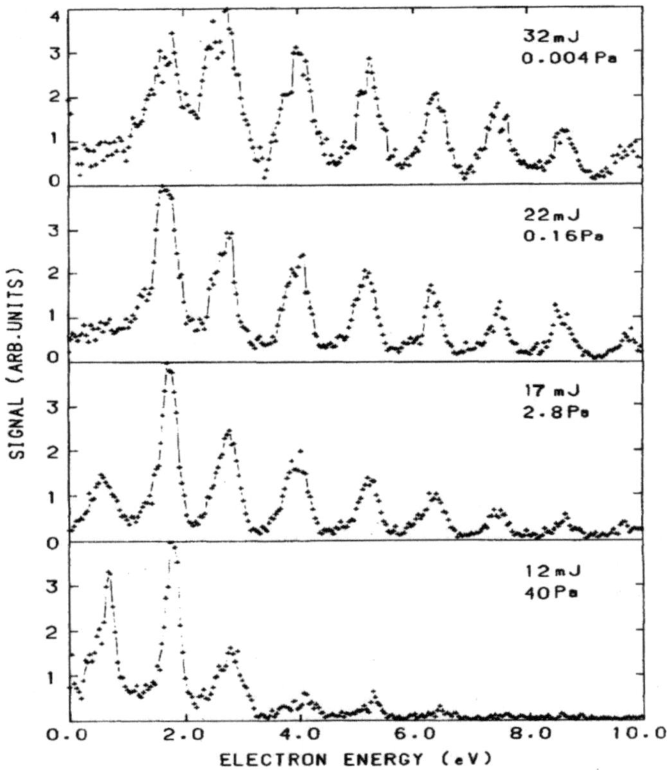

Fig. 1.9 A sequence of well-developed ATI spectra observed by Kruit and coworkers [6]

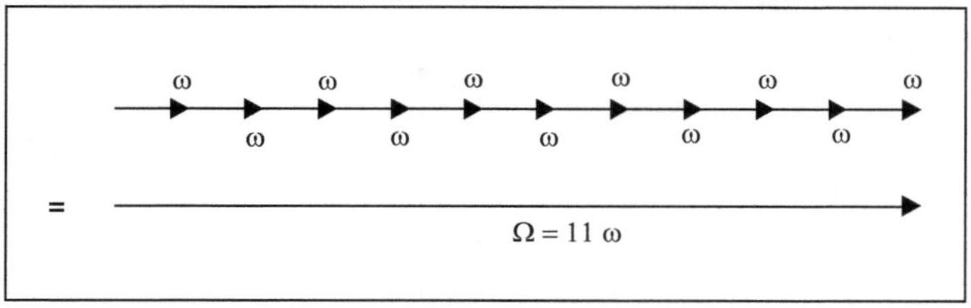

Fig. 1.10 A schematic diagram defining high harmonic generation (HHG)

According to the perturbation theory, the yield of the nth harmonic should change as the nth power of intensity I, since n laser photons are absorbed (and 1 spontaneously created photon is emitted) in the process.

However, the intensity of the observed peaks for high-order harmonics exhibited a region of a *plateau* with comarable strengths for the different order harmonics and beyond it a sharp cut-off appeared (Fig. 1.11).

These properties were subsequently well understood from the now well-known *Lewenstein model* [7] developed on the basis of the strong-field KFR-ansatz.

1.7 The Volkov State and KFR-Theory

One of the building blocks of intense-field theories, specifically the S-matrix theories, in which the KFR model of ionization is the simplest example, is the wave function of a free electron in an intense electromagnetic field. First, we consider classically an electron interacting with a plane wave electromagnetic field. By applying Newton's classical equation of motion, one easily obtains a trajectory exhibiting a simple sinusoidal motion with a drift. The electron drifts along a straight line if there are no forces other than that from the electromagnetic field; however, the presence of the nuclear center may make the straight trajectory to bend into a curved one. Typically, an electron rapidly oscillates at the laser frequency with the maximum amplitude called the *quiver radius* of the electron (see Fig. 1.12).

The plane wave solution of the Schrödinger equation (in the dipole approximation) for an electron in an electromagnetic field is widely known as the (nonrelativistic) Volkov solution (see Figs. 1.13 and 1.14), and is the quantum counterpart of the classical trajectory mentioned above. In this state, the electron has a wave vector $\boldsymbol{k}$ and an entire sequence of virtual energies, above and below its initial energy, separated by the photon energy. It oscillates in both space and time. The normalized amplitudes are given by the *generalized Bessel functions*. They depend on two parameters, namely, the *quiver radius*, and an energy parameter, called the *ponderomotive energy*, U_p, that is associated with the mean energy of the oscillation of the quivering electron.

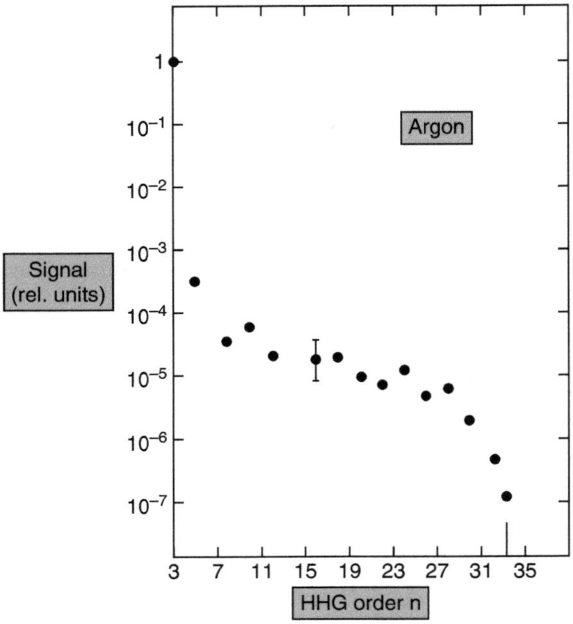

Fig. 1.11 A spectrum HHG obtained by Ferray and coworkers [8]

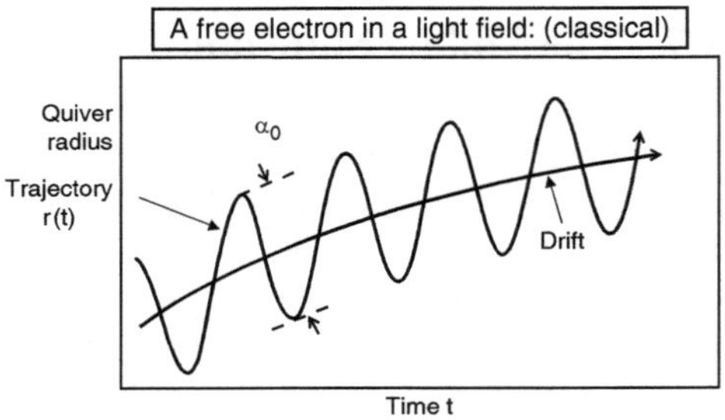

Fig. 1.12 A schematic diagram showing a classical free-electron trajectory in a laser field and the quiver radius

Note that the initial energy is enhanced or reduced, not only by the multiple of the photon energy, but also by U_p. Thus, when the field is applied, the energy of the free electron develops into a *Volkov spectrum* of virtual energy (Fig. 1.14), and the rapid transitions among them correspond to the classical oscillations associated with the quiver motion. A free electron cannot really emit or absorb a photon by itself, since it cannot simultaneously satisfy the energy–momentum conservation condition. Therefore, the Volkov spectrum of the electron is entirely virtual. However, in the ATI process these virtual states become really occupied by the electron when the energy–momentum conditions are satisfied through the available distribution of momentum of the initial bound state, due to the presence of a *third body*, i.e., the atomic nucleus.

Fig. 1.13 Volkov wave function with nonlinear parameters

Volkov Wavefunction:

(D.V.Wolkow, Z.Phys., 94, 240 (1935)

$$\Psi_k(r, t) = \sum_n e^{-i(E+n\hbar\omega+U_p)t/h}\, J_n\,(k.\alpha_0; U_p/2\hbar\omega)\, e^{ik.r}$$

$E = \hbar^2\kappa^2/2m$ = free energy; k = wave vector

$U_P = e^2E_0^2/4m\omega^2$ = quiver energy

$\alpha_0 = eE_0/m\omega^2$ = quiver radius

E_0 = peak field strength

$J_n(a, b) = \sum_m J_{n+2m}(a)J_m(b)$

= generalized Bessel function of two arguments

$$a = k.\alpha_0 \;;\quad b = U_p/2\hbar\omega$$

Fig. 1.14 Volkov energy spectrum

Distribution of Energy of a Free Electron in a Laser Field

(Energy levels)	(probability)
.	
.	
.	
$E+U_P+6\hbar\omega$,	$J_6^2(a, b)$
$E+U_P+5\hbar\omega$,	$J_5^2(a, b)$
$E+U_P+4\hbar\omega$,	$J_4^2(a, b)$
$E+U_P+3\hbar\omega$,	$J_3^2(a, b)$
$E+U_P+2\hbar\omega$,	$J_2^2(a, b)$
$E+U_P+1\hbar\omega$,	$J_1^2(a, b)$
$E+U_P+0\hbar\omega$,	**$J_0^2(a, b)$**
$E+U_P-1\hbar\omega$,	$J_{-1}^2(a, b)$
$E+U_P-2\hbar\omega$,	$J_{-2}^2(a, b)$
$E+U_P-3\hbar\omega$,	$J_{-3}^2(a, b)$
$E+U_P-4\hbar\omega$,	$J_{-4}^2(a, b)$
$E+U_P-5\hbar\omega$,	$J_{-5}^2(a, b)$
$E+U_P-6\hbar\omega$,	$J_{-6}^2(a, b)$

The first-order KFR amplitude for ionization is essentially the matrix element of the laser–atom interaction between the initial unperturbed bound state and the final Volkov state of the electron. The four factors that appear in the simple ionization rate formula (see Fig. 1.15) describe the essential physical ingredients needed for the ATI process to occur: (1) the initial momentum distribution, (2) the probability of occupying the *n*th Volkov state, (3) the amount of the energy transferred to the electron from the field, and (4) the density of the final continuum states.

Note in passing that it provides a good example of the generalized Fermi golden rule, where the $T(n)$-matrix can easily be identified by a comparison; it depends on

KFR-Model

[L.V.Kedysh, Zh.Ekps.Theor. Phys. 47,1945 (1964); JETP 20,1307 (1965)
F.H.M.Faisal, J.Phys. B5, L89 (1973)
H.R.Reiss, Phys.Rev. A22, 1986 (1980)]

Probability of N-photon transition to the continuum (in a.u., e = ħ = m =1):

$$dW^{(N)}/d\Omega = 2\pi\,|(N\omega - \underbrace{U_p}) \, J_N(\underbrace{\alpha_0 . k}; \underbrace{U_p/2\omega}) \, \phi_i(k)|^2 \, k \qquad \text{(a.u)}$$

$N\omega = E_B + U_p + k^2/2$,

E_B = binding energy,

$k^2/2$ = kinetic energy,

$U_p = I/4\omega^2$ = 'quiver energy' = poderomotive energy,

$\alpha_0 = I^{1/2}/\omega^2$ = 'quiver radius',

$\boxed{J_N(a; b) = S_m \, J_{n+2m}(a) J_m(b)}$ = generalized Bessel function of two arguments,

$\phi_i(k)$ = Fourier transform of the initial state.

Fig. 1.15 First-order KFR-rate formula for ionization

the field parameters through the generalized Bessel function, which is a highly non-linear function of the laser field intensity and frequency. Note also that the ionization threshold is lifted by an amount equal to the ponderomotive energy U_p, so that, in intense fields, the Einsteinian photoelectric equation, which determines the kinetic energy of the ejected electron, is as discussed already by: K.E. $= n \times$ (photon energy) – $I_p - U_p$ – a formula which has been amply confirmed experimentally.

Figure 1.16 shows schematically the physical contents of this KFR model. We note that the unperturbed threshold shifts upward by U_p. Therefore, in the case illustrated, it is necessary to absorb more than ten photons for the ionization to occur, although, in a very weak field, when U_p is negligible, the absorption of only eight photons would be sufficient for the ionization to occur. In other words, in a sufficiently strong field the lower order channels for the ionization could be closed by the ponderomotive shift of the threshold – this phenomenon is known as *channel-closing*; its evidence can be seen already in the earliest fully developed ATI spectra, taken for a sequence of intensities by the FOM-Amsterdam group in 1981 (see Fig. 1.9).

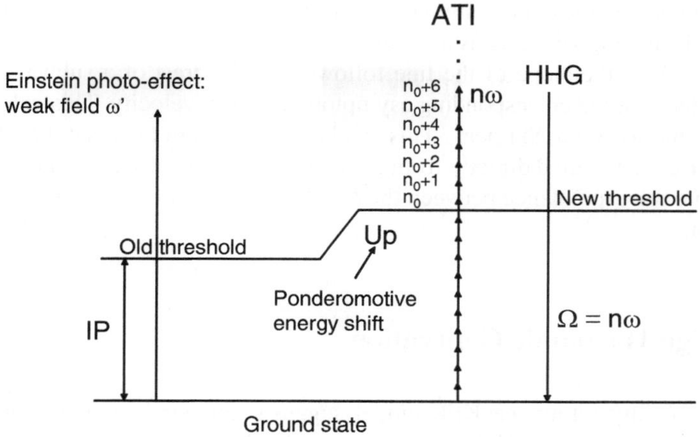

Fig. 1.16 A schematic diagram illustrating the KFR model of ionization, with ponderomotive potential (shift $U_{\rm p}$), ATI, and HHG

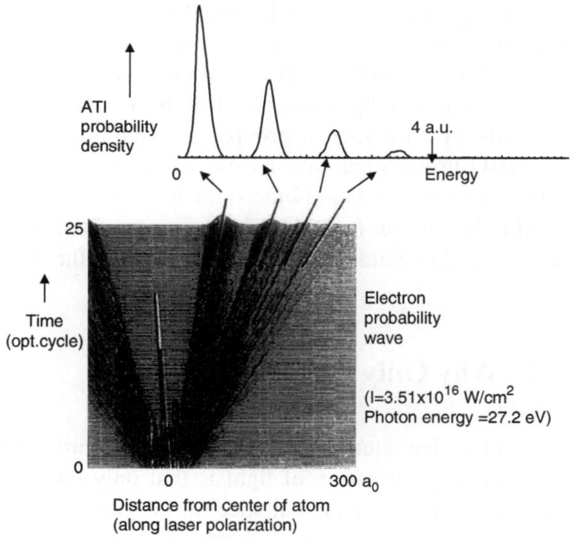

Fig. 1.17 Schrödinger evolution of ATI in space–time U. Schwengelbeck and F.H.M. Faisal [9]

Figure 1.17 shows the temporal evolution of the ATI process generated by a direct simulation, in which the probability wave for the ejected electron is plotted along the polarization direction x in units of Bohr radius and along the evolution time axis, t, in units of the laser cycles (vertical axis).

We observe that from a sharp initial distribution at $t = 0$, it quickly evolves into a complex distribution. After this, the wave packet is seen to split into discernible multiple wavelets with independent moving crests. We observe here a purely quantum evolution that emerges into a number of linear classical-like crest-trajectories, each of which is moving with a constant velocity.

In fact, from the slope of the line followed by the crest of a subwave one can easily determine the corresponding asymptotic electron velocity. This velocity also gives the energy of an ATI peak. This can be verified from the inset that shows the ATI spectrum computed directly from the entire wave packet seen in the figure. The one-to-one correspondence between the crest velocities and the ATI peak positions is evident.

1.8 High Harmonic Generation

Figure 1.16 illustrating the KFR-model suggests another interesting and important process, namely the HHG in intense fields. Suppose that the bound electron absorbs seventeen photons as in the illustrative figure and attains the corresponding excitation energy in the continuum; then it may either leave the atom, resulting in ionization, or it may emit a single photon of exactly the same energy as the excitation energy and go back to the initial bound state. In the latter case, the frequency of the emitted photon would be the seventeenth harmonic of the laser frequency.

This idea of HHG was proposed first in connection with the electrons that emerge at the ATI peaks, by Shore and Knight in 1987 [10]; it was soon observed experimentally by a number of groups, as noted above. As one can see from the figure, the HHG process can occur from virtual excitations to both above and below the ionization threshold, and indeed the lower harmonics may arise only from the latter condition, without requiring the presence of the ATI electrons as such, and in fact the above-threshold harmonics compete with the ATI.

1.9 Why Only Odd Harmonics?

One of the characteristic features of the high-order harmonic generation from atoms with linear polarization of light is that only the odd harmonics were found to be present in the experimental spectra, i.e., no even harmonics were observed at all. This we show is due to the following four general factors, (1) conservation of parity in electrodynamics, (2) central-symmetry of atoms, (3) a given parity of the ground state, and (4) the odd parity of the dipole interaction.

The nth harmonic emission is associated with the net absorption of n laser photons and emission of one harmonic photon. Therefore, the net change of parity must be $(-1)^{n+1}$. This must match the change of parity between the initial and the final states in the transition. For the coherent HHG process the final state must be the same as the initial state, and therefore no change of parity: $(-1)^0 = 1$. Thus, for the nth harmonic to be emitted one requires $(-1)^{n+1} = 1$, and hence $n =$ odd, q.e.d.

Corollary 1. If the initial state is a mixed-parity state, both odd and even order parity changes would be allowed in the transition (from the mixed-parity initial state to itself) and therefore, simultaneous emissions of both odd and even order harmonics would be permitted.

One way to create a mixed-parity initial state is to use a static electric field. A better way is perhaps to introduce a second laser field with a frequency that is tuned to be resonant or near resonant with an optical transition, to couple a pair of bound states of different parity, before applying the probe pulse for generating the HHG signal. This mixed-parity state can generate harmonics of both odd and even orders simultaneously.

How can one evaluate the probability of HHG radiation? A useful theory of HHG that gives the relative strengths and many other properties of the generated high harmonics has been developed by Lewenstein et al. [7] who employed the strong-field KFR-ansatz to obtain the expectation value of the dipole operator. The intensity of the nth harmonic signal is calculated from the popular semi-classical rule that the strength of the harmonic is proportional to the nth Fourier transform of the dipole expectation value.

We point out in passing that a subject of much current research interest is to observe the electronic motions in atoms and molecules. One way to do so is to couple initially two different (or more) electronic states by a short "pump pulse," next to apply an attosecond "probe pulse," after a delay Δt, and monitor the HHG signal as a function of Δt, to track the electronic motion.

1.10 Tests for the KFR-Model via the Floquet Theory and Experiments

In view of its usefulness, it is worthwhile to test the consequences of the KFR-Ansatz against possible "exact" calculations and/or experimental data. For example, to test the KFR model for ionization, we may compare the total ionization rate given by this model with the ionization rate of the hydrogen atom, obtained from the "exact" Floquet calculation.

The Floquet method assumes that the laser field be a periodic function of time with the periodicity of the laser cycle. According to the Floquet theorem, the solution of the Schrödinger partial differential equation can be generally expanded as a Fourier series in time with a Floquet characteristic exponent E, which need not be (and generally is not) commensurate with a multiple of the photon energy (see, e.g., [1]).

The characteristic exponent is essentially the perturbed energy, often called the *quasi-energy* of the atomic system in the presence of the laser interaction. Its difference with the unperturbed energy is a good measure of the so-called *light shift*, which is also known as the *AC Stark shift*, $\Delta E = E(I) - E(0)$, where the Floquet quasi-energy $E = E(I)$ is a function of the light field intensity, I, and $E(0)$ corresponds to the unperturbed or bare energy of the atomic state.

The usefulness of Floquet transformation lies in the fact that one can use a discrete, hence countable, Fourier expansion of the solutions with a quasi-energy exponent, rather than a Fourier integral transformation, which is continuous, and hence, nondenumerable. A Floquet expansion can thus convert the time-dependent Schrödinger equation of the periodic Hamiltonian system into a set of coupled stationary (time-independent) equations. The practical advantage of the Floquet method is the greatest when the photon energy is comparable with the typical electron energy or larger. Otherwise, in an intense field, the number of relevant Fourier components in the wave function would be too many to calculate, without loss of significant accuracy.

Before comparing the KFR ionization rates with the Floquet rates for an H atom, we note that the KFR model is directly applicable to the case of short-range potentials where there is no asymptotic Coulomb potential, as in a photo-detachment process. For the ionization problem in intense fields, one needs also to account for the joint effect of the Coulomb and the laser interactions in the final Volkov state. This can be crudely but simply accounted for by a WKB-type correction, which gives a simple multiplicative correction factor to the ionization rate (see, e.g., [11]).

The Floquet rates show (see Fig. 1.18, upper dotted line) certain peaks that are associated with the *Freeman resonances* mentioned earlier, superimposed on the relatively smooth variation of the ionization rate with the light field intensity. There are also certain rather sharp, cusp-like, changes of slopes at certain intensities. These are the signature of the *channel-closing* effect on the total rate, due to the upward shift of the ionization threshold by the ponderomotive potential.

It can be seen from the comparison in the figure that over the wide range of laser field strengths, the Coulomb-corrected KFR rate is in good agreement with the

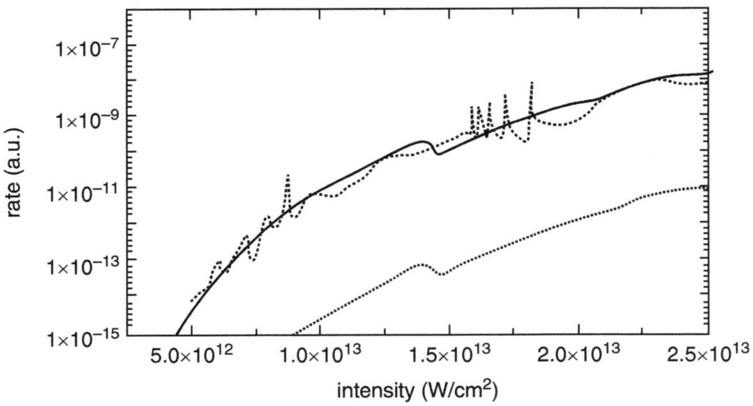

Fig. 1.18 Ionization of H atom vs. intensity. Comparison of Coulomb-corrected KFR rates (*solid line*) with "exact" Floquet rates (*upper dotted line*). Plane-wave KFR rates (*lower dotted line*) [12]. (For more details, see references cited in [11].)

Floquet rates when the effect of the resonances are averaged out. Note that the KFR rates do not wash out the effect of the channel-closing phenomenon.

Regarding the experimental data of ionization of atoms in intense laser fields, we first note that an extensive set of experimental data is available covering a wide range of intensities for different polarizations and wavelengths. The ionization rates given by the KFR model have been compared with 36 different sets of experimental data for the ionization yields for various noble gases, including He, and were found to fit well [12]. In a figure there we show an example for the theoretical yields calculated from the rate equation for ionization with basic rates given by the KFR model, in which Gaussian space–time laser pulse profiles was adopted that mimic the laser pulses used in the experiments with the yields that have been experimentally obtained for the ionization of Xe at the wavelengths of 586, 800, and 1053 nm. Note that the resonances that were found to be present in a typical Floquet calculation of the ionization rates mentioned earlier, seem to be smoothed out in both of the experimental (and the theoretical KFR yields) because of the averaging over the various intensities that are present in the space-time profile of the laser pulse.

As the last topic of this first part of the presentation, I shall very briefly discuss the limiting quasi-static case of tunneling, when the intensity becomes sufficiently high and the frequency becomes very low. If the frequency should approach zero, then, clearly, the electric field of light would be a static electric field. Quantum-mechanical study of ionization of atoms in a static electric filed has a long history starting with the work by Oppenheimer, followed by a large number of researchers (cf., e.g., [11]).

Keldysh [3–5] was the first to consider the ionization rate when an intense light field approached the static limit. He showed that the ionization rate in an intense field in the low-frequency limit is essentially given by the static field formula with the characteristic nonanalytic field dependence in the exponent, which is typical of the tunneling phenomena, and he found that the first correction due to the finite frequency is rather small in the *tunnel regime*, i.e., the now well-known *Keldysh parameter* $\gamma = \sqrt{(I_p/4U_p)} < 1$.

We note here, parenthetically, that the above condition is necessary for *tunneling*, but is not sufficient. A condition for a proper *tunnel regime* that is both necessary and sufficient is: $n_o > 1/\gamma > 1$; where n_o is the minimum number of photons required to overcome the threshold energy; when the second inequality is not obeyed one has the well-known "multiphoton-regime", and when the first inequality is violated one has above-the-barrier ionization.

The usefulness of the tunneling formula for low-frequency intense-field processes has been investigated experimentally by Corkum et al. [13, 14] and others. Some recent high-resolution momentum and energy measurements indicate that even in the tunneling domain, *quantum remnants* of the discrete photon character of the light field are present.

We may point out that the limiting procedure of replacing the discrete sum over the allowed photon numbers by a continuous integration, might result in the smooth Gaussian variation in the momentum dependence of the discrete photon absorption rates at the tunneling regime, and that this turns out to be is too drastic for

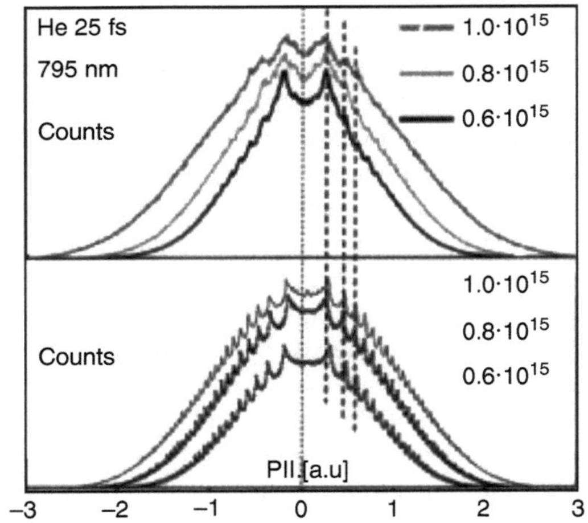

Fig. 1.19 Breakdown of the *tunnel* model for $\gamma < 1$. Discrete photon signatures (*quantum remnants*) showing a central minimum with a *two-horn* distribution for the parallel momentum (in place of a smooth Gaussian with a central maximum predicted by the tunneling model) has recently been observed [15, 16] and interpreted [17] (courtesy, R. Moshammer)

determining the momentum (and energy) distributions even at the *tunneling limit*, $\gamma = \sqrt{(I_p/4U_p)} < 1$.

In fact when the discrete sums over the photon numbers are properly accounted for, and the Coulomb effect is included, all the discrete structures reappear, signaling the breakdown of the Gaussian momentum distribution predicted by tunnel ionization model. Thus, in high-resolution experiments [15, 16] one obtains instead the double-horn central distribution with many side peaks along the flanks [17] (see Fig. 1.19).

1.11 Many-Electron Atomic Systems in Intense Light Fields

In this second part of my presentation, I shall discuss how we approach more complex problems involving many-electron systems subjected to intense laser fields.

Analysis of such problems may require addressing the following two effects simultaneously: (1) many-body correlation effect, and (2) nonperturbative field effect. As a matter of fact, all systems, except the hydrogenic or a one-electron system, require us to account for both of them together. However, in the processes involving single-electron excitation or ionization of many-body systems, an approximate simplicity can occur.

Such a process is dominated by coupling of the radiation field to an active electron of the system. This leads to the so-called *single active electron* (SAE) ansatz, according to which the active electron moves in its own effective potential and is affected independently of the other electrons by the field. This ansatz has been successfully employed in analyzing many single-electron processes in intense laser fields, such as single-ionization and high-order harmonic generation from atoms and molecules. Why is this possible? The reason behind this is that the fundamental light–matter interaction operator is a sum of independent one-electron operators:

$$V(t) = \sum (i = 1, N)\ [-e\boldsymbol{E}(t) \cdot \boldsymbol{r}(i)],$$

where $\boldsymbol{E}(t)$ is the electric field strength and N is the number of electrons. This is true even without the dipole approximation, as can be easily seen from the interaction in the minimal coupling Coulomb gauge (e.g., [1]):

$$V(t) = \sum (i = 1, N)\ [e/mc\, \boldsymbol{A}(\boldsymbol{r}(i), t) \cdot \boldsymbol{p}(i) + e^2/2mc^2\, \boldsymbol{A}(\boldsymbol{r}(i), t)],$$

where $\boldsymbol{A}(\boldsymbol{r}(i), t)$ is the vector potential at the position $r(i)$, and $p(i)$ is the momentum operator of the ith electron. This is why a light field influences the electrons primarily independently. However, the presence of the *e–e* correlation (Coulomb interaction between all pairs of electrons) can and does couple the electrons along with the interactions of the light field and the individual electrons. This can even be dominant, under special circumstances, for two-(or many-)electron processes, e.g., in the process of so-called nonsequential double-ionization of noble gas atoms in intense near-infrared laser fields, as we discuss below.

Several possible approaches can be taken to face the general challenge posed by the interaction of intense laser light with many-electron systems when the correlation effect cannot be ruled out a priori. These include direct simulation of the total wave packet of the many-electron Schrödinger equation. For one-electron processes in the SAE ansatz [18], this approach has been very successful in the past. Since, in that case, the problem reduces to the direct numerical solution (simulation) of a 3D partial differential equation, and this can now be routinely performed. It has the merit of offering an "exact" evolution of the total wave packet in the laser field, which contains all possible information about the behavior of the system. However, to perform an equally "exact" simulation beyond a few-electron system (say, a three-electron atom, requiring a 9D space–time simulation in a linearly polarized field) is currently hardly possible due to the limitations of the fastest/largest available computers. Nevertheless, interesting progress has been made recently for two-electron systems (in 6D simulations by Taylor and coworkers [19] or by reduction to four dimensions [20]).

An alternative to direct simulation is an extension of the Floquet R-matrix approach [21] from the current one electron ionization problem to the two-electron ionization problem. However, the Floquet method for the double ionization is difficult to implement for the low-frequency and high-intensity, e.g., for the currently available intense infrared Ti-sapphire laser fields (e.g., 800 nm, $I = 5\,\mathrm{PW/cm^2}$). In this case, the presence of a large number of photon blocks and the double-ionization partial continua would have to be treated properly. Going to many-electron processes (e.g., multiple ionization) would be impractical at present.

The exact solutions are always important (whenever feasible). Nevertheless, the insights gained into various types of processes by approximate means are often very useful. What is needed is to develop a systematic approximation method, wherein problems are not attempted to be solved "exactly"; instead, attempts are made to extract step-by-step as much information as may be needed to understand a mechanism behind a process, or phenomena, of interest, albeit by testing the outcome, whenever possible, against the available experimental data, or by predicting new phenomena and effects, which could be tested subsequently in the laboratory. Such an approach is given by the IMST [22, 23], which we shall discuss next.

1.12 Intense-Field Processes in Many-Body Systems

The development and applications of the IMST is reviewed recently [11]. Here, we may first note that in the last several years it has been fruitfully applied to a series of problems, such as NSDI [24–26], correlated momentum distribution [27, 28] and coincident energy distribution [29, 30], multiple ionization of noble gas atoms [31–33], two-slit interference phenomena and suppressed ionization of molecules [34], many-fold ionization to high-charge states of fullerenes [35], inelastic vibronic ionization [36, 37], adiabatic alignment of molecules [38, 39], dynamic HHG from coherently rotating molecules [40], etc. I shall introduce the IMST approach, and illustrate the use of this versatile method, by discussing its application to the double ionization of He, a system that has been studied extensively experimentally, and interpreted quantum-mechanically by IMST.

1.13 Correlations: Static and Dynamic

In the context of double ionization we shall be concerned with the effect of *e–e* correlation on intense-field dynamics. It is this *e–e* or Coulomb correlation that makes all electronic systems such as complex atoms and molecules "nonseparable," and hence difficult to tackle genuine quantum many-body systems.

We note in passing that in two- or many-body quantum systems there is another correlation present. For example, when two particles are in an *entangled state*, there need not be any electron–electron or any kind of interaction at all, and yet the two particles would behave as correlated. Indeed, they may, under free motion, remain so correlated for an arbitrary separation between them! The electron–electron correlation can be *static* or *dynamic*. An example of a static correlation effect can be seen if one computes the ground-state energy of a helium atom, and finds that there is a contribution coming from the *e–e* interaction, which can be small compared to the total energy. Another example is the polarizability of an atom or a molecule, which also arises from the *e–e* correlation. These *static* effects of the *e–e* correlation appear in the stationary states, e.g., in the ground state, of the system.

However, there is another class of effects of the *e–e* correlation, called the *dynamic* correlation, which appears in *short-lived* or transient virtual states of the system, e.g., in the doubly or multiply excited states of atoms in the real continua, or in the virtual excited states. It becomes active during the interaction of the system with external perturbations, e.g., with intense laser fields, and manifests itself by influencing the outcome in the finally observed channel(s).

1.14 Intense-Field Many-Body *S*-Matrix Theory

Before we apply the IMST to the concrete illustrative case, we shall briefly motivate the approach itself, indicate its derivation, and point out some of its particular advantages. Heisenberg introduced the *S*-matrix in 1943. Wheeler had considered a similar idea before that. Heisenberg, in line with his early philosophy of sticking only to observable quantities for the formulation physical theories, considered a typical experimental measurement as a transition process, in which a system is at first "prepared" in an initial state, |in>. Then a certain interaction is brought to bear on it that transforms, or "scatters," the initial state of the system into all possible states. This is described by a transformation or "scattering operator" *S*. Then, finally, the system is "detected" in an outgoing state, <out|. This completes the basic experiment, which is repeated many times to determine the probability of the occurrence of a specific transition process.

Theoretically, the probability amplitude for the process is to be given by the matrix element of the operator *S* taken between the "in" and the "out" states, i.e., by <out|*S*|in>. The question is what is the true *S*? Despite many attempts by many researchers over decades, a complete specification of the "*S*-matrix," using the fundamental invariance principles, causality, analyticity and other requirements, *S* could not be determined completely, without knowing the system more

IMST(1996):

$$(S-1)_{fi}(t) = S_{fi}^{(1)}(t) + S_{fi}^{(2)}(t) + S_{fi}^{(3)}(t) + \ldots :$$

$$S_{fi}^{(1)}(t) = -\frac{i}{\hbar}\int^{t} < \phi_f | V_i | \phi_i >$$

$$S_{fi}^{(2)}(t) = -\frac{i}{\hbar}\int^{t} < \phi_f | V_f G^0 V_i | \phi_i >$$

$$S_{fi}^{(3)}(t) = -\frac{i}{\hbar}\int^{t} < \phi_f | V_f G^0 V_0 G^0 V_i | \phi_i >$$

....

(Note: $\int^{t}$ stands for integrations over the intermediate times, from the initial up to the final time t.)

Ref.: F.H.M. Faisal and A. Becker, in *Selected Topics in Electron Physics,* Eds. D.M. Campbell and H. Kleinpoppen, (Plenum Press, New York, 1996), p. 397.

Fig. 1.20 Intense-field many-body *S*-matrix theory (IMST)

specifically, e.g., by knowing the Hamiltonian of the system or the equations of the motions.

For our purpose, we shall derive it from the Schrödinger equation of the many-body atomic or molecular systems interacting with intense laser fields. The mathematical form S is formally hardly more transparent than an engineer's "black box" that is connected to an externally accessible "input" and "output" channels. However, one can use Feynman-like diagrams to "visualize" the processes that go on inside it! It turns out that the S-matrix formulation, e.g., IMST (see Fig. 1.20), together with the Feynman-like diagrams it generates, can provide not only a quantitative estimate of the probability of a transition process of interest, but also it can help to intuitively understand the significant "mechanisms" that might lay behind a process.

Consider a general reaction process, $\mathrm{A} + \mathrm{BC} + V(i) \rightarrow \mathrm{AB} + \mathrm{C} + V(f)$, in which the atoms or the molecules are initially "prepared" in a stationary state of the unperturbed "reference" Hamiltonian of the "reactants" (A + BC), and it is finally "detected" in a state of the reference Hamiltonian of the "products" (AC + B).

In general, these reference Hamiltonians in the initial and the final states are not the same. The usual perturbation expansion of the transition amplitude, which depends only on one of these interactions, (e.g., the standard Dirac or Dyson perturbation series developed with respect to the power of $V(i)$ or the initial interaction) is highly inconvenient for calculating the transition amplitude for the final product state, since the two reference Hamiltonians are different, and hence, should be modified. IMST effectively provides a systematic method for generating the modified series that also covers such processes as well. The mathematical procedure for constructing the IMST series is to partition the total Hamiltonian of the system, independent of whether it is a stationary or a time-dependent Hamiltonian, in three different ways: (1) an initial partition of the total Hamiltonian as a sum of the initial reference Hamiltonian and the initial interaction, (2) a final partition as a sum of the final reference Hamiltonian and the final interaction, and (3) an intermediate partition with an intermediate reference Hamiltonian and an intermediate interaction.

Of course, all the three partitions add up to the same total Hamiltonian, but they separately account for the relevant interactions and the states, that is, the interactions in the initial and in the final states, as well as, in the potentially dominant intermediate states (variously known as the *doorway*, *virtual fragment*, or *transition* states). In fact, the IMST accounts for all the three factors simultaneously and in a systematic way. It provides not only a method of calculation of the transition probability of interest, but it also opens up the possibility of investigating, and testing in a systematic way, various "mechanisms" that may be considered a priori or intuitively as those governing the complex process.

To be concrete, we consider here an atom or a molecule interacting with a laser field, for example, a He atom in an intense Ti-sapphire laser pulse. Initially there is no interaction and the He atom will be in the ground state. If it is being ionized in the presence of the intense pulse, the final state would be more like two Volkov electrons, rather than an eigenstate of the neutral He atom. Therefore, the

Fig. 1.21 The conventional "prior" and "post" forms of S-matrix series

PRIOR-form:

$$(S-1)_{fi} = S_{fi}^{(1)} + S_{fi}^{(2)} + S_{fi}^{(3)} + \ldots :$$

$$S_{fi}^{(1)} = -\frac{i}{\hbar} \int < \phi_f | V_i | \phi_i >$$

$$S_{fi}^{(2)} = -\frac{i}{\hbar} \int < \phi_f | V_i G_i^0 V_i | \phi_i >$$

$$S_{fi}^{(3)} = -\frac{i}{\hbar} \int < \phi_f | V_i G_i^0 V_i G_i^0 V_i | \phi_i > \ldots$$

POST-form:

$$(S-1)_{if} = S_{fi}^{(1)} + S_{fi}^{(2)} + S_{fi}^{(3)} + \ldots :$$

$$S_{fi}^{(1)} = -\frac{i}{\hbar} \int < \phi_f | V_f | \phi_i >$$

$$S_{fi}^{(2)} = -\frac{i}{\hbar} \int < \phi_f | V_f G_f^0 V_f | \phi_i >$$

$$S_{fi}^{(3)} = -\frac{i}{\hbar} \int < \phi_f | V_f G_f^0 V_f G_f^0 V_f | \phi_i > \ldots$$

final reference Hamiltonian is now different from the initial reference Hamiltonian, and hence, the need for the two different partitions arise, which take care of the two different reference Hamiltonians and the end-interactions. Furthermore, the passage to the double ionization might be through the set of doorway states in which there is one virtually ionized Volkov electron and a virtually occupied ionic electron of He^+; clearly, such a set of product eigenstates does not belong either to the initial or to the final reference Hamiltonians.

If you expand the integral equation for the Schrödinger equation in terms of only the initial interaction with the help of the initial reference Green's function, you get the traditional "prior" expansion of the S-matrix in a perturbation series, which emphasizes the initial interaction and the initial reference Hamiltonian in the intermediate states. Similarly, an expansion using the final interaction and the final reference Green's function generates the so-called "post" expansion of the S-matrix, which emphasizes the final interaction and final Green's function in the intermediate states (see Fig. 1.21). Obviously neither of them can take account of both the interactions on equal footing in the end states, nor can they account for the virtual doorway states of the third kind mentioned above except perhaps by evaluating the series, if at all, to infinite orders! IMST overcomes such difficulties in a simple and systematic way, as shown below.

First, an integral equation is constructed from the Schrödinger equation using the total Green's function or propagator. This gives a time-dependent version of the stationary *Lippmann–Schwinger equation*, well known in the stationary S-matrix theory [41]. Second, the total Green's function is expressed in terms of the initial Green's function, the initial interaction, and itself. Third, the total Green's function is expressed, but now in terms of the final Green's function, the final interaction, and itself. This gives formally a closed form (two terms) expression for the total wave function, in which the total Green's function appears in the middle of the second term. Finally, the total Green's function in the middle of the second term is expanded for the third time, now in terms of the intermediate Green's function and the intermediate interaction in an infinite series. This gives a systematic expansion of the IMST wave function of the total system, in which all the above-mentioned crucial interactions and propagators appear systematically in the appropriate places.

Note that the initial state and interaction are present on the extreme right-hand side. And the projection to any states of the final reference Hamiltonian from the left-hand side can be done, to obtain the desired S-matrix element, now most conveniently by extracting the corresponding single term from the final state Green's function through ortho-normality alone. The door-way propagator appears from the second leading term onwards. The result is the expansion of the desired transition amplitude within the IMST (Fig. 1.21).

In view of its structure, it is clear that the IMST series can be used to provide both quantitative information and physical insights into the transition process even within the lowest few terms of the series. This is of considerable practical importance, since they often are also the only terms that may be evaluated without great difficulty in any form of the S-matrix series. At present, like virtually all other forms of S-matrix series, the exact convergence properties of IMST are not known. However, it should be noted that powerful techniques exist, e.g., the Shank's transformation [42], which may be used to obtain estimates of the sum of the series using the first few terms of the series, if it exists or at least if it is an asymptotic series, or to construct a transformed series with a faster approach to the sum.

Finally, we may note that it is not difficult to show that the conventional "prior" and "post" forms of the S-matrix series are special cases of the IMST series (Fig. 1.22). They can be obtained from the IMST, respectively, by restricting the final and the intermediate interactions and Green's functions to be equal to the initial ones, or by restricting the initial and the intermediate interactions and Green's functions to be equal to the final ones. Note also that IMST differs qualitatively from either the "prior" or the "post" series, from the second term onwards.

As an illustration of the application of IMST, we may consider the problem of double ionization of atoms in intense laser fields, especially the NSDI of the He atom in both near-infrared and the ultra-violet wavelengths.

1.15 Nonsequential Double Ionization

It was often thought in the past that double ionization of atoms in intense laser fields proceeds *stepwise*, i.e., single ionization of the neutral atom followed by the ionization of the ion. However, the experiments performed for the double ionization of He with intense lasers of near-infrared wavelengths during the first half of the 1990s [24, 25] clearly demonstrated that this could not be the case. For example, the yield of doubly ionized He ions, in the intensity domain between 10^{14} and $10^{15}\,\mathrm{W/cm^2}$ with 780 nm Ti-sapphire lasers turned out to be many orders of magnitude greater than that calculated on the assumption of the stepwise process. It was therefore clear that the correlated motions of the electrons play a role for the breakdown of the stepwise mechanism. The phenomenon was called nonsequential double ionization (NSDI).

Intense-field Many-body S-matrix Theory 4

The **'post' form**:

$$G_0\,(t, t') = G_i^0\,(t, t') \qquad \text{and} \qquad V_0(t) = V_i(t). \tag{14}$$

Add and Subtract:

$$-i\int_{t_i}^{t_f} dt_1 < \phi_f(t_1)|V_f(t_1)|\phi_i(t_1) >, \tag{15}$$

Then,

$$\begin{aligned}(S-1)_{fi}(t_f, t_i) = {} & \Delta\,(t_i, t_f) \\ & -i\int_{t_i}^{t_f} dt_1 < \phi_f(t_1)|V_f(t_1)|\phi_i(t_1) > \\ & -i\int_{t_i}^{t_f}\int_{t_i}^{t_f} dt_2 dt_1 < \phi_f(t_2)|V_f(t_2) \\ & \qquad \times G_i^0\,(t_2, t_1)\ V_i\,(t_1)|\phi_i(t_1) > \\ & -i\int_{t_i}^{t_f}\int_{t_i}^{t_f}\int_{t_i}^{t_f} dt_3 dt_2 dt_1 < \phi_f(t_3)|V_f(t_3) \\ & \qquad \times G_i^0\,(t_3, t_2)\ V_i\,(t_2)\ G_i^0\,(t_2, t_1)\ V_i\,(t_1)|\phi_i(t_1) > \\ & i\ldots \\ = {} & \Delta\,(t_i, t_f)\ -i\int_{t_i}^{t_f} dt_1 < \phi_f(t_1)|V_f(t_1)|\psi_i(t_1) > \\ = {} & -i\int_{t_i}^{t_f} dt_1 < \phi_f(t_1)|V_f(t_1)|\psi_i(t_1) >\end{aligned} \tag{16}$$

(The difference integral

$$\begin{aligned}\Delta\,(t_i, t_f) \equiv {} & -i\int_{t_i}^{t_f} dt_1 < \phi_f(t_1)|V_i(t_1) - V_f(t_1)|\phi_i(t_1) > \\ = {} & -i < \phi_f(t_1)|\phi_i(t_1) > |_{t_1=t_i\to-\infty}^{t_1-t_f\to\infty} = 0, \qquad (\phi_f \neq \phi_i)\end{aligned} \tag{17}$$

and, does not contribute to the transition *rate*.)

Fig. 1.22 "Post" S-matrix series as a special case of IMST

IMST automatically generates two sets of suggestive diagrams, in the first and the second order (Fig. 1.23).

One suggestion for its origin was the so-called *shake-off* (SO) mechanism (Fig. 1.24) that was known to dominate the double ionization of atoms by a high-frequency synchrotron photon (in weak fields). In this mechanism one electron absorbs the field energy and leaves the atom so fast that the other electron cannot adiabatically adjust its motion, but it can, like any transient process, "jump" to a state

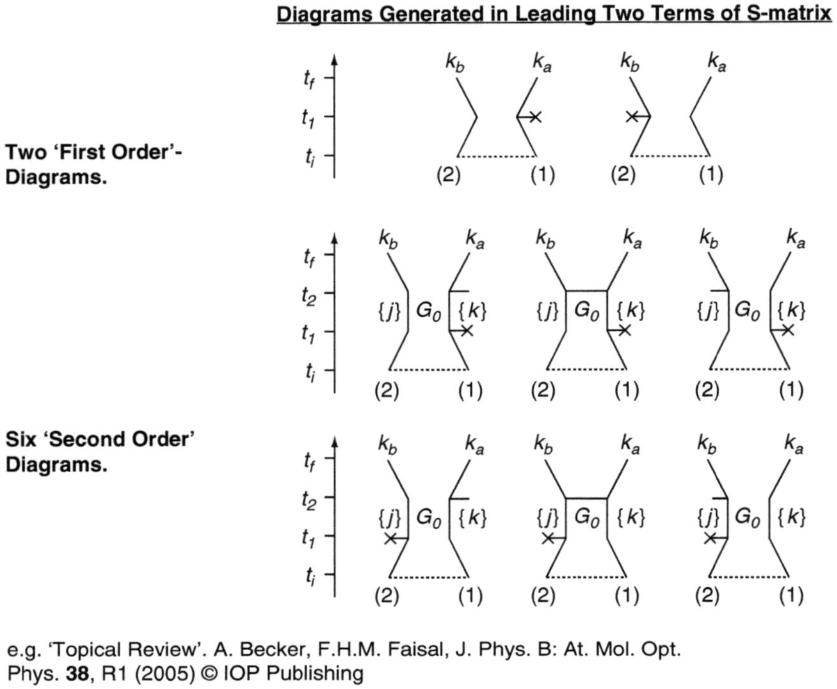

Fig. 1.23 First- and second-order IMST diagrams. Note that each diagram represent also a "mechanism"; hence, we have the useful formula: Diagram = mechanism

Fig. 1.24 Shake-off diagram (*left*) and mechanism (*right*) [30]

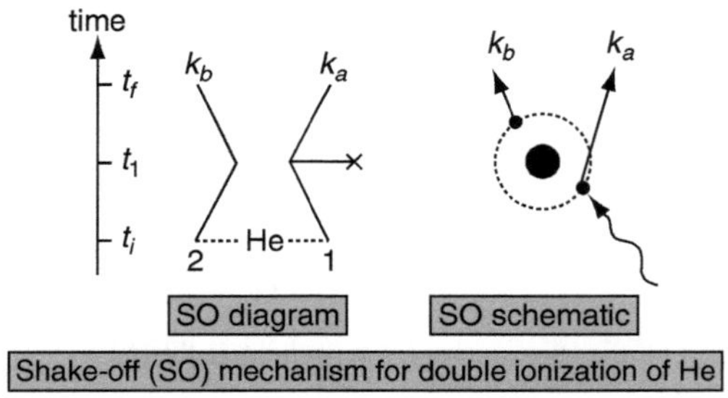

of the residual ion. The probability of the latter jump occurring into the continuum of the ion determines the probability of the double ionization by the SO mechanism.

Another possibility is the so-called *rescattering* mechanism as discussed by Corkum [13, 14]. In this case the first electron is ionized effectively by quasi-static tunneling with zero kinetic energy, then it is accelerated and driven back to the ion core by the reversal of the direction of the laser field (after a half-cycle) and collides (rescatters) with the second electron to make the ion release it as well to complete the double-ionization process.

1.16 The "CES" Diagram and "Mechanism" of Double Ionization

A systematic quantum derivation, identification, and evaluation of the *correlated energy sharing* (CES) diagram (Fig. 1.25), that incorporates the quantum analog of the classical rescattering mechanism [13, 14] (shown on the right), was made using the IMST [11, 22, 23, 26]. It also includes a direct "on the way out" scattering (middle sketch), in a unified manner. The actual calculation of the double ionization rates for He and the comparison with the experimental data for the double ionization yields, at, e.g., 780 nm wavelength, as we shall discuss soon, fully confirmed the nonsequential nature of the process for the near-infrared wavelength.

In contrast, the data for the double ionization of He at a ultraviolet wavelength (248 nm) are in remarkably good agreement [11, 26] with the results obtained only from the stepwise mechanism. In fact, the calculation of the CES contribution to the process at this short wavelength turned out, consistently, to be negligibly small. Clearly, therefore, the dominant role of the CES (or rescattering) mechanism in intense-field double ionization of He breaks down at the ultraviolet wavelength.

In order to see the nature of the correlation in this context, consider first the top two diagrams in the list of diagrams generated by the first two terms of the IMST series for double ionization. These two arise from the first term of the series. Here, 1 and 2 are the two electrons of the target He atom. The two upward-running lines refer to the evolution of their states with increasing time. At the initial time the two elections of the He atom are weakly joined together, as represented by the dotted line connecting them, by the *e–e* correlation in the unperturbed ground state. Then, at a latter time, one of them interacts with the intense laser pulse, picks up the field energy and is emitted with a wave-vector $\boldsymbol{k}_a$.

The other electron, which did not interact directly with the laser pulse, however, does not fail to respond, since the rapid change of the state of the first electron can

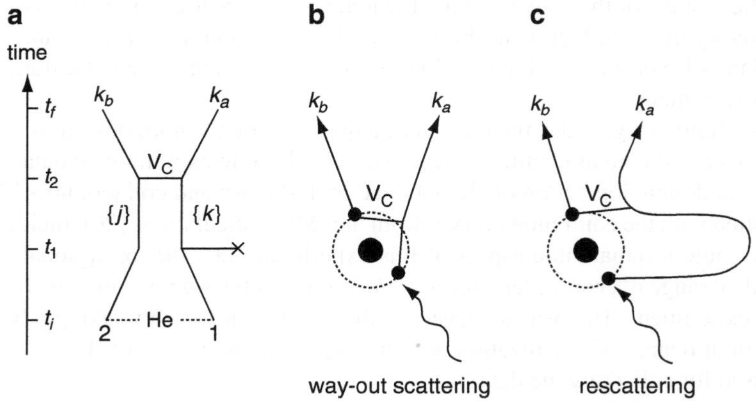

Fig. 1.25 CES diagram (*left*) and mechanisms (*middle* and *right*) [30]

cause the other electron to jump to any other state, specially to a free state of the ion, and proceed freely with the wave-vector $\boldsymbol{k}_b$, thereby causing the atom to be doubly ionized. This is the so-called shake-off, or SO, mechanism mentioned above, which is already clear from the SO diagram.

Note that, because of the identity of the electrons, there is always a symmetric extra diagram in which the role of one and the other electron is interchanged; they refer to the same mechanism due to the indistinguishable nature of the electrons. The corresponding amplitudes must, therefore, be added coherently together before computing the probability.

If we take only the SO diagram into account, the computed double-ionization probability at near-infrared laser wavelengths (e.g., from Ti-sapphire lasers) falls greatly short of the observed probability by many orders of magnitude [11]. Clearly, the SO mechanism is not the mechanism representing the observed NSDI process.

Naturally, we are forced to consider the second-order diagrams as the next set that arises from the second leading term of IMST. In these diagrams there are always two interaction events, and between them the two electrons propagate in the intermediate (virtual) states created by the interaction with the laser pulse. This propagation is given by the Green's function that refers to the intermediate Hamiltonian in which one electron that interacts primarily with the laser pulse is in the virtual Volkov states and the other electron is in the virtual ion states.

Note also that, in the list (Fig. 1.23), there are two kinds of diagrams, that is the *disjoint* diagrams in which the two electron lines are unconnected and the *conjoint* diagrams in which they are connected. If the correlation is important, the conjoint diagrams should be dominant, since it is the *e–e* correlation interaction that alone can connect the electron lines. Neither the laser interaction nor the nuclear Coulomb interaction can do so because both of these interactions are given by single-electron operators.

In fact, the numerical calculations [26] of the yields of the double ionization observed in the experiment at the Ti-sapphire wavelength [24, 25] fully confirm this expectation. Note that, for the comparison with the experimental ion yields, the theoretical rates for the single and double ionization were used in the rate equations for deriving the populations in the focal region of a Gaussian laser beam, with a Gaussian pulse profile, having the same beam waist and the same pulse duration as in the experiment.

In the figure (Fig. 1.26), the intensity increases along the horizontal axis, and the ion yields are shown along the vertical axis. The discrete experimental data of both single and double ionization of He obtained by DiMauro and collaborators [24, 25] are plotted with the continuous curves from the IMST calculations [26]. Incidentally, we may note a remarkable aspect of this experiment, i.e., the exceptionally large dynamical range over some ten orders of magnitude of the signal values are recorded in this experiment. This was achieved by the experimentalists by taking advantage of the high degree of stabilization and the high repetition rates of the Ti-sapphire laser used for collecting the data.

The theoretical double ionization curve corresponds to the conjoint diagram in the middle of the list of diagrams. Reading it (cf. Fig. 1.25) from the bottom to the top, we see that one electron interacts with the laser, and is goes into the virtual Volkov states. Then, it interacts with the other electron via the *e–e* correlation, shares its energy with it until they both have enough energy to escape together from the binding force of the atom and proceed to the detector(s) with the respective wave-vectors. This is the quantum picture of the CES mechanism.

As already mentioned, for the given values of the laser parameters in the experiment, this diagram, representing its associated (CES) mechanism, gives the overwhelming contribution to the double-ionization rates. Note that in the analytic expression of the diagram, the time spent by the system in the intermediate virtual states must be integrated for the initial to the final time of observation of the system. This implies that all values of the accumulated phase contribute to the NSDI probability. Nevertheless, a stationary-phase estimation of the intermediate time integration, carried out by many authors (see [11]), suggests that most of the contributions to the intermediate time integral comes from the time range of the order of the classical return time to the origin of an electron emitted initially with zero energy. This picture corresponds to the classical rescattering mechanism proposed earlier [13, 14].

The quantum CES diagram in effect includes both the classical rescattering as well as the on-the-way-out scattering trajectories. Finally, it can be seen from the figure (Fig. 1.26) that the calculated results (full lines) are in excellent agreement with the experiment data (crosses) over the entire intensity and dynamic range. Also, as shown clearly by the dotted curves, the theoretical results of the contributions from the stepwise double ionization process fall off too rapidly by many orders of magnitude, over the most part of the intensity domain, but, interestingly, they approach the experimental data at the highest intensities employed in the experiment.

The reason for this approach is that, at the highest intensities, the single ionization in the laser focus becomes saturated (as can be seen in Fig. 1.26) and, therefore, the

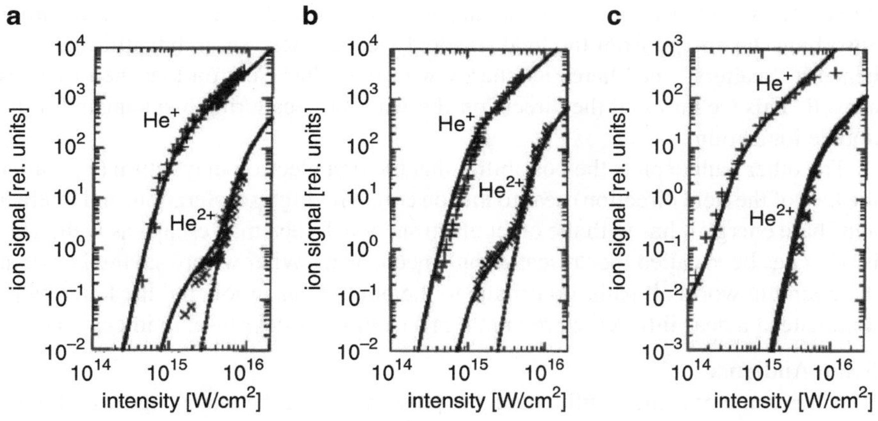

Fig. 1.26 NSDI yields in He; comparison of experimental data vs. IMST

residual singly ionized ions at the laser focus can be further singly ionized. This is clearly a stepwise double ionization process under the condition of saturation of single ionization of the neutral atom. However, when the field intensity grows toward the saturation and before the full saturation region for the single ionization is reached, there is a competition between the two processes, namely, the NSDI and the stepwise double ionization. This leads to the characteristic "knee" structure of the NSDI yield curve.

We may add that the CES rescattering mechanism has been found to hold at other wavelengths as well (see Sect. 1.5 in [11]). However, this is not always the case. Let me give a concrete example where the rescattering mechanism becomes negligible as a mechanism for intense-field double ionization, namely, in the case of an experiment at a ultraviolet wavelength of 248 nm (cf. [26]).

Here unequivocally the results for the stepwise ionization, calculated using IMST, agree well with the experimental data, demonstrating that the rescattering mechanism is not significant at this ultraviolet frequency. In fact, as indicated earlier in the evaluation of the contribution from the CES diagram (Fig. 1.25), the rescattering mechanism is found to be negligible compared to the contribution of the stepwise mechanism at this wavelength. In fact the contribution from the rescattering or CES diagram is nondiscernible in the scale of the graph that compares the experimental and the theoretical results for the double ionization yields in logarithmic scales. This result warns us that a universal expectation of the prevalence of NSDI in intense laser fields, independent of the wavelengths, can be easily misleading.

Let me stop here for discussions.

Professor Kono
You have mentioned two types of mechanisms, namely, the scattering and rescattering. What is implied by rescattering, especially large-amplitude rescattering?

Professor Faisal
The two classical paths in the figure represent an artist's depiction of two extreme "trajectories" in the sense of "Feynman paths." In one path, the first electron – after absorbing the energy from the field – quickly leaves the system, but still interacts, hence it "scatters" and shares its energy with the other electron that then escapes as well. This we may call the direct "on-the-way-out" scattering mechanism for the double ionization.

The other path depicts the possibility that the first electron may return (e.g., after the turn of the field direction) near to the ion core with high acceleration, and interact and share energy it has with the other electron, which then may escape as well. This is what has been called the "rescattering" mechanism. What we are saying is that in the quantum world all paths contribute to the double ionization, but the latter paths dominate at a near-infrared wavelength, in a semi-classical sense, in intense fields.

From Audience
Which parameter is more influential, the pulse width, or the wavelength? Although the wavelength structure is expected, what investigation has been made regarding the width of the light pulse?

Answer
All our calculations for the double ionization of He were made with the same pulse widths as in the experiments, e.g., 160 fs for 780 nm, 200 fs for 670 nm, and 500 fs for 248 nm, etc. In all the cases the agreement with the experimental ion yields, both for the single- and double-ionization yields, were remarkably good. Nevertheless, the dependence of the ion yields on the pulse length is only of secondary significance. Let me briefly say why. The theory gives the basic rates, that is, the "rate constants" of the single ionization and double ionization. They are essentially constants in the adiabatic timescale of the slowly varying pulse envelope, and therefore depend only parametrically, not dynamically, on time.

The direct experimental measurements of the ionization rates has not been performed by any body yet; instead, the ion yields have been measured as a function of the mean intensity of the laser pulse. The connection between the time-dependent yields and the adiabatically constant rates is given by the rate equations for the ion yields with in the laser focus. Therefore, the yields contain merely the averaged information of the pulse duration which is accounted for in the calculations by integrating the rate equations over the space-time intensity profile within the laser focus.

When does the adiabatic rate concept break down? We have not looked into this question specifically for the double-ionization rates, but have investigated it for the ionization of an H atom; we calculated an exact ionization probability, by direct simulation in 3D, with successive pulses of shorter durations. The results were rather surprising to us. We found an effectively constant rate of ionization for the pulse durations longer than only three cycles (e.g., for the 800-nm laser). Therefore, for pulses on the order of or shorter than one or two cycles, it is may be necessary to consider the ionization probability directly (instead of the "rates") as a function of time.

From Audience
The rescattering can be either quantum or classical, but they are in fact disjoined. In other words, the timescales are significantly different because the quantum timescale is only approximately 10^{-17} s, whereas many lasers in the laboratories have a scale of 10^{-15} s. A factor of a difference of 100 implies that the quantum process must always terminate before we consider the classical process. Furthermore, if the perturbation theory fails, it does not lend itself to the summation of an infinite number of diagrams. This is because with regard to convergence of the series, the summation of any finite number of terms has no meaning when it fails.

Answer
These are two different questions. Let me take them one after another. First, I have pointed out that the quantum CES diagram actually sums over all the paths (and hence times) including the rescattering and the direct-scattering (or on-the-way-out) trajectories. Qualitatively speaking, the motion of the first kind lasts at least on the order of the laser cycle, that is, the field should reverse at least once, while the motion of the second kind may take much shorter time. Semi-classical or stationary-phase integration over the entire intermediate times in the CES amplitude suggests

that, for double ionization of He at near-infrared wavelengths, the rescattering paths contribute the most per cycle to the CES amplitude. In principle, of course, quantum motion may occur in all scales of time, short or long. An example of the latter is the motion in a quantum stationary state that may last for an indefinitely long time.

Now, for the second question: Your point relates to a truly divergent series. In that case, of course, the sum does not exist. However, this does not apply to asymptotic series which may be summed in a generalized sense of, e.g., Cesàro, Borel, Padé, or Shank, among others.

From Audience
A classical problem, which was a hot topic in the 1950s, has not yet been resolved. In 1952, Feynman–Dyson showed that quantum electrodynamics is generally not convergent, and therefore, the series must be asymptotic. In 1962, it was demonstrated that it does have convergence. Furthermore, not only is the radius of convergence not zero, there is also an upper limit and everything is intensity dependent such that even such external field theories do not apply if the intensity is very high.

Answer
The essential point is that there is an important distinction between the divergent series and the asymptotic series. The sum of the latter series can be estimated by various techniques, for example, by Shank's transformation [42].

1.17 Comments on Sum-Momentum Distributions

Let me briefly comment on the momentum distributions of the electrons measured in NSDI of He atom. Consider the experimental data [27, 28] for the distribution of the sum-momentum of the two outgoing electrons, parallel to the laser polarization direction (Fig. 1.27). We see the double-hump distribution with a dip in the middle. Note that the corresponding distribution in the perpendicular direction reveals no dip. Moreover, the parallel momentum distribution observed is very broad; such a broad momentum distribution is very unusual in ionization by collision with atoms or electrons and/or in weak-field photo double-ionization processes, and has not been observed before.

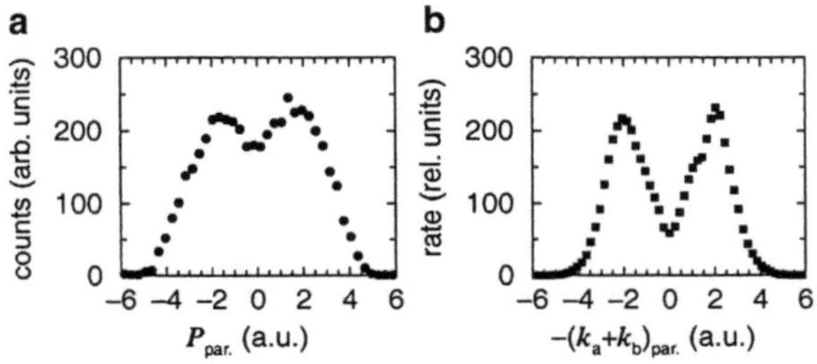

Fig. 1.27 Comparison of parallel sum-momentum for NSDI of He [28]: (**a**) experiment (**b**) theory, IMST

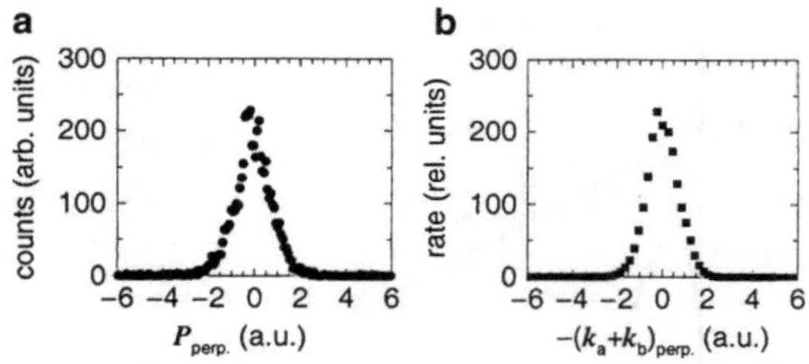

Fig. 1.28 Comparison of perpendicular sum-momentum in NSDI of He [28]: (**a**) experiment, (**b**) theory, IMST

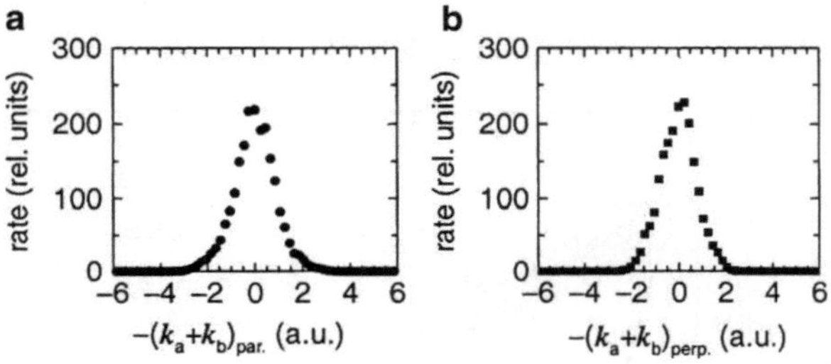

Fig. 1.29 Sum-momentum distributions calculated [28] by removing the "field dressings" of the two electrons in the final state: (**a**) parallel case, (**b**) perpendicular case. Note the drastic change in the parallel case with and without the field dressings, and the little change in the perpendicular case

The theoretical results shown here are obtained from the same CES diagram (Fig. 1.25) as was used for the calculation and interpretation of the ion yields earlier [26]. It can be seen from the comparison that the theory also gives a two-hump distribution with a dip, albeit sharper, in the middle. Moreover, the broad widths are also in good agreement with each other. The theoretical perpendicular distribution is also very similar to the corresponding experimental distribution. It is rather narrow with a single peak in the middle (Fig. 1.28).

What is the origin of the big difference between the parallel and perpendicular sum-momentum distributions in Fig. 1.27? This can be found out as follows. The S-matrix theory allows us to study easily two types of final states of the ejected electrons. In the first case (Fig. 1.27), the electrons are both "dressed" by the field, i.e., they are taken to be in the plane wave Volkov states, while in the second case (Fig. 1.29), the field dressing is removed and they are taken to be in the ordinary plane wave states, but with the same momenta as before.

The difference between the distributions calculated with and without the Volkov dressing in the final state, is rather drastic in the parallel case; in contrast, the perpendicular case hardly changes. The broad double-hump parallel distribution collapses to the narrow single hump distribution in the absence of the final state Volkov dressing. Therefore, the observed feature is clearly a consequence of the presence of the laser field in the final state. Why does the perpendicular distribution remain the same when the field dressing in the final state is absent? This is because, classically speaking, in the direction perpendicular to the field polarization, there is no component

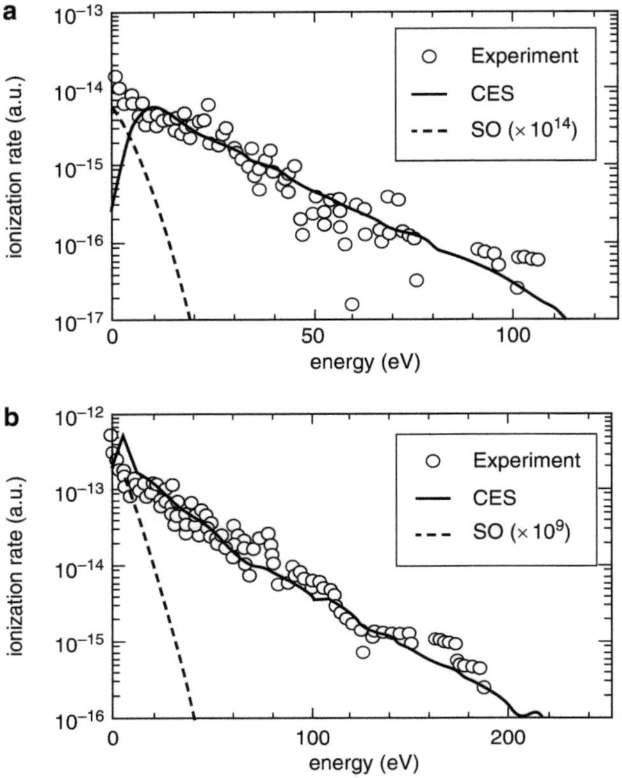

Fig. 1.30 Comparison of experimental data vs. SO and CES mechanisms from IMST [30]

of the electric force present to affect the distribution in this direction. The large change in the parallel distribution is precisely due to the opposite effect, namely, the entire electric force of the field is directed along the polarization. We point out that the theory also gives a simple analytical formula for estimating the actual width of the parallel distribution in NSDI, which compares rather well with the experimental data. We note also that results of coincidence energy measurements (cf. [29, 30]) of the electrons could be well reproduced (Fig. 1.30) by the same CES amplitude.

1.18 Comments on Multiple Ionization

Next, I comment briefly on the difficult problem of multiple-ionization in intense laser fields. At present a full numerical calculation of the probability of multiple ionization is practically impossible.

However, IMST suggests a qualitative mathematical modeling for the process, which is quite simple. Taking the hint from the double-ionization CES diagram, it is not unlikely that a similar diagram in which the first electron absorbs the laser energy, and then, ionizes the second, third, fourth, etc., electrons by collision with the residual ion, could be the dominant mechanism also for the triple, quadruple, or multiple ionization in intense near-infrared fields (Fig. 1.31).

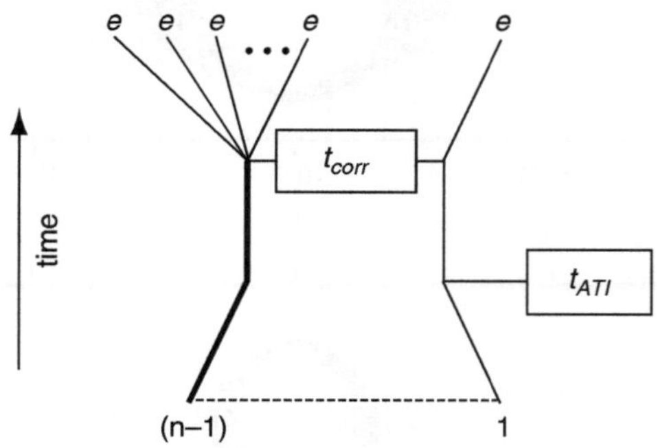

Fig. 1.31 An IMST diagram showing the mechanism of nonsequential multiple ionization [26]

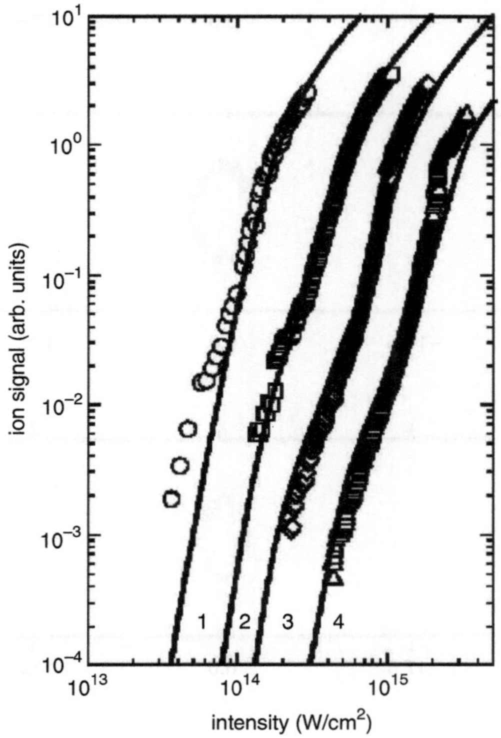

Fig. 1.32 Comparison of experimental data and IMST for multiple ionization of Kr.; Maeda et al. [32]

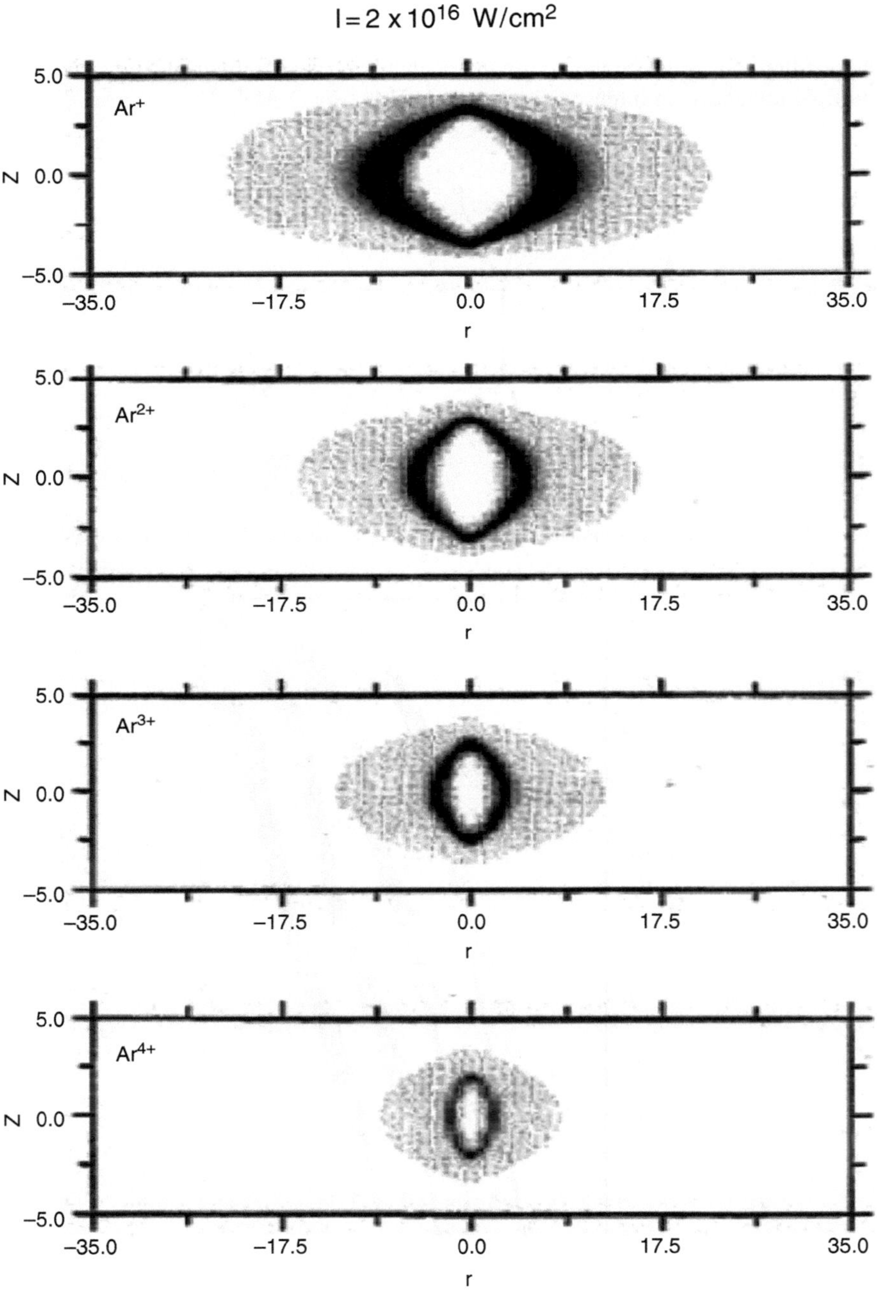

Fig. 1.33 Calculated spatial distribution of the charge-states of Ar-ions formed from neutral Ar-atoms in a Gaussian laser focus of an intense Ti-sapphire laser field [43]

In fact, a crude estimate made by an approximate formula [33] obtained from this analogy, shows that the result agrees satisfactorily with the observed yields [31, 32] of multiply charged ions (e.g. Fig. 1.32).

Figure 1.33 illustrates the calculated spatial distribution of the multiply charged ions created in the laser focus. It shows that the laser focus acts like a *reaction vessel* that lasts as long as the pulse lasts (e.g., 30 fs) and the various reaction products (i.e., the charge states of the ions) distribute themselves with different concentrations along the different intensity domains in the laser focus. At the center where the intensity is extremely large, we observe that there are no singly ionized ions because these ions are already doubly and triply ionized in this intense region of the beam. However, some singly ionized ions can be observed at the edge of the beam where the intensity is low and the singe ionization is not saturated.

References

1. F.H.M. Faisal, *Theory of Multiphoton Processes* (Plenum, New York, 1987)
2. P. Agostini, F. Fabre, G. Mainfray, G. Petite, N.K. Rahman, Phys. Rev. Lett. **42**, 1127 (1979)
3. L.V. Keldysh, Sov. Phys.-JETP **20**, 1307 (1964)
4. F.H.M. Faisal, J. Phys. B **6**, L89 (1973)
5. H.R. Reiss, Phys. Rev. A **22**, 1786 (1980)
6. P. Kruit, J. Kimman, H.G. Muller, M.J. van der Wiel, Phys. Rev. A **28**, 248 (1983)
7. M. Lewenstein, Ph. Balcau, M.Yu. Ivanov, A. L'Huillier, P.B. Corkum, Phys. Rev. A **49**, 2117 (1994)
8. M. Ferray, A. L'Huillier, X.F. Li, L.A. Lompre, G. Mainfray, C. Manus, J. Phys. B: At. Mol. Opt. Phys. **21**, L31 (1988) © IOP Publishing
9. U. Schwengelbeck, F.H.M. Faisal, Phys. Rev. A **50**, 632 (1994)
10. B.W. Shore, P.L. Knight, J. Phys. B **20**, 413 (1987)
11. A. Becker, F.H.M. Faisal, J. Phys. B, **38**, R1–R56 (2005), sec. 4.4
12. A. Becker, L. Plaja, P. Moreno, M. Nurhuda, F.H.M. Faisal, Phys. Rev. A **64**, 023408 (2001)
13. P.B. Corkum et al., Phys. Rev. Lett. **62**, 1259 (1989)
14. P.B. Corkum, Phys. Rev. Lett. **71**, 1994 (1993)
15. R. Moshammer et al., Phys. Rev. Lett. **91**, 113002 (2003)
16. A. Rudenko et al., J. Phys. B **37**, L407 (2004)
17. F.H.M. Faisal, G. Schlegel, J. Phys. B **38**, L223 (2005)
18. K.C. Kulander, Phys. Rev. A. **38**, 778 (1988)
19. J.S. Parker, B.J.S. Doherty, K.G. Meharg, K.T. Taylor, J. Phys. B, **36**, L393 (2003)
20. C. Ruiz, L. Plaja, L. Roso, A. Becker, Phys. Rev. Lett. **96**, 053001 (2006)
21. C.J. Joachain, M. Doerr, N.J. Klystra, Adv. At. Mol. Opt. Phys. **42**, 225 (2000)
22. F.H.M. Faisal, A. Becker, in *Selected Topics on Electron Physics*, ed. by D.H. Campbell, H. Kleinpoppen (Plenum Press, New York, 1996)
23. F.H.M. Faisal, A. Becker, *Multiphoton Processes*, ed. by P. Lambropoulos, H. Walther. IOP conference series, vol. 154 (Institute of Physics, 1996)
24. D.N. Fittinghoff, P.R. Bolton, B. Chang, K.C. Kulander, Phys. Rev. Lett. **69**, 2642 (1992)
25. B. Walker, B. Sheehy, L.F. DiMauro, P. Agostini, K.J. Schafer, L.C. Kulander, Phys. Rev. Lett. **73**, 1227 (1994)
26. A. Becker, F.H.M. Faisal, Phys. Rev. A, **59**, R3182 (1999)
27. T. Weber et al., Phys. Rev. Lett. **84**, 443 (2000)
28. A. Becker, F.H.M. Faisal, Phys. Rev. Lett. **84**, 3546 (2000)
29. R. Lafon et al., Phys. Rev. Lett. **86**, 2762 (2001)

30. A. Becker, F.H.M. Faisal, Phys. Rev. Lett. **89**, 193003 (2002)
31. S. Larochelle, A. Talebpour, S.L. Chin, J. Phys. B, **31**, 1201 (1998)
32. H. Maeda et al., Phys. Rev. A **62**, 035402 (2003)
33. A. Becker, F.H.M. Faisal, Phys. Rev. A **59**, R3182 (1999)
34. J. Muth-Boehm, A. Becker, F.H.M. Faisal, Phys. Rev. Lett. **85**, 2280 (2000)
35. A. Jaron-Becker, A. Becker, F.H.M. Faisal, Phys. Rev. Lett. **96**, 143006 (2006)
36. X. Urbain et al., Phys. Rev. Lett. **92**, 163004 (2004)
37. A. Requate, A. Becker, F.H.M. Faisal, Phys. Rev. A **73**, 033406
38. A. Jaron-Becker, A. Becker, F.H.M. Faisal, Phys. Rev. A **69**, 023430 (2004)
39. A. Jaron-Becker, A. Becker, F.H.M. Faisal, J. Phys. B **36**, L375 (2003); [7] sec. (6.3) (2006)
40. F.H.M. Faisal et al., Phys. Rev. Lett. **98**, 143001 (2006)
41. C.J. Joachain, *Quantum Collision Theory* (North-Holland, Amsterdam, 1984)
42. C.M. Bender, S.A. Orszag, *Advanced Mathematical Methods for Scientists and Engineers* (McGraw-Hill, New York, 1978)
43. A. Becker, F.H.M. Faisal (2000, unpublished)

Chapter 2
Foundations of Strong-Field Physics

Howard R. Reiss

Abstract The general structure of the nonperturbative theory of laser-induced reactions is developed from first principles, including the means by which quantum transition amplitudes are formulated. The possibility of treating strong-field ionization of atoms and molecules by analytical approximations leads naturally to the *strong-field approximation* (SFA). Some widespread misconceptions are addressed, particularly that strong-field ionization constitutes a simple tunneling process. Such misconceptions underlie the separation of strong-field processes into a *tunneling* domain and a *multiphoton* domain. The entire framework of a tunneling description of ionization is based on the notion of the laser field being treated as if it were a quasi-stationary electric field, whereas a laser field is a plane-wave field. Although there is a gauge transformation connecting the two concepts when the dipole approximation is applicable, this cannot avoid the fact that a laser produces plane waves, and treating a plane wave as a quasi-stationary electric field is responsible for fundamental misconceptions. This overview of strong-field foundations is continued to the intensity domain where the dipole approximation fails altogether due to the onset of magnetic-field effects. Further increase in intensity brings on relativistic effects, for which illustrations are given.

2.1 Introduction

There are certain inconsistencies in the way strong-field physics is presently constituted, and this chapter will question some of those inconsistencies that appear to be widely accepted for inappropriate reasons. The starting point is a description of the features that are unique to strong-field physics, since physicists and chemists are educated very differently in the discipline of perturbation theory. When physicists and chemists consider strong fields, it is difficult to forget the misleading

---

H.R. Reiss
American University, Washington, DC, 20016-8058, USA Max Born Institute,
12489 Berlin, Germany

K. Yamanouchi (ed.), *Lectures on Ultrafast Intense Laser Science 1*, Springer Series in Chemical Physics 94, DOI 10.1007/978-3-540-95944-1_2,

perturbation theory background. This is the likely cause of a number of the inconsistencies that have arisen. To ensure that this problem is perceived correctly, a general quantum transition amplitude is derived that is independent of perturbation theory. Perturbation theory can be derived as a limiting case of this formalism. However, the general transition amplitude has special properties that are essential. This transition amplitude is usually not discussed in lectures. It is generally regarded as something that should be well known, and does not require repeated mention.

One of the reasons why this matter is being addressed here is because of a recent movement taking place in the strong-field community to view the length gauge, where the interaction is $\boldsymbol{r} \cdot \boldsymbol{E}$ – the scalar product of the radius vector $\boldsymbol{r}$ and the electric field $\boldsymbol{E}$ – as a fundamental gauge [1]. It is often referred to as the *correct* gauge, or as a *gauge-invariant* gauge. These are descriptions that are self-contradictory. Somehow, this notion about the priority of the length gauge has gained a real foothold in the community. There are several recent papers that adopt this point of view.

Further, particular attention will be paid to the *strong-field approximation* (SFA) [2], which is the leading method for the calculation of strong-field effects by analytical approximations. This is often viewed as equivalent to the *Keldysh–Faisal–Reiss* (KFR) method [3]. This causes conceptual problems since K [4], F [5], and R [2] differ in fundamental ways. The SFA is the same as the R of the KFR.

Under the heading of strong-field approximation, the applications treated are rates, spectra, angular distributions, and momentum distributions. The SFA itself provides many of the basic properties of momentum distributions that are causing much puzzlement and debate at this time.

The next item under the SFA concerns the separation of phenomena into the putative tunneling domain and the multiphoton domain. This is fundamentally inconsistent and self-contradictory [6].

The issues of time domains and rescattering are very interesting, and have required some recent clarification [7]. The last items under this topic are some new phenomena that come into action when the fields are strong enough that magnetic and relativistic effects become significant.

Quantum mechanics is required if strong fields and the presence of bound states are to be considered. Quantum mechanics will have to be employed in a very substantial way. Today, everybody recognizes that strong-field phenomena require nonperturbative methods. However, this was questioned during the 1980s, when there were some researchers who believed strong-field phenomena to be nothing more than a manifestation of higher-order perturbation theory. Strong-field physics is considerably more than that. The signature characteristics of a strong-field process are considered first.

2.2 Special Features of Strong-Field Problems

In a strong-field problem, there are several energies: the photon energy $\hbar\omega$; the binding energy of the bound electron E_{B}; the electron rest energy mc^2; and the ponderomotive energy U_{p}, which is just the interaction energy of a charged particle with the field in which it is immersed. Dimensionless intensity parameters can be formed from ratios of these energies [8]. That is, three independent ratios can be formed from four energies to give three dimensionless parameters: the nonperturbative intensity parameter (z); the bound-state intensity parameter (z_1); and the free-electron intensity parameter (z_f). In each of these, the ponderomotive energy is in the numerator, and this interaction energy of the electron with the field is compared with other fundamental energies. These three different ratios imply that it should not be expected that strong-field phenomena can be scaled with one parameter.

One example of how these intensity parameters characterize physical properties is nondipole behavior, which is commonly overlooked. Its onset is measured by the ratio of the intensity I divided by the cube of the frequency: I/ω^3. A second example is the onset of relativistic effects, where the relevant intensity parameter is proportional to I/ω^2. Two independent types of behavior have already been found, so it is clear that scaling cannot be described by only one parameter. Moreover, it does not allow sufficient flexibility to discuss all possible strong-field effects. The Keldysh parameter is widely considered to be sufficient. It is useful, but it is not sufficient.

Figure 2.1 gives a basic illustration of strong-field phenomena. The lower x-axis gives field frequency in atomic units, and the upper x-axis gives the wavelength. These axes are correlated. The intensity is given in atomic units on the left y-axis, and in watts per square centimeter on the right y-axis. The relativistic domain is shown here as bounded by the slanted solid line. The line represents the conditions at which the ponderomotive energy of an electron in a field is one-tenth of mc^2. The quantity mc^2 is the rest energy of an electron, which is a very fundamental parameter. When the interaction energy of the electron with the field is so strong that it approaches mc^2, relativistic effects are expected.

The more steeply slanted dashed line marks the boundary of an intensity domain where intensity-caused nondipole effects begin to appear. In the strong-field case, the low-frequency limit on the onset of nondipole effects has nothing to do with a ratio of wavelength to atomic size. Textbooks in quantum mechanics employ the criterion that, if the wavelength of an applied field is much larger than the size of the atom, then the dipole approximation can be used. In fact, the dipole approximation fails when this blue line is approached, which is far more of a limitation than the comparison of wavelength to atomic radius. It fails as a function of intensity.

Next, one more element in the figure is examined: the horizontal line along which the electric field is equal to one atomic unit. If one looks first to the right side of Fig. 2.1, which is the high-frequency (or short-wavelength) region, 1 a.u. is not a particularly strong field. If one looks to the left side, which is the low-frequency (or long-wavelength) region, 1 a.u. of electric field is extremely strong and deeply into the relativistic domain.

However, in both regions the electric fields are equal to 1 a.u. Generally, it is said: "Strong-field effects occur when the electric field is of the order of one." That may be true in the case of the widely used Ti:sapphire frequency, but it is not a general result. In fact, the electric field seems to have very little significance. There

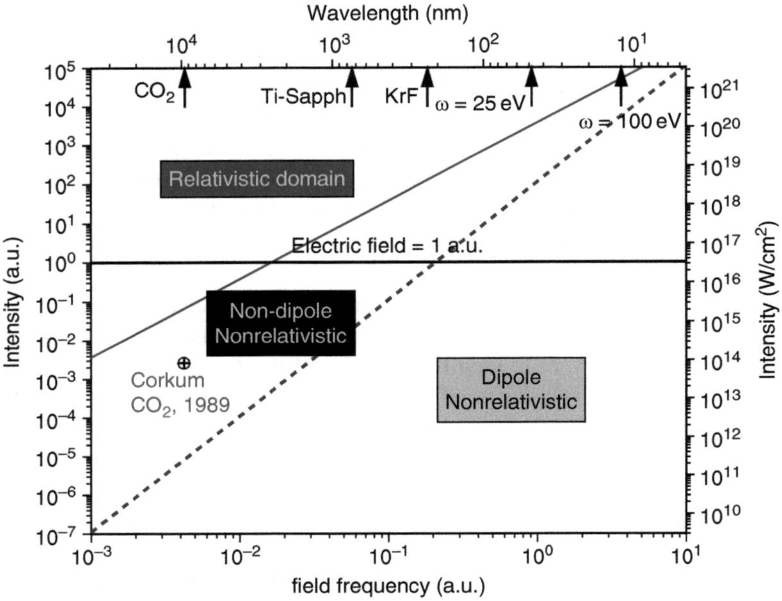

Fig. 2.1 This figure shows the frequency–intensity domain encompassed by laser physics. The division of this domain into regions characterized by dipole-approximation nonrelativistic phenomena; nondipole, nonrelativistic phenomena; and the fully relativistic domain, is basic. Nevertheless, it is commonly overlooked. The figure makes clear that the magnitude of the electric field of the laser has no particular significance. This is true despite the fact that a quasi-static-electric view of laser fields depends fundamentally on the magnitude of the electric field

is a reason for this. The Schrödinger equation is an energy equation. It is not stated in terms of electric or magnetic fields. At one time, there was a concerted effort to rewrite the Schrödinger equation so that it depends only on the fields instead of on the potentials [9–14]. All those attempts were unsuccessful. The motivation for the attempts to make the Schrödinger equation depend on fields rather than potentials comes from the near-universal attitude that fields are fundamental and potentials are just auxiliary mathematical quantities. This is a standard statement in textbooks. In fact it is not true, and there are many indications of its inadequacy that originate not only from strong-field physics, but from many other fields of physics. The electromagnetic potentials are more fundamental than the field strengths. This matter will be revisited later.

One further piece of information needs attention in the frequency–intensity illustration that is being examined. This refers to experiments done in Canada in 1988 by a group from Université Laval [15] and in 1989 by Paul Corkum [16] with a CO_2 laser, which is a very low frequency laser. The experiment is well below an electric field of one atomic unit and yet it is a very strong-field experiment. It is in the region in which the dipole approximation has already failed and it is approaching the limit where the relativistic domain begins.

In these experiments, the atom that was used was ground-state Xe. Because the energy of each photon is so small for the CO_2 laser, it takes about 100 photons just to provide the binding energy. This is a very nonlinear process, and to provide the ponderomotive energy requires an additional 9,000 photons. This number is so large that it may no longer be worth counting the photons. This is an illustration of what a strong-field experiment really is like. Of the three intensity parameters mentioned above, the z parameter of 9,000 coincides with the number of photons for a very good reason: that is one way to view the physical meaning of z. The z_1 parameter is almost 200. The z_f, which is the relativistic parameter, is still small, but it is getting close to the relativistic domain. Despite its early date, this is a very strong-field experiment; so strong that the dipole approximation is not valid and it is nearly relativistic. On the other hand, the electric field is only 0.053 a.u. In this intense-field experiment, the electric field hardly matters. The electric field cannot be used as a criterion; it does not tell us very much.

2.3 General Quantum Transition Amplitude

2.3.1 Preliminaries

Our next topic is a derivation of the general quantum transition amplitude. An essential aspect of this is to define an operational procedure for how an experiment is done in the laboratory, because laboratory experiments have to be conducted in order to verify theories and to observe vital phenomena. It will be found shortly that it is not necessary to describe how a state evolves in time. In fact, there are certain dangers involved in using the time evolution operator. It can be misleading. Also, it is possible to avoid the need for *adiabatic decoupling*. This is an important point as it has become a standard statement that adiabatic decoupling is required in the definition of a transition amplitude, which means that as time goes to minus infinity or plus infinity somehow the field has to be turned off in a gradual way. It is not necessary to say that. All that needs to be said is that the experiment is done in a region that is bounded in space and time. Almost all experiments are done in that way. The dynamics are all in the transition amplitude itself, because in order to derive the transition amplitude the Schrödinger equation has to be used, and so all the dynamics are automatically included.

Some of the points addressed here are in the nature of a general criticism of the AMO (atomic, molecular, and optical) community of researchers. In traditional atomic, molecular, and optical physics, the dipole approximation is used almost universally. There is very little experience with anything that goes beyond the dipole approximation. Electromagnetic phenomena are also very common in the nuclear physics community. However, in nuclear physics, the dipole approximation is nothing more than a special case. People use magnetic multipoles and electric multipoles of all orders. Hence, when a lot of inconsistencies are found, it can be assumed that it originates from the fact that the education of the AMO community focuses on the dipole approximation and on perturbation theory.

For example, it is assumed in the AMO community that the quantum transition amplitude transforms as a unitary transformation under gauge transformation, which is not true. A time-dependent gauge transformation – such as the gauge transformation between the length gauge and the velocity gauge – does not transform the Hamiltonian unitarily. Many in the AMO community employ gauge transformations assuming that the Hamiltonian is always a unitary operator. It is not. It is a matter of great importance that physical interpretations are gauge-dependent. This appears to be a source of confusion throughout quantum physics with electromagnetic fields, but it should not be because it is also true in classical physics.

In electromagnetic theory textbooks where gauge transformations are treated, one can find cases in which a perfectly simple physical situation in the laboratory is considered, a gauge transformation is carried out, and the end result is something that has no physical interpretation whatsoever. It could be a total mess, but it will have the same solution as the original problem. When it comes to physical interpretations, of course they are going to be changed. Sometimes the gauge transformation changes the problem to a physical situation that can be recognized; but it will, in general, be quite different from the original problem.

The currently fashionable matter of the length gauge being special is simply incorrect. On the face of it, a special quality attached to the length gauge does not make any sense, but unfortunately it is gaining some currency right now. There are a lot of papers that claim to show that. There is also the matter mentioned a little earlier about electromagnetic potentials being more than just auxiliary quantities. It appears that they are more fundamental than the fields.

2.3.2 History of the S-Matrix

A brief history of S-matrices is necessary. Wheeler [17] was apparently the first one to think of that approach, although Heisenberg [18] and Stückelberg [19] are the names most often associated with S-matrices in the early days. All applications were to scattering problems and that is why it is called the S-matrix. The S stands for scattering. On the other hand, in the 1960s there was a need for a general nonperturbative formulation to use to replace the standard perturbation theory for strong-field problems. The S-matrix concept appeared to be a suitable candidate, and it was shown [20] that it can be applied for any quantum process, not just scattering.

Scattering is a free–free process; in other words, there is a free particle, it scatters under the influence of some interaction, and goes off as a free particle with altered properties. But the S-matrix can also be used for bound–free, free–bound, or bound–bound processes. That means it can be used to describe ionization, recombination, or excitation and de-excitation transitions. The demonstration of this universality was done in 1970 [20].

2.3.3 Derivation of the Transition Amplitude

In a rigorous derivation of the S-matrix, the only condition that is imposed is that the transition-causing interaction occurs only in a domain bounded in space and time. This is almost always the case. Certainly for transitions in the focus of a laser pulse it is true.

For this derivation, the physical situation hypothesized is that of an atomic electron that is being exposed to a focused laser beam. This is a convenience, not a necessity. This is a very general derivation that applies to almost all processes.

Nonrelativistic conditions will be assumed, so that the Schrödinger equation is applicable. Everything done here can also be done for the Dirac equation [8]. It is not a fundamental constraint to limit this development to the Schrödinger equation. If there is a bound electron and it is put in the focus of a laser beam, when something happens to it, that electron can be described by a Hamiltonian H, which may be a rather complicated quantum operator.

The Schrödinger equation with this Hamiltonian will have a complete set of solutions. In the laboratory, measuring instruments must be consulted to find out what happened. These instruments never see a laser field; they do not "know" what a laser field is. A deduction has to be made as to what has happened in the interaction region based on readings taken outside the interaction region. So there is another Schrödinger equation and another complete set of solutions describing what the laboratory instruments observe. This Hamiltonian H_0 contains everything except the transition-causing interaction due to the laser.

A summary of the derivation thus far is as follows: We use the language of an atomic electron subjected to a pulsed, focused laser beam. This is a convenience, not a requirement.

There will be a complete set of states $\{\Psi_n\}$ that satisfy the Schrödinger equation describing the atomic electron that may be undisturbed or in interaction with a laser beam:

$$\mathrm{i}\partial_t \Psi = H\Psi. \tag{2.1}$$

The outcome of any experiment will be measured by laboratory instruments that never experience a laser field. As far as the laboratory instruments are concerned, there is a complete set of states $\{\Phi_n\}$ that satisfy the Schrödinger equation describing an atomic electron that does NOT experience the laser field:

$$\mathrm{i}\partial_t \Phi = H_0\Phi. \tag{2.2}$$

The laser pulse is constrained to be finite in time as well as in space, so that the difference between the complete Hamiltonian and the noninteracting laboratory Hamiltonian will vanish at both plus and minus infinity in time because the field is not present there. There are then two complete sets of solutions: solutions for the H_0 Hamiltonian and another set of solutions for H. At time minus infinity – in other words, before the laser pulse arrives at the atom – a one-to-one identification will be made between the elements of these two sets, so that the starting point is well-defined. Then, after the laser interaction has occurred, the laboratory instruments have to deduce what happened. They examine what comes out the interaction region. Suppose it is an ionized electron. The instruments can measure its energy and the angle at which it emerged, and other quantities that can be measured in terms of these noninteracting states. The way to evaluate what has happened is to take probability–amplitude overlaps onto each of the possible noninteracting states to gain full information about what happened in the interaction. This is the S-matrix.

As time goes to plus infinity, which means after the interaction is over, wave function overlaps are formed to deduce what happened. For our discussion, time going to minus infinity or to plus infinity, simply means before and after the interaction happens. These limits may be femtoseconds apart; they do not have to be some macroscopic time apart. They only need to be before the interaction and after the interaction. If the amplitude that no interaction has occurred is subtracted, what remains is $S - 1$. An overlap is being formed at time equal to plus infinity that is needed in order to assess what has happed, and the corresponding overlap at minus infinity is subtracted from it. At minus infinity, unless these two indices are the same the result is zero; if they are the same the result is unity. That is the Kronecker delta. It is written this way because of the symmetry when both these terms are of the same form, with one applicable at plus infinity in time and the other at minus infinity.

These points are summarized as: By hypothesis, the laser pulse is finite, so

$$\lim_{t \to \pm\infty} [H(t) - H_0] = 0. \tag{2.3}$$

We can organize the two complete sets of Φ states and Ψ states so that they correspond at $t \to -\infty$:

$$\lim_{t \to -\infty} [\Phi_n(t) - \Psi_n(t)] = 0. \tag{2.4}$$

After the laser interaction has occurred, the only way for the laboratory instruments to discover what has happened is to form overlaps of all possible final Φ_f states with the state that began as a particular Ψ_i state. This is the S-matrix:

$$S_\mathrm{fi} = \lim_{t \to +\infty} (\Phi_\mathrm{f}, \Psi_\mathrm{i}) . \tag{2.5}$$

Subtract the amplitude that no transition has occurred:

$$M_\mathrm{fi} \equiv (S - 1)_\mathrm{fi} = \lim_{t \to +\infty} (\Phi_\mathrm{f}, \Psi_\mathrm{i}) - \lim_{t \to -\infty} (\Phi_\mathrm{f}, \Psi_\mathrm{i}) . \tag{2.6}$$

We now have the form of a perfect differential:

$$M_\mathrm{fi} = \int_{-\infty}^{+\infty} \mathrm{d}t \frac{\partial}{\partial t} (\Phi_\mathrm{f}, \Psi_\mathrm{i}) . \tag{2.7}$$

Carry out the time derivatives in the integrand, and use the Schrödinger equations to remove the time derivatives.

The final result is:

$$M_{\mathrm{fi}} = -\mathrm{i} \int_{-\infty}^{\infty} \mathrm{d}t \, (\Phi_{\mathrm{f}}, H_{\mathrm{I}} \Psi_{\mathrm{i}}) \, , \tag{2.8}$$

$$H_{\mathrm{I}} \equiv H - H_0, \quad \lim_{t \to \pm\infty} H_{\mathrm{I}} = 0. \tag{2.9}$$

An alternative form is:

$$M_{\mathrm{fi}} = -\mathrm{i} \int_{-\infty}^{\infty} \mathrm{d}t \, (\Psi_{\mathrm{f}}, H_{\mathrm{I}} \Phi_{\mathrm{i}}) \, . \tag{2.10}$$

This is exactly the result if a perfect differential were to be integrated. Within the integrand, the time derivative can be carried out on each of the states, which yields two terms. Then the relevant Schrödinger equations are introduced, which gives the information about what happens with the time derivatives of the states. That leads to the final result where the H_{I} – the interaction Hamiltonian – is just the difference of the two Hamiltonians H and H_0. This is a general quantum transition matrix element. No perturbation theory is involved here at all. Everything is based on what is observed in the laboratory; this is very important to remember.

First, it should be noted that there is an alternative form. One form is called *post*, the other *prior*; or one is *direct time*, the other *time reversed*. They are equivalent. The difference between them is that instead of corresponding one set of solutions to the other at minus infinity, it is done at plus infinity, and then the inquiry is made as to what initial states could have given rise to this final state. It is just a time-reversed version of the first procedure. This time-reversed transition amplitude turns out to be very convenient for ionization problems. It is the basis of the SFA.

2.4 Gauge Transformations

Now a gauge transformation of this transition amplitude is carried out. The standard statement of what happens in a gauge transformation is that the vector potential is modified by adding the gradient of the generating function of the gauge transformation, and the scalar potential has subtracted from it the time derivative of that same generating function. When the gauge transformation is done, the quantum transition amplitude is not the same as it was originally. The interacting states transform, but the states describing the laboratory instruments do not transform. They do not "know" anything about the laser field. They cannot possibly contain any information about what gauge that laser field is expressed in.

The noninteracting states and the noninteracting Hamiltonian are independent of gauge. The consequence is that the interaction Hamiltonian H_{I} changes, and the interacting state Ψ changes, but the noninteracting state Φ does not change, so that the transformed amplitude is not the same as the initial amplitude. This is essential. It cannot be the same as the initial one because when a gauge transformation is performed, the physical interpretation is different. Were the gauge transformation to leave the transition amplitude unchanged, there would be no possibility of a different physical interpretation. This is really the foundation of most of the mistakes that are found in the AMO literature, continuing right through to the present day. People gauge-transform everything in the transition amplitude. That is not the correct way to do it. That improper procedure is part of the basis for the claim that the length gauge is an invariant gauge [1, 21]. The transition amplitude must change under a gauge transformation.

These points are summarized in the following discussion. In a gauge transformation:

$$\begin{aligned} &\vec{A} \to \vec{A}' = \vec{A} + \vec{\nabla}\chi, \quad \varphi \to \varphi' = \varphi - \frac{1}{c}\partial_t \chi \\ &\Rightarrow \quad H \to H', \quad H_{\mathrm{I}} \to H_{\mathrm{I}}', \quad \Psi \to \Psi' = U\Psi \\ &\Phi, \quad H_0 \quad \text{unchanged.} \end{aligned} \tag{2.11}$$

It is clear and unambiguous that the states $\{\Phi_{\mathrm{n}}\}$ and Hamiltonian H_0 for the world of the laboratory instruments have nothing to do with the laser field. A change of gauge of the laser field can have no possible effect.

The transition amplitude changes as:

$$\begin{aligned} &M_{\mathrm{fi}} = -\mathrm{i}\int_{-\infty}^{\infty} \mathrm{d}t \ \left(\Psi_{\mathrm{f}}, H_{\mathrm{I}}\Phi_{\mathrm{i}}\right) \to \quad M_{\mathrm{fi}}' = -i\int_{-\infty}^{\infty} \mathrm{d}t \ \left(\Psi_{\mathrm{f}}', H_{\mathrm{I}}'\Phi_{\mathrm{i}}\right) \\ &\Rightarrow M_{\mathrm{fi}} \neq M_{\mathrm{fi}}'. \end{aligned} \tag{2.12}$$

When a gauge transformation is carried out in the completely interacting Hamiltonian, an extra time derivative term arises beyond that which would be the case for a simple unitary transformation of an operator. The fact that this time derivative exists means that if the gauge transformation function has time dependence, then the Hamiltonian does not transform like a unitary operator. Standard textbooks do not state this. This is a real problem.

A famous example of an apparent gauge dilemma noted by Willis Lamb in 1952 [22], was resolved by Zoltan Fried [23]. He was one of the early members of a strong-field group that I set up in Washington in 1960, about 4 years before the Keldysh paper. One of Fried's major contributions was to resolve the paradox where Lamb purportedly showed that the length gauge was more fundamental than the velocity gauge. The way Lamb came to that conclusion was when he calculated a line shape for the Lamb-effect transition by second-order perturbation theory. He calculated this by length-gauge theory and got reasonable results. He did the calculation again in the velocity gauge and obtained a result that was implausible. The velocity-gauge line shape he found was asymmetrical, whereas in the laboratory it was symmetrical. The length-gauge result was symmetrical. And so, rather reluctantly, Lamb concluded there was something special about the length gauge. He did not understand what happened to gauge invariance, but this was the only way he could think of to solve that problem.

Fried realized there was something wrong with the Lamb velocity-gauge calculation. In the length gauge, Lamb knew that the intermediate states in a second-order perturbation theory calculation were such that only those states close in energy to initial or final states mattered. He dealt with the problem of having to sum over an infinite number of intermediate states by retaining only those states close in energy to the initial or final states. Fried pointed out that in the velocity gauge the physical picture is different. It is not just nearby energy states that matter; all intermediate states matter, and if the calculation is done over again with that point of view in mind then the right answer is obtained from the velocity gauge. This is a very direct demonstration that physical interpretations change with gauge.

Here is another example of a gauge-caused change in physical interpretation that is much more recent. The velocity gauge – used here since it is the generic gauge for plane-wave fields – agrees with the high-frequency approximation (HFA), an approximation being used by Mihai Gavrila [24]. It is based on an assumption of high frequency, so it is mostly of formal interest because all laboratory experiments with current lasers are at low frequency. However, it had some real significance. Using this HFA, Gavrila was able to show that there is such a thing as stabilization.

The HFA is the dashed line in Fig. 2.2. It shows a transition rate for ionization as a function of intensity [25]. As the intensity goes up, it might be expected that the rate increases also. That is not the case. There is a maximum, and if the intensity is increased beyond that point, then the rate starts to fall. This seems to be a fairly general phenomenon. It is not widely known, because until strong-field processes came along, people did not look at what happened at extremely high intensities. Stabilization is, in fact, more the rule than the exception. The same calculation as Gavrila was done with the SFA [26] instead of the HFA. There is no adjusting of scales in Fig. 2.2. The results are very close for eight orders of magnitude in intensity.

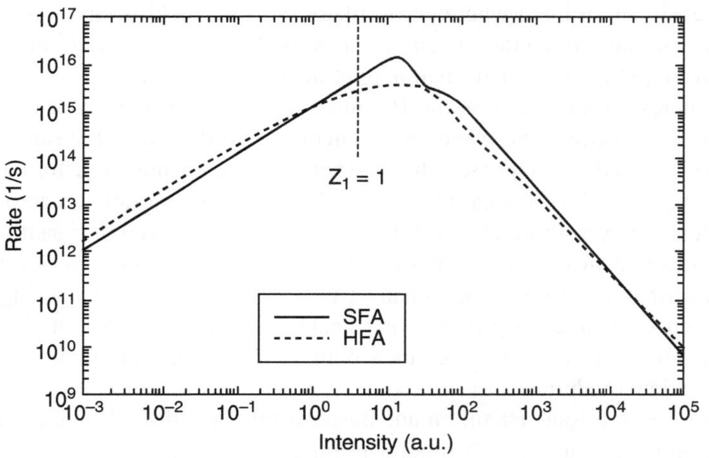

Fig. 2.2 Velocity-gauge SFA at high frequency. Ionization of ground-state hydrogen, $\omega = 2$ a.u. The comparison shown here is between stabilization predictions that follow from two different gauges providing different physical explanations for the same phenomenon. SFA from [26]; HFA from [25]

The calculations by these two different methods agree also in that they show that the maximum point is in much the same spot. The difference between these two calculations is that when Gavrila does the HFA, he removes the A^2 term (the square of the vector potential) by a gauge transformation.

In the dipole approximation this is purely a time-dependent function, so it can be removed by a simple gauge transformation. That is not done in the strong-field approximation, and the physical interpretation given for the occurrence of this peak is the fact that this marks the intensity at which ionization is no longer possible with a single photon. It is necessary to supply so much ponderomotive energy to the ionized electron that there is a threshold at which the single-photon process becomes energetically impossible, and more than one photon is needed. That makes a major difference in the transition rate and the rate declines after that.

In the Gavrila theory the A^2 term has been gauged-transformed away, the ponderomotive potential is gone, and the channel-closing physical interpretation makes no sense whatsoever. It is necessary to recognize that physical interpretations are gauge-dependent. To my knowledge, Gavrila never actually gave a qualitative explanation for stabilization in the HFA, but it has to be different from that in the SFA. It is necessary to think of some other physical reason why this maximum happens.

The point to be emphasized here is that the SFA and the HFA give essentially the same result because of gauge invariance, but physical interpretations are completely different.

It was previously noted that physical interpretations change even in classical electromagnetism. It is possible to find all kinds of problems where a simple gauge transformation is made and everything changes. An example worked out for self-knowledge is one that can be solved in closed form classically. It is a one-dimensional problem where an oscillatory electric field is superimposed on a harmonic oscillator potential. It can be solved with or without the A^2 term because there is a simple gauge transformation to remove that A^2 term.

The interesting thing is that the Hamiltonian for the problem looks totally different in the two cases, but when the problem is solved, exactly the same classical trajectory is found in each case. This is what a classical gauge transformation protects. In the case of a classical physical problem it is the particle trajectory that is unchanged, but everything else is different. This corresponds to a transition amplitude in quantum mechanics changing under gauge transformations. Even though gauge-invariant quantities do not change, the overall problem does. Gauge invariance is sort of a magical property that enables us to say: "Two totally different problems are being analyzed, but they both give the same prediction for certain gauge invariant quantities."

Figure 2.3 is a total Hamiltonian. Because this is a time-dependent problem, the Hamiltonian changes with time. The time axis in the graph covers one complete period. With the A^2 term present, the solid line can be called the laboratory gauge because the A^2 term really is there in the laboratory. If the A^2 term is gauged away, what happens to the Hamiltonian and the total energy is given by the dotted line; something completely different. It could be said if this were being observed classically, where it is possible to follow time-dependent changes, that these two

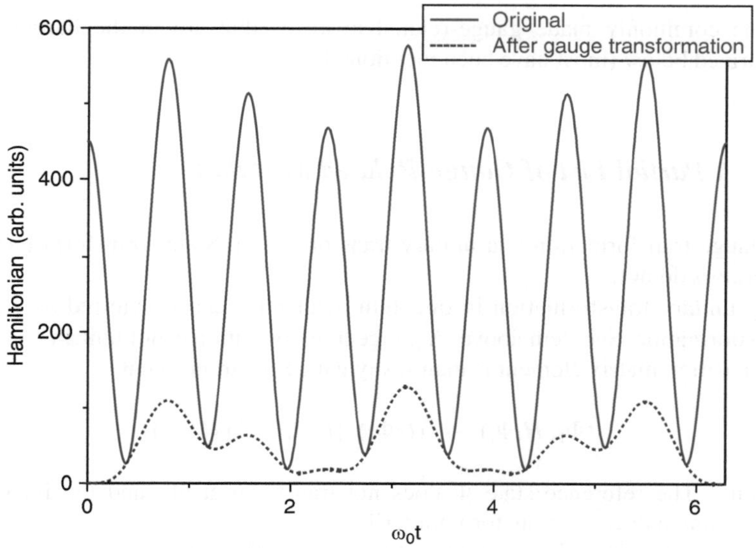

Fig. 2.3 This figure illustrates the large discrepancies that can occur after a gauge transformation when a quantity is examined that is not represented by a quantum-mechanical Hermitian operator. The potential energy, exhibited here, is not a laboratory observable, and so it can change in a major way after a gauge transformation is introduced

Gedanken experiments (thought experiments or conceptual experiments) could look entirely different. However, the particle follows the same trajectory in both cases because they are gauge-equivalent.

One way to think about this is that electric and magnetic fields identify the classical problems, but they do not define all aspects of those problems. Within any given classical problem identified by the electric and magnetic fields, there are many different electromagnetic potentials that can give rise to those same fields. These can be thought of as elements of a larger set. The large set is identified by the fields, and the elements within that set are all the possible potentials that will give rise to the same fields. This is saying basically that the potentials are more fundamental than the fields. That is a statement that has been in the literature again and again over the years and has been strangely ignored.

The question can be asked about what emerges from a laser: fields or potentials? The answer would have to be that it is both: the effect of the fields is dependent on the context, and the context requires knowledge of the potentials. That is, if the fields are truly fundamental, then all possible associated potentials will have to provide the same detailed results, and they do not. There are many different potentials that can give rise to the field, and it makes a difference in the classically measurable time-dependent electromagnetic energy. Therefore, the potentials are more fundamental. A gauge transformation may amount to a change to a coordinate system moving in a possibly complicated path with respect to the original system, and the potentials must be known to be able to understand the results.

Some commonly made gauge-related mistakes that are in the literature are summarized below (most have been mentioned above).

2.4.1 A Partial List of Gauge-Related Mistakes

1. A gauge transformation is a unitary transformation. States transform unitarily, operators do not.
2. Any unitary transformation in quantum mechanics can be viewed as a gauge transformation. See item above. A gauge transformation is not unitary.
3. A transition matrix element is manifestly gauge-invariant. That is:

$$(\Phi_\mathrm{f}, H_\mathrm{I}\Psi_\mathrm{i}) \to ((U\Phi_\mathrm{f}), \left(UH_\mathrm{I}U^{-1}\right)(U\Psi_\mathrm{i})).$$

   Wrong! The reference state Φ does not transform at all, and the interaction Hamiltonian does not transform unitarily.
4. The interaction Hamiltonian transforms unitarily. There is an elementary counter-example: $H_\mathrm{I} = \boldsymbol{r}\cdot\boldsymbol{E}$ transforms to

$$H_\mathrm{I}' = UH_\mathrm{I}U^{-1} = \boldsymbol{r}\cdot\boldsymbol{E} \text{ instead of to } H_\mathrm{I}' = \boldsymbol{A}\boldsymbol{p}/c + \boldsymbol{A}^2/2c^2.$$

5. Any purely time-dependent phase factor can be extracted from the transition matrix element without consequences. This is based on the classical theorem that any purely time-dependent function can be added to the Hamiltonian without changing the equation of motion. In QM:

$$\begin{aligned}(\Psi_\mathrm{f}, H_\mathrm{I}\Phi_\mathrm{i}) &= \left(\mathrm{e}^{\mathrm{i}f(t)}\Psi_\mathrm{f}', H_\mathrm{I}\Phi_\mathrm{i}\right) = \mathrm{e}^{-\mathrm{i}f(t)}\left(\Psi_\mathrm{f}', H_\mathrm{I}\Phi_\mathrm{i}\right) \Rightarrow \\ |(\Psi_\mathrm{f}, H_\mathrm{I}\Phi_\mathrm{i})|^2 &= \left|\left(\Psi_\mathrm{f}', H_\mathrm{I}\Phi_\mathrm{i}\right)\right|^2.\end{aligned} \tag{2.13}$$

   However, $\int \mathrm{d}t$ must be done *before* the absolute square.
6. The length gauge is a "special" gauge, or a "gauge-invariant" gauge, or even the "correct" gauge. This is incorrect for numerous reasons, among them that the length gauge approximates a propagating laser field (a transverse field) by treating it as a scalar field (longitudinal field). It cannot be a fundamental gauge.

In classical physics, there is a very general, very important theorem showing that if any function depending only on the time is added to the Hamiltonian or Lagrangian functions, then the equations of motion are unchanged. This theorem cannot be used directly in quantum mechanics, because measurements are not done directly. They are done by analyzing results in terms of laboratory states, and the laboratory reference states are not gauge-transformed.

For instance, a dipole-approximation Volkov solution, which has the A^2 term in the exponential, is a function of time only. The inner product that is involved in the transition matrix element represents integration over spatial coordinates. It then might appear that the time-dependent exponential can be extracted, and when an absolute square is done, it seems to go away. However, when a transition matrix element is formed, an integration over time has to be performed, and that integration has to be done before an absolute square is done. Therefore, the time-dependent exponential does not go away.

Does a laboratory gauge exist? If it is accepted that physical interpretations change when the gauge is changed, then it could be asked: "What physical interpretation is closest to what the laboratory instruments know about?" This is a very old question and has been discussed many times. In fact, the best discussions are to be found in classical electrodynamics books, not in quantum mechanics books. The situation is summarized in the following.

2.4.2 Does a Laboratory Gauge Exist?

It is clear that there are gauges that might have mathematical meaning, but do not have any obvious physical significance. We first express the gauge that does seem to have a clear meaning:
Basic Hamiltonian:

Canonical momentum: p
Kinetic momentum: $p - qA/c$
Kinetic energy: $(1/2m)(p - qA/c)^2$
Scalar potential (Coulombic): $V(r)$

If $A(r,t)$ is the vector potential of the laser field with no dipole approximation, then the radiation gauge (or Coulomb gauge, or velocity gauge) Hamiltonian is:

$$H = \frac{1}{2m}\left[\vec{p} - \frac{q}{c}\vec{A}\left(\vec{r},t\right)\right]^2 + V\left(\vec{r}\right). \tag{2.14}$$

The expression of (2.14), where the vector potential includes all transverse fields, like propagating laser fields or radio waves, carries an implication that is not often discussed. When the radiation gauge is used (also called the Coulomb gauge, or, in the dipole approximation, it is called the velocity gauge) it is the vector potential that represents all transverse fields. The word *transverse* has the meaning that such a field has an electric field in one direction, a magnetic field perpendicular to it, and then a direction of propagation perpendicular to both of those fields.

A propagating plane-wave field is a very special kind of field. Once established, it can propagate to infinity with no further input of energy. That cannot be described by a scalar potential because a transverse field has to exist in three dimensions and the scalar potential does not have any vector properties. That is what is meant when it is called a *scalar* field. On the other hand, static potentials like the Coulomb potential are longitudinal fields, and they are most appropriately described by a scalar potential; only a scalar function is needed, not a vector function.

When the transverse fields are described by a vector potential and longitudinal fields are described by a scalar potential, this is called the radiation gauge or the Coulomb gauge. Classical electrodynamics people view this as the most fundamental gauge available. Some reservations have to be expressed here, because when the problem is formulated relativistically, a difficulty appears that is not obvious in what has been written above with (2.14). The problem exists because of the fact that a static field implicitly has an infinite velocity of propagation, so it is difficult to accommodate this adequately in a relativistic theory. That has been the stumbling block, but there are ways to get around it; the well-known textbook by J.D. Jackson on classical electrodynamics discusses some of them.

So it appears that if there is a laboratory gauge, meaning that our laboratory instruments, never exposed to the fields, have to deduce what has caused the results that they measure, it is most consistent to describe the problem in the Coulomb gauge. The Coulomb gauge holds true when there is spatial dependence as well as time dependence. If the dipole approximation is being used, and the spatial dependence of propagating waves no longer appears, then this radiation gauge or Coulomb gauge is called the velocity gauge in AMO physics. This would appear to say that the velocity gauge gives us something closest to what would be in the everyday experience of our laboratory instruments. It is difficult to do anything more than that.

One more point needs to be made about the A^2-removed gauge because this is a gauge that is physically impossible, even though the gauge-invariance principle means that useful information can nevertheless be extracted. If the dipole approximation is used, and the kinetic momentum expression with the vector potential is a function of time only, squaring this yields an A^2 term that is purely a function of time. As was already mentioned, that means that there is an elementary gauge transformation that can remove this term. The gauge transformation that accomplishes this removal is generated by a function that depends on time only, and such a gauge transformation alters the scalar potential but leaves the vector potential unchanged. These points are summarized below:

There is a gauge transformation that removes the dipole-approximation $A^2(t)$ term by introducing a new potential energy that subtracts it from H:

$$H' = \frac{1}{2}\left[\vec{p} - \frac{q}{c}\vec{A}(t)\right]^2 - \frac{1}{2}\frac{q^2}{c^2}\vec{A}^2(t) + V(r). \tag{2.15}$$

This can be done *only* in the dipole approximation.

The subtracted A^2 term has no physical meaning.

What is the meaning of this subtractive term that now appears in the scalar potential in (2.15)? It is nonphysical. It is not there in the laboratory. It is just a result of the gauge transformation that yields some mathematical simplifications, but it is not real. Making physical interpretations after having subtracted that $A^2(t)$ term would only be misleading. This is exactly what happened in the stabilization problem mentioned earlier. In the HFA, the $A^2(t)$ term was gauged away, making it appear that there could be no channel closings. This was not done in the SFA, so channel closings were part of that theory.

In ordinary laser experiments, A^2 can be a very large number even when the dipole approximation is applicable. It would appear to be possible to remove it. Nevertheless, if a completely relativistic treatment is being done, A^2 is a function of $\mathbf{r}$ as well as t, so there is no simple transformation that removes A^2. It has to be retained, and everything should be done in the Coulomb gauge/radiation gauge. So working by extrapolation from a relativistic case to the nonrelativistic case is another way to see that it is not justifiable to remove the A^2 term and expect to get a physically plausible result.

That the laser field is transverse and that the Coulomb field is longitudinal are reliable interpretations, and they are violated in the length gauge. Something more needs to be said about this. The length gauge is an approximation that treats the laser field as if it were a slowly varying electric field. A constant electric field can be treated as a scalar field. If the field varies very slowly, then it is not so different from the constant field and so a scalar potential can be used. This is why the length gauge is well-liked. It is conceptually easier to work with just a one-component scalar potential.

Now let us consider again the subject of the laboratory gauge. The laboratory instruments do not know anything about gauges, so if you're going to put a physical interpretation on what happens in the field, it should be consistent with what the laboratory instruments know about. In a laser problem the Hamiltonian is explicitly time dependent. In classical electrodynamics or classical mechanics, it is known that if the Hamiltonian is time dependent, energy conservation is not a valid principle. It is also known that it is then not justifiable to interpret these different terms as kinetic and potential energies. However, the laboratory instruments respond according to a Hamiltonian that is not time dependent, so they can make meaningful energy measurements. This is very important, because in the laboratory it is reasonable to make energy measurements, even though in principle a time-dependent Hamiltonian exists where energy is not well defined. It is the correspondence between what can be measured and what measurements are being sought, that is really important to this matter of physical interpretations.

2.5 SFA (Strong-Field Approximation)

The strong-field approximation, customarily referred to as the SFA [2, 8, 27] will now be examined. The SFA is nonperturbative; it is based on a time-reversed transition amplitude where the interacting state is the final state, not the initial state; and it is so designed that when the laser field is dominant over the Coulomb potential, then it is a good approximation for calculation of the properties of strong-field ionization.

How is the importance of the laser field measured? It is measured by the ponderomotive energy of an electron in the field. How is the importance of the Coulomb field measured? It is measured by the binding energy of the electron in the Coulomb potential. So just that ratio of energies will be examined. The previously defined intensity parameter $z_1 = 2U_\mathrm{p}/E_\mathrm{B}$ is just such a ratio.

2.5.1 SFA Rates

The SFA transition amplitude in atomic units (a.u.) is written as:

$$M_{\mathrm{fi}}^{(\mathrm{exact})} = -\mathrm{i} \int_{-\infty}^{\infty} \mathrm{d}t \left(\Psi_{\mathrm{f}}, H_{\mathrm{I}} \Phi_{\mathrm{i}} \right), \tag{2.16}$$

$$M_{\mathrm{fi}}^{(\mathrm{SFA})} = -\mathrm{i} \int_{-\infty}^{\infty} \mathrm{d}t \left(\Psi_{f}^{\mathrm{Volkov}}, H_{\mathrm{I}} \Phi_{\mathrm{i}} \right). \tag{2.17}$$

In the velocity gauge (dipole-approximation Coulomb gauge), H_{I} and Ψ^{Volkov} are written as:

$$H_{\mathrm{I}} = \frac{1}{c} \vec{A} \cdot \left(-i \vec{\nabla} \right) + \frac{1}{c^2} \vec{A}^2, \tag{2.18}$$

$$\Psi^{\mathrm{Volkov}} = \frac{1}{V^{1/2}} \exp \left[i \vec{p} \cdot \vec{r} - i \frac{p^2}{2} t - \int_{-\infty}^{t} \mathrm{d}\tau H_{\mathrm{I}} \left(\vec{p}, \tau \right) \right]. \tag{2.19}$$

The exact transition amplitude that was presented before in (2.8) and (2.9) looks this way if the time-reversed form is used. The final state is the interacting state. It is in this final state where, in an atomic ionization process, the electron has become detached from the atom and finds itself in a really intense laser field. The ponderomotive energy can be huge, as in the example shown earlier. The ponderomotive energy may exceed the binding energy by many factors of 10. In that CO_2 laser experiment (Fig. 2.1), U_{p} was of the order of a hundred times as big as the binding energy. Then it can be said: "The Volkov solution will be applicable, because the Volkov solution is an exact solution of the quantum equation of motion for a free electron in this very strong plane-wave field." That means the SFA. The sole approximation is to neglect Coulomb corrections to the wave function of the ionized electron. The quantity being neglected is the remnant effect of the Coulomb field on an electron whose motion is dominated by the laser field. It is possible to write in closed form all the correction terms to this approximation. They are given in one of the appendices in the 1980 paper [2]. That is, the exact transition amplitude can be written in the usual form for a time-reversed transition amplitude. The SFA is the leading term when the exact interacting final state is expanded in powers of the Coulomb potential energy. They are hard to calculate, but they can still be written. Equation (2.19) gives just one way to express the Volkov solution.

When the SFA transition amplitude is written, there are certain things that immediately occur that are summarized as follows:

With a laser field inserted, many algebraic steps can be accomplished in general. An energy-conserving delta function occurs, that also identifies a minimum photon order n_0.

$$n_0 \geq \left(E_{\mathrm{B}} + U_{\mathrm{p}} \right) / \hbar\omega \tag{2.20}$$

$$p^2/2 = n\hbar\omega - E_{\mathrm{B}} - U_{\mathrm{p}}. \tag{2.21}$$

In words:

Kinetic energy of the photoelectron = energy contributed by n photons minus the energy required to supply the binding and potential energies.

The general velocity-gauge SFA transition matrix element can be written with certain identifiable components.

There always seems to be a multiphoton process. It does not matter what the Keldysh parameter is; everything is multiphoton. The photon number serves as the summation index in the sum over n, and the photon number has to have a minimum value, n_0. What is this minimum value? The product of n_0 and $\hbar\omega$ is the minimum energy supplied by the laser. The field, at the minimum, must be able to overcome the binding energy, and also to supply the ponderomotive energy to the ionized electron. Then, to find the kinetic energy of that ionized electron, it is necessary to know how many photons have been absorbed. The n index has to be at least n_0, but it can be much greater. This is the above-threshold ionization (ATI) phenomenon. Then it is necessary to subtract the binding energy and the ponderomotive energy. It is the remainder that is available for kinetic energy of the ionized electron.

There is a simple way to write the overall SFA quantum transition amplitude. First, it will be expressed in spherical coordinates. The differential transition rate as a function of solid angle, $\mathrm{d}W/\mathrm{d}\Omega$, can be written as:

$$\frac{\mathrm{d}W}{\mathrm{d}\Omega} = \frac{1}{(2\pi)^2} \sum_{n_0}^{\infty} p \left(\frac{p^2}{2} + E_\mathrm{B} \right)^2 \left| \hat{\varphi}_\mathrm{i} \left(\vec{p} \right) \right|^2 \left[J_n \left(\alpha_0 p \cos\theta, -\beta_0 c \right) \right]^2 \tag{2.22}$$

and each element in this form has a meaning. It is necessary to recall again the kinetic energy expression of (2.21), which defines what is meant by the momentum p. The differential transition rate can be written in other terms, but it is written most transparently with the momentum. The first multiplicative factor p in (2.22) comes from doing the phase–space integration.

When calculating the total transition amplitude for producing photoelectrons it is necessary to integrate over all the possible final states that can be present. That is where the p factor comes from. The next term is just the square of the net energy transferred to the ionized electron. It is necessary to transfer the kinetic energy and to overcome the binding energy. Then there is the momentum-space wave function of the initial state. There is a specific atom in the experiment, and it is the momentum state properties of that atom that enter into the final expression.

Finally, in (2.22) the generalized Bessel function has been written in a somewhat nonstandard way. In the first of the arguments in the generalized Bessel function, α_0 is the amplitude of motion of a free particle in an oscillatory electric field. This is a standard intensity-dependent quantity. This first argument also has the factor $p\cos\theta$. In this case, θ represents the angle from the electric field direction, so this is basically just the component of the momentum projected onto the electric field. That is, it is what is called the *parallel momentum*. The β_0 in the second argument of the generalized Bessel function will have to be defined.

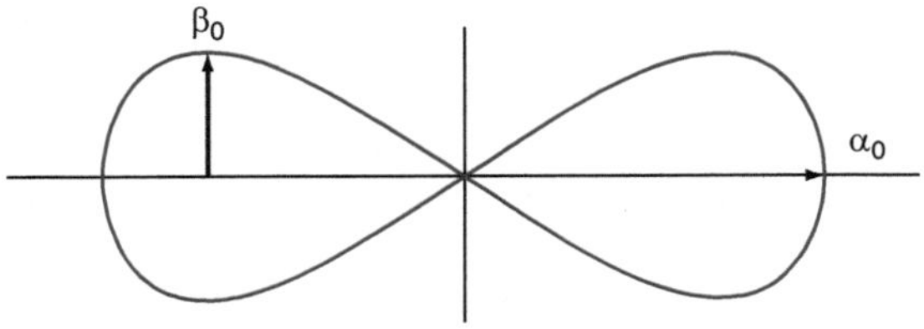

Fig. 2.4 Classical free electron in a plane-wave field. Electron follows a figure-8 trajectory resulting from the combined action of the electric and magnetic fields. α_0 = amplitude along electric field direction; β_0 = amplitude along direction of propagation. A free electron in a plane-wave field follows a figure-8 pattern when viewed in the simplest frame of reference. The amplitudes along the electric field direction and along the field propagation direction play basic roles in qualitative understanding of strong-field phenomena

In order to do that, it is necessary to introduce Fig. 2.4 with the famous figure-8 diagram from classical electromagnetic theory. If a free charged particle – not bound to anything – is put in a very strong, transverse, propagating electromagnetic field, and viewed in the simplest possible coordinate system, the result is a figure-of-8 motion. In that figure-of-8 motion, α_0 is the length of one of the lobes parallel to the electric field direction. That is a standard quantity, almost universally designated by α_0. At relatively low intensities, these loops collapse almost to a straight line, and it looks like a linear back-and-forth motion. The quantity β_0 is the amplitude of this figure-of-eight configuration in the direction of plane-wave propagation. The nonzero magnitude of β_0 indicates that motion is not straightforward back-and-forth.

$$\alpha_0 = \frac{c}{\omega}\sqrt{2z_\mathrm{f}}, \quad \beta_0 = \frac{c}{\omega}\frac{z_\mathrm{f}}{4}, \quad \frac{\beta_0}{\alpha_0} = \frac{1}{4}\sqrt{\frac{z_\mathrm{f}}{2}}, \quad z_\mathrm{f} \equiv \frac{2U_\mathrm{p}}{mc^2}.$$

There is a persistence of this plane-wave behavior in the generalized Bessel function even though the expression written here is in the dipole approximation. This is not present in any length-gauge theory.

2.5.2 SFA Spectra

The next topics are the properties of the photoelectron spectrum as predicted by the SFA, and then the momentum distributions. These can be examined in very general terms, and that is sufficient to explain many of the things that are observed in the laboratory at present.

The SFA works better at higher intensities and thus a good prediction of the spectrum can be expected when the field intensity is high.

A circular polarization experiment that was conducted by Bucksbaum et al. in 1986 [28] revealed a sharply defined ATI spectrum, which had the highest peak near the middle of the spectrum (Fig. 2.5). In the laser community at that time, the ATI phenomenon of well-defined peaks occurring in a spectrum at orders larger than

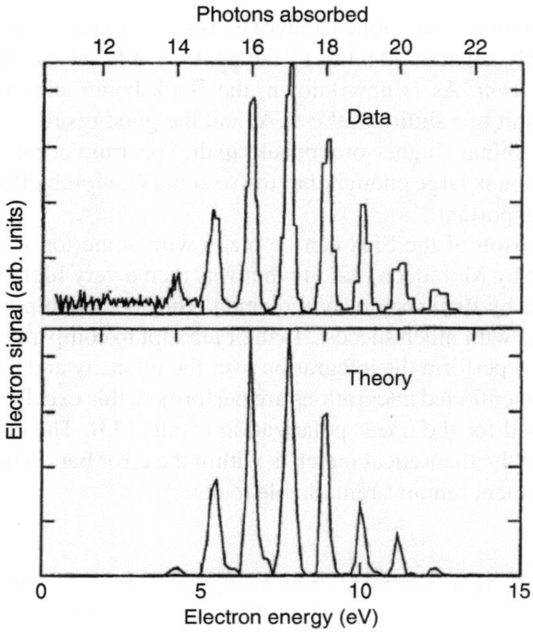

Fig. 2.5 The first successful quantitative explanation for an ATI spectrum is shown here in a figure reproduced from an article by Bucksbaum [30]. The experimental result in the upper panel is for the ionization of Xe by circularly polarized 1,064 nm light [28], and the lower panel shows the theoretical replication of the spectrum using what is now called the SFA. The reference to Keldysh is now known to be inappropriate, since that is a tunneling theory, and a tunneling mechanism is incapable of predicting individual spectral peaks

the minimum perturbative order was not well-understood, but it was thought to be probably just a higher-order perturbative correction. The experimental findings of Bucksbaum et al. contradicted that, since it is impossible in perturbation theory to have higher-order terms be dominant over lower orders.

The original SFA paper [2], though published in 1980, had not yet come to the notice of the community, and the experimental results created quite a stir, since they seemed to be inexplicable. It was shown in a 1987 paper [29] that the SFA gives a very good prediction of the spectrum. Bucksbaum later applied the SFA to his results using a more sophisticated computer than was available for [29], and incorporated the effects of the intensity distribution in the laser focus. The Bucksbaum calculation was never published in open journals. It was published [30] in the proceedings of a conference held in 1988, and it marked the first time that it could be said without ambiguity that nonperturbative laser-induced phenomena had been measured in the laboratory and replicated by theory. Figure 2.5 is a reproduction of the figure from the Bucksbaum paper in [30].

Clearly separated multiple peaks in the spectrum, at an energy spacing of $\hbar\omega$, were observed for the first time in 1979 by Agostini et al. [31] with linear polarization, and then (as mentioned above) by Bucksbaum et al. with circular polarization. The clearly separate identity of the peaks would later come to be called a *multiphoton spectrum*. As is now known, the Bucksbaum experiment was at the lower intensity limit of validity of the SFA, and the good results can be ascribed to the fact that the dominant higher-order peaks in the spectrum occur where the energy of the photoelectron is large enough that the residual Coulomb effects neglected by the SFA are not important.

Now a comparison of the SFA can be made with some ionization experiments with helium done by Mohideen [32] at what was then a very high intensity. He was part of a group led by Bucksbaum and Freeman. They had attempted to find a match to the SFA, but not with much success. In their attempt to compare theory and experiment, they did not perform the integration over the intensity and time distributions. When the above-mentioned integrations are performed, this excellent comparison of Fig. 2.6 is obtained for the linear polarization results [33]. The experimental error bars are small, but the theoretical match is within the error bars. This close match of theory and experiment remains remarkable to date.

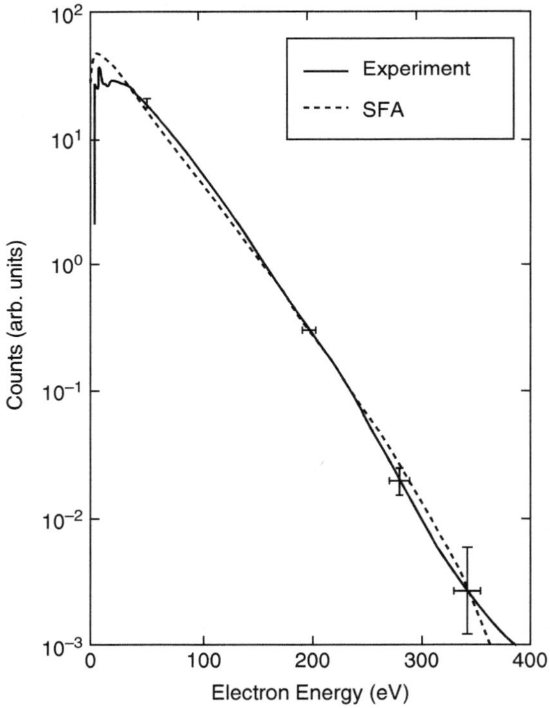

Fig. 2.6 A comparison of the SFA theory [33] and a 1993 experiment [32] for ionization of helium by linearly light shows agreement to within the experimental error bars. The 7×10^{15} W/cm^2 cited in [32] is reduced to 3.6×10^{15} W/cm^2 in the theory. The adjustment that had to be made in the peak intensity is typical for that era. Absolute measurements of intensity, while much better now, are still difficult to make

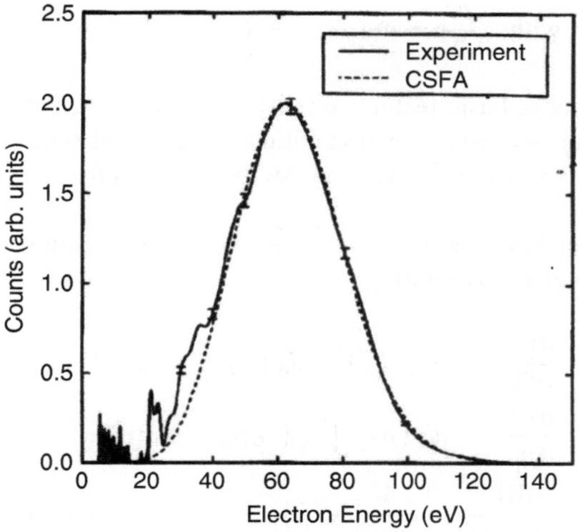

Fig. 2.7 Circular polarization spectrum for He^+ compared with [32]. Error bars are as given in [32]. The $6 \times 10^{15}\ \mathrm{W/cm^2}$ cited in [32] is reduced to $1.275 \times 10^{15}\ \mathrm{W/cm^2}$ in the theory. The match shown here between a measured circular polarization experiment [32] and the SFA [33] is nearly perfect, since the strange features at low energy are known to be artifacts, and the error bars at higher energies are very small. The seemingly large adjustment that had to be made in the reported peak intensity is strongly supported by the fact that, after the adjustment, the peak in the circular polarization spectrum occurs at about 80% of the maximum U_p, the ponderomotive energy. This is just about what would be expected, whereas the maximum would be at less than 20% of U_p for the reported peak intensity. Such a low proportion of U_p is not possible for strong fields

Mohideen also obtained [32] a circular polarization spectrum shown in Fig. 2.7. According to Mohideen, the strange features at low energy are experimental artifacts and should be ignored. However, in the remainder of the spectrum, the error bars are very small. The motivation for the experiment was to obtain precise spectral measurements, and the SFA-calculated spectrum [33] matches the experimental spectrum within very small error bars.

2.5.3 SFA Momentum Distributions

The SFA is represented in terms of cylindrical coordinates, suited for calculating momentum distributions as:

$$\frac{\mathrm{d}W}{\mathrm{d}p_{\parallel}} = \frac{1}{2\pi} \sum_{n=n_0}^{\infty} \left(\frac{p^2}{2} + E_\mathrm{B} \right)^2 \left| \hat{\varphi}_\mathrm{i}(p) \right|^2 \left[J_n \left(\alpha_0 p_{\parallel}, -\beta_0 c \right) \right]^2 , \tag{2.23}$$

$$\frac{\mathrm{d}W}{p_{\perp}\mathrm{d}p_{\perp}} = \frac{1}{2\pi} \sum_{n=n_0}^{\infty} \frac{1}{p_{\parallel}} \left(\frac{p^2}{2} + E_\mathrm{B} \right)^2 \left| \hat{\varphi}_\mathrm{i}(p) \right|^2 \left[J_n \left(\alpha_0 p_{\parallel}, -\beta_0 c \right) \right]^2 , \tag{2.24}$$

$$\text{with} \quad \frac{p^2}{2} = n\omega - E_\mathrm{B} - U_\mathrm{p} = \frac{1}{2}\left(p_{\|}^2 + p_{\perp}^2\right). \tag{2.25}$$

There are a number of basic features on display here: the square of the generalized Bessel function; the square of the momentum space, initial-state wave function; and the square of the energy transferred, consisting of the kinetic energy plus the binding energy.

Expressions for both parallel and perpendicular momentum distributions have momentum dependence as follows:

$$\frac{\mathrm{d}W}{\mathrm{d}\Omega}: \quad p\left|\hat{\varphi}_\mathrm{i}\left(\vec{p}\right)\right|^2\left[J_n\left(\alpha_0 p_{\|}, -\beta_0 c\right)\right]^2. \tag{2.26}$$

$$\frac{\mathrm{d}W}{\mathrm{d}p_{\|}}: \quad \left|\hat{\varphi}_\mathrm{i}\left(\vec{p}\right)\right|^2\left[J_n\left(\alpha_0 p_{\|}, -\beta_0 c\right)\right]^2. \tag{2.27}$$

$$\frac{\mathrm{d}W}{p_{\perp}\mathrm{d}p_{\perp}}: \quad \frac{1}{p_{\|}}\left|\hat{\varphi}_\mathrm{i}\left(\vec{p}\right)\right|^2\left[J_n\left(\alpha_0 p_{\|}, -\beta_0 c\right)\right]^2. \tag{2.28}$$

These distributions have many features in common with each other and with the energy spectrum. The differences arise from the kinetic momentum p. The energy spectrum is proportional to p. This means that spectra calculated with the SFA tend to approach very closely to zero at the low end of the spectrum. This result is in strong contradiction to that obtained from tunneling theory. In tunneling theory, the spectrum always has a maximum at zero energy.

For the parallel momentum distribution, the SFA predicts either zero or near zero in the forward direction. The reason why "near zero" is said is that the total energy conservation expression given here has to be satisfied, and an exactly zero result for the momentum is not normally obtained. This is in contrast with tunneling theory, which always predicts a maximum in the parallel momentum distribution at zero momentum, in contradiction to most of the experimental measurements.

In the total rate written in spherical coordinates, there is a factor p that originates from the phase space integration in spherical coordinates, and this factor always forces the low energy end of the spectrum to be at zero or near zero. The momentum-space distributions are expressed in cylindrical coordinates, and have different dimensionality than the spherical coordinates used up to this point. For example, the parallel-momentum distribution $\mathrm{d}W/\mathrm{d}p_{\|}$ lacks the factor p that is present in $\mathrm{d}W/\mathrm{d}\Omega$. The perpendicular-momentum distribution $\mathrm{d}W/p_{\perp}\mathrm{d}p_{\perp}$ has an extra factor of momentum in the denominator, which appears in the result as division by $p_{\|}$. This makes possible cusp-like behavior at small momenta. Since the parallel component of the momentum never becomes exactly zero (except at a single point in a two-dimensional space), the value of $1/p_{\|}$ does not diverge. This general dependence on p is directly attributable to the geometry of phase space.

Specifically momentum-dependent factors have been singled out in the energy spectrum, the parallel-momentum distribution and the perpendicular-momentum distribution. The role of the multiplicative factor of p in the energy spectrum has already been mentioned, as well as the division by $p_{\|}$ in the perpendicular momentum distribution. The small-p behavior in the parallel momentum distribution is determined by the momentum-space initial-state wave function, which appears in all three of the distributions, but is the primary governing factor controlling dependence on p in the parallel-momentum distribution. The especially interesting feature of the momentum-space wave function is that it can be quite different for different types of initial states.

For hydrogenic states, the momentum-space wave functions are presented in closed form as:

$$\varphi(p) \propto \frac{(np)^l}{\left[(np)^2 + 1\right]^{l+2}} \tag{2.29a}$$

$$\to (np)^l\,,\; np << 1 \tag{2.29b}$$

$$\to \frac{1}{(np)^{l+4}},\; np >> 1. \tag{2.29c}$$

The momentum-space wave function is proportional to $(np)^l$, with the quantity np appearing again in the denominator. Here, n is the principal quantum number for the hydrogenic state and l is the angular momentum quantum number; therefore, these are basic atomic parameters. It is not necessary to postulate rescattering to explain unusual features.

For low-lying states where np is small, then the momentum-space wave function behaves as np to the l-th power, so that it depends fundamentally on the angular momentum. If high-lying states are considered, then the momentum-space wave function behaves as $1/(np)^{l+4}$, which is a totally different behavior. Thus a variety of qualitatively different behaviors are expected in parallel-momentum distributions.

A useful example can be treated, with the principal quantum number maintained at 3. The number $n = 3$ is selected to be able to show *s*, *p*, and *d* angular momentum states. In Fig. 2.8 the *s* state is seen to have a peak at zero parallel momentum. However, local minima are observed in both the *p* and *d* states when the parallel momentum is zero. In the *d*-state case, the dip is very strong. Thus, the behaviors are totally different. The widths of the distributions are also quite different for the *s*, *p*, and *d* states. All of these qualities are consequences of the angular momentum of the initial state.

What has been shown is the result of performing the SFA calculation for a fixed intensity. It does not include the intensity distribution in the focus.

For a real atom as used in an experiment, a standard procedure has been to introduce analytical Hartree–Fock wave functions for the initial state. This means that there is a sum over several angular distribution factors. Any real atom other than hydrogen will be described by a complicated mixture of different hydrogenic states. The momentum-space wave functions of each of those Hartree–Fock components have to be added together to get the overall wave function. This is a fairly complicated procedure, but it should give an excellent representation of an actual atom.

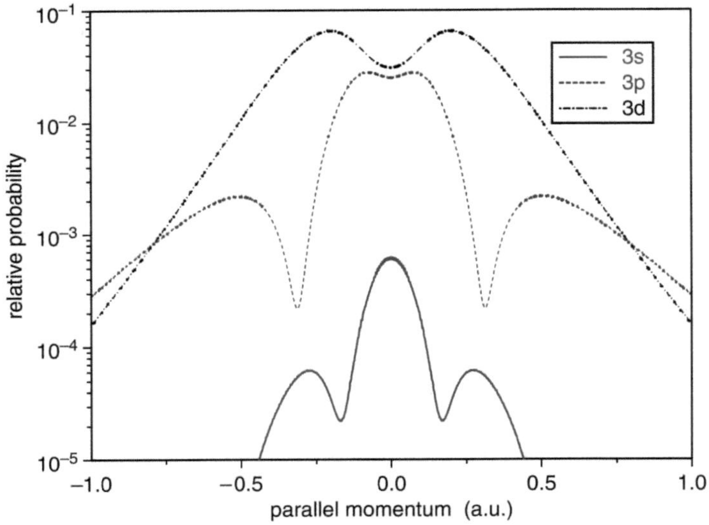

Fig. 2.8 A clear way to visualize the effect of the initial state of an atom on its momentum distribution properties can be seen in this example. A hydrogenic atom with a principal quantum number $n = 3$ can have s, p, d substates. They have significantly different momentum distributions even though the atom is the simple hydrogen atom. The light pulse characteristics are $\lambda = 10.6\,\mu\text{m}$, $I = 3.5 \times 10^{12}\,\text{W/cm}^2$, and Gaussian shape with 250 fs width

It was mentioned earlier that a cusp, that is, a sharp point has been observed at the origin in the perpendicular momentum distribution. This has led to hypotheses about Coulomb final-state corrections and rescattering processes. None of that is necessary or appropriate. The cusp arises because of the $p_{||}$ that occurs in the denominator in the expression for the perpendicular momentum distribution. Since the value of $p_{||}$ depends upon the discrete photon order index n, a value of $p_{||}$ that is exactly zero will occur only "on a set of measure zero"; in other words, it will never occur. Nevertheless, $p_{||}$ can be quite small, giving a cusp-like behavior.

The initial-state momentum distributions that have been discussed enter in a very interesting way into an evaluation of tunneling behavior. It has been mentioned that tunneling theory always gives a maximum in the parallel-momentum distribution in the forward direction. A tunneling limit can be found for the SFA [34], and it has been emphasized that the SFA permits a variety of behaviors in the parallel-momentum distribution. It is necessary to reconcile these apparently contradictory behaviors for the complete SFA and for its tunneling limit.

Figure 2.9 presents the parallel-momentum distribution in the ionization of helium by a laser field of peak intensity $3.5 \times 10^{15}\,\text{W/cm}^2$ in a Gaussian pulse of 25 fs duration ($\lambda = 800\,\text{nm}$). The solid curve is the complete SFA result, which appears slightly flat on top. The SFA tunneling limit is shown by the dashed curve. Tunneling behavior can be identified by an exponential factor of the form $\exp(-C/E)$, where C is a constant that depends on parameters of the atom, and E

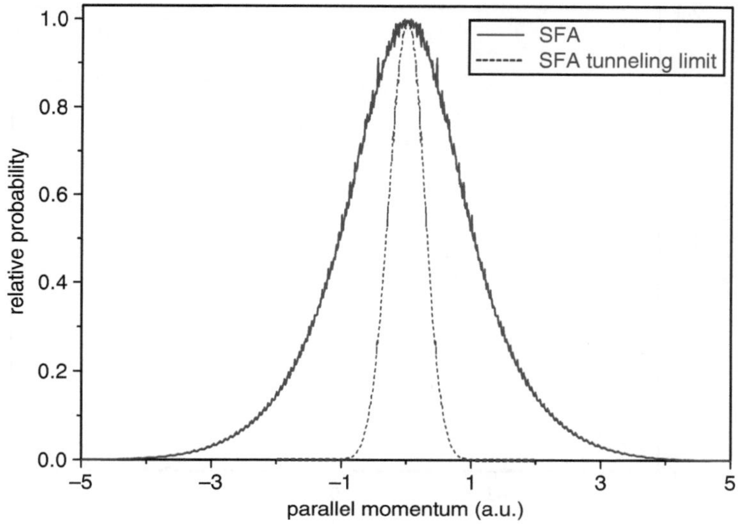

Fig. 2.9 An attempt can be made to find a "tunneling limit" of the SFA by confining attention to small momentum values. This is an artificial and fruitless goal, as shown here for helium. There are no amplitude adjustments in this figure, so the small-momentum part matches as expected, but higher components of the momentum distribution are importantly underestimated in this "limit"

is the electric field magnitude. This exponential can be extracted from the velocity gauge SFA. However, the tunneling part is only a fraction of the overall SFA result, and the remaining part is also important. The SFA and the tunneling limit match near the zero value of the parallel component of the momentum; they do not match further than this. In other words, the tunneling behavior is always narrow and the complete SFA behavior is much broader.

The parallel-momentum distribution for the ionization of 1s hydrogen can be calculated, as shown in Fig. 2.10 [35]. The ground state is an s state as in the preceding case, and the behavior is similar. The correspondence between the full SFA and the tunneling limit of the SFA is limited to that portion of the parallel-momentum distribution that is right at the center, near $p_{\|} = 0$.

Figure 2.11 shows totally different behavior. The parallel-momentum distribution shown is for ionization of the ground state of neon [35]. The neon ground state is a p state, and this is responsible for the large wings, with a very deep minimum between them. Again, the tunneling limit matches at zero parallel momentum. In this case, the tunneling limit constitutes only a small part of the overall ionization process. Therefore, in the SFA, the results obtained are strikingly different in a comparison between the velocity gauge and its tunneling limit.

If an ab initio tunneling theory is constructed, such as the Keldysh theory [4], PPT (Perelomov, Popov, and Terent'ev) [36], or ADK (Ammosov, Delone, and Krainov) [37], then the small result at zero value of the parallel momentum distribution is larger than for the SFA, but it will always be peaked there. It is the velocity-gauge

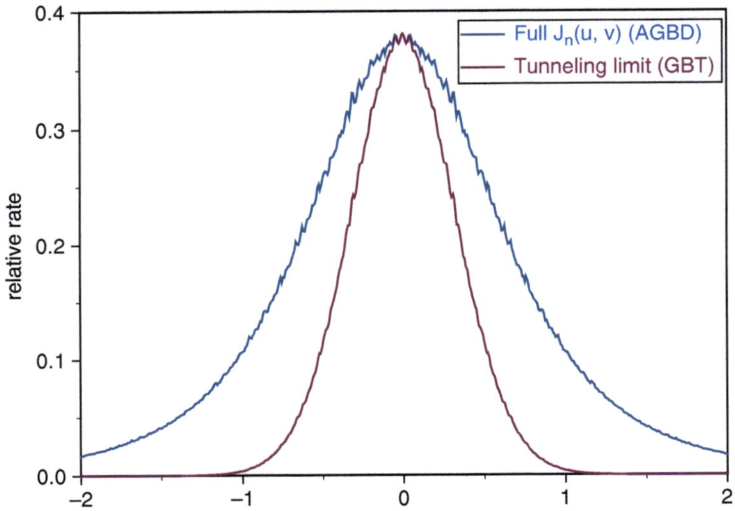

Fig. 2.10 Ionization of 1 s hydrogen (light pulse characteristics are $\omega = 0.057$ a.u., $I_{\text{peak}} = .13$ a.u., Gaussian pulse of 25 fs). The comparison between the "tunneling limit" and the full SFA for hydrogen is not much better than for helium in the preceding figure

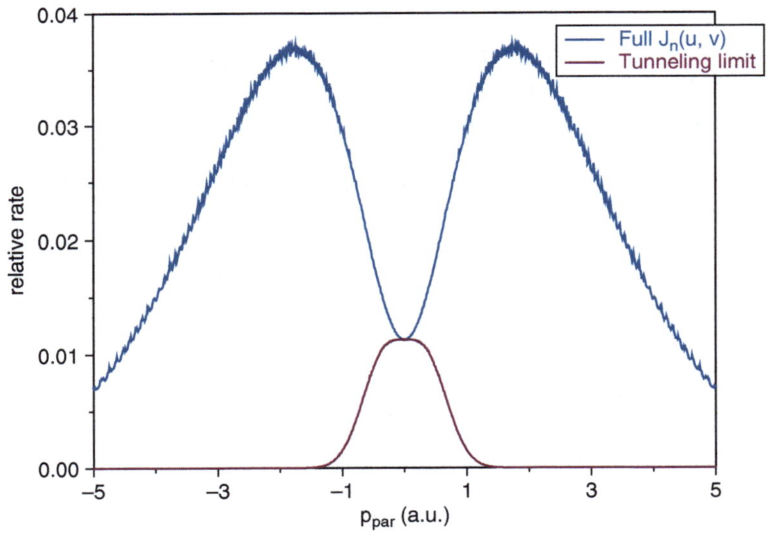

Fig. 2.11 Ionization of ground state of Ne (light pulse characteristics are $\omega = 0.057$ a.u., $I_{\text{peak}} = .216$ a.u., Gaussian pulse of 25 fs). When the atom has an initial p state, the putative "tunneling limit" is completely inadequate because the initial p state has such an important component of parallel momentum

SFA that predicts the great variety of shapes for the parallel-momentum distributions that are observed in experiments. The velocity gauge SFA provides more substantial information than the tunneling theory.

The SFA itself is gauge dependent. The length-gauge SFA and velocity-gauge SFA give different results. (The term *SFA* was originally intended [27] to apply only to the velocity gauge. However, this was evidently not stated with sufficient clarity, and other authors [38] began to employ the same procedure in the length gauge, and also calling it SFA.)

2.6 Tunneling/Multiphoton Misconception

Tunneling is exclusively a length-gauge concept. This is because ionization is represented on the basis of a sum of scalar potentials such that a Coulomb potential will be depressed on one side by a quasi-static electric field potential, resulting in a depressed barrier as shown in Fig. 2.12. The Coulomb potential is indicated by the heavy solid line. The electric field is given by the straight-line potential. The addition of the Coulomb potential and the electric field increases the Coulomb potential on one side and decreases it on the other. The ground state of atomic hydrogen has

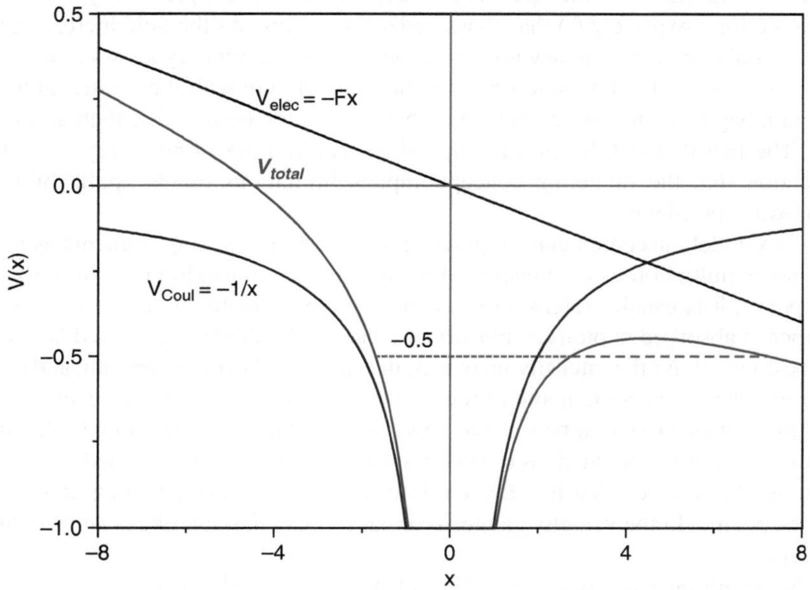

Fig. 2.12 In a tunneling approach, the laser field is represented by a scalar potential, so it can be added directly to the scalar Coulomb potential, creating the possibility of barrier penetration. A laser field, however, is a transverse propagating field that cannot be adequately represented as if it were a quasi-static field

a binding energy of -0.5 a.u. The diagram is drawn to scale for ground-state hydrogen. Thus, there is a possibility of tunneling through the barrier at the depressed side of the Coulomb potential. This is the concept of tunneling ionization.

This can be done specifically because the Coulomb and the laser fields are represented by scalar potentials. If the velocity gauge is used, the laser field is represented by a vector potential, and this diagram does not exist.

There are fundamental problems associated with the concept of tunneling. (Note: Subsequent to the delivery of these lectures, a paper was published [6] on the subject now about to be discussed. Very recently, another paper [39] gave explicit limits to applicability of tunneling methods.) The tunneling treatment of laser effects is limited to the dipole approximation. If the laser field is sufficiently strong, the dipole approximation will fail entirely as a function of intensity. This happens because the magnetic field, always present as a component of the laser field, begins to exert a significant force on the photoelectron as the intensity becomes high. This force is in a direction parallel to the propagation direction of the laser field. That is, it is perpendicular to the direction of the electric force. This will result in a displacement of the electron away from the simple back-and-forth linear oscillation associated with the electric field. When this magnetically induced displacement approaches an amplitude that is a significant fraction of an atomic unit, then the dipole approximation is no longer valid, and the tunneling diagram that has been discussed is no longer applicable.

Another limitation of the tunneling concept arises from the previously mentioned algebraic form $\exp(-C/E)$ that characterizes tunneling. As the field increases, this exponential approaches an asymptote. Almost every other theory shows a stabilization effect, where the ionization rate reaches a maximum with increasing intensity, and then begins to decline. Stabilization behavior is the norm rather than an exception. The fact that stabilization is impossible within a tunneling theory is another indication that the tunneling concept implies limitations on its applicability to plane-wave problems.

It is a widely accepted current practice to classify ionization problems as being in either a multiphoton or a tunneling domain. If a spectrum exhibits distinguishable ATI peaks, it is usually referred to as being in a multiphoton domain, because the number of absorbed photons responsible for each individual peak, spaced $\hbar\omega$ apart, can be counted. As the intensity increases, the spectrum becomes smooth, and peaks corresponding to specific photon orders cannot be distinguished. However, at very high intensities, large numbers of photons can contribute to a spectrum. In the SFA, photons are always counted even when a smooth distribution is obtained.

It should be noted that the division into multiphoton and tunneling domains is gauge-specific. In the velocity gauge, as already stressed, tunneling is an ill-defined concept.

The definition and properties of the well-known Keldysh parameter, γ, will now be presented.

2.6.1 Tunneling and the Keldysh Parameter

A common practice in recent years has been to describe ionization problems as being either in the *multiphoton regime* or in the *tunneling regime*, with the *Keldysh parameter* γ as the sole deciding criterion:

$$\gamma = \sqrt{E_B/2U_p}.$$

Physically, γ is the ratio of the time required to tunnel through the barrier as compared to the period of the laser field.

If $\gamma < 1$, the process is said to be by tunneling.

If $\gamma > 1$, the process is said to be multiphoton.

This entire scheme is contradicted by the following diagrams, for the case of ground-state hydrogen in a laser field of 800 nm.

In fact, this is the inverse square root of the z_1 intensity parameter, but the Keldysh parameter is often defined in terms of the *tunneling time*. In a tunneling theory, one can find a property identifiable as the time required for a particle to tunnel through a classically opaque barrier. The ratio of this time to the period of the laser is one way to define the Keldysh parameter. Tunneling takes place when the Keldysh parameter is small, and the spectrum lacks individual photon peaks. That domain of intensity where identifiable peaks occur is called the multiphoton domain, and the Keldysh parameter is large in such a region. This scheme is inconsistent, as will now be shown.

The example considered is ground-state hydrogen in a laser field of 800 nm. An electric field is selected of sufficient magnitude so that the top of the barrier exactly aligns with the ground state in the presence of this laser (Fig. 2.13). For hydrogen, this corresponds to a Keldysh parameter γ of 0.91, or roughly 1. This could be regarded as the dividing line between the tunneling and multiphoton regimes. If it is desired to enter further into the tunneling regime, then the field strength has to be increased.

If the field strength is increased, then it is observed that there is no tunneling at all, because if the initially bound electron can escape over the barrier, then that is the most probable pathway to ionization. Ionization by means of tunneling through the barrier is exponentially damped in comparison to ionization over the top of the barrier as Fig. 2.14 shows. The value of the Keldysh parameter γ in this case is 0.3. This is supposed to be in the tunneling domain; however, there is no tunneling.

If the Keldysh parameter γ is reduced further to 0.1, this is supposedly deeply into the tunneling domain, but Fig. 2.15 shows that ionization takes place far over the barrier. Therefore, the whole tunneling vs. multiphoton scheme is backwards.

As the Keldysh parameter decreases, ionization is nominally supposed to be more and more deeply into the tunneling regime. Instead, ionization is increasingly far above the barrier. If the Keldysh parameter becomes larger than 0.91, so that the problem enters into the nominal multiphoton domain, then the hydrogen ground state is below the top of the barrier, and the only way ionization can occur is by

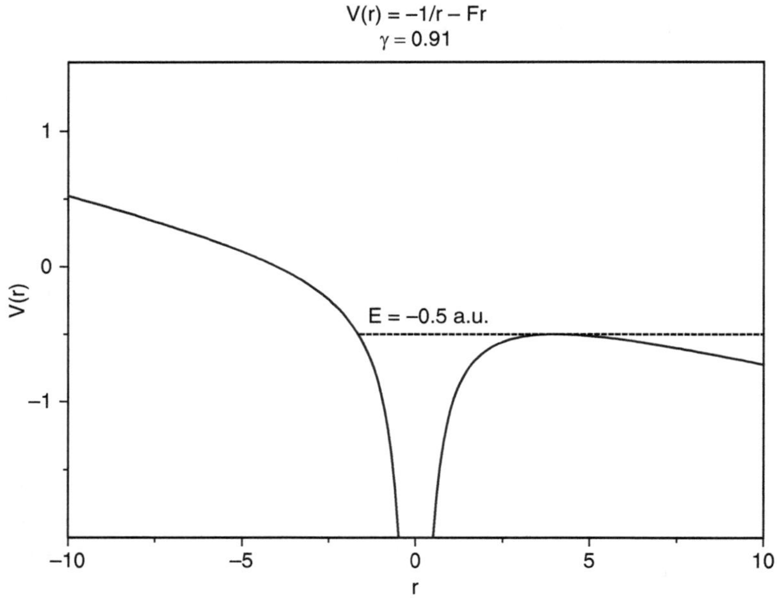

Fig. 2.13 This diagram, for hydrogen in a field of 800 nm wavelength, is given for the top-of-the-barrier case that nominally separates the tunneling and multiphoton regimes. The Keldysh parameter has the value $\gamma = 0.91$

tunneling through the barrier. This means that tunneling is required for ionization in the supposedly multiphoton regime.

The tunneling concept is therefore self-contradictory for atomic ionization. If ionization is treated using the SFA, The Keldysh parameter γ does not appear, but the z_1 parameter is applicable, which is one of the ratios of the ponderomotive energy to some other energy quantity. (Remember that the connection is $z_1 = 1/\gamma^2$.) In this case, it is the binding energy, and this parameter increases for a strong field and decreases for a weaker field.

In the velocity gauge, everything is stated in terms of sums over photon orders. This may sound like the Keldysh concept of a multiphoton domain, but both multipeaked and continuous spectra can arise from the velocity-gauge SFA. When the intensity is high, the sum over individual photon orders is still necessary, but the spatial and temporal intensity distributions in the laser focus cause individual photon contributions to overlap so extensively that the spectrum appears to be smooth and continuous. Hence, it is no longer possible to identify individual photon orders because of this overlapping, and a smooth result is obtained.

Therefore, it can be said that all strong-field processes are multiphoton. A multipeaked ATI spectrum for circular polarization was shown (Fig. 2.5), as well as a continuous-appearing spectrum (Fig. 2.7). Both those spectra arise from a

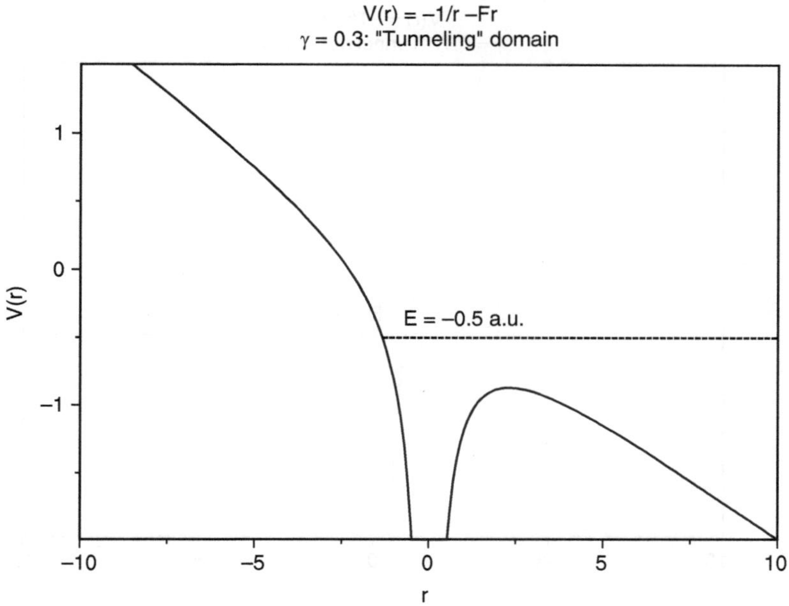

Fig. 2.14 The field is now much stronger than in the previous diagram. Here, $\gamma = 0.3(<< 0.91)$, so the problem should represent an example of tunneling. However, the diagram (drawn to scale) shows that ionization will occur far above the barrier. It cannot be described as tunneling ionization

calculation that is multiphoton, but account is taken of the intensity distribution in the laser focus and the time distribution in the laser pulse.

The spectrum of Fig. 2.5, from Bucksbaum [30], is in the so-called multiphoton regime. Within the velocity-gauge SFA, each of these peaks can be identified with a specific photon order. However, the photon order is always low. That is, it does not exceed 10 or 12.

In the spectrum of Fig. 2.7, from Mohideen [32], the individual multiphoton peaks cannot be seen; they merge together. In the velocity-gauge SFA, the concept of tunneling does not arise, since that is a physical interpretation specific to a length-gauge theory, where the scalar Coulomb potential has superimposed on it an approximation of the laser field as another scalar potential. As has now been seen, these length-gauge potential energy diagrams are backwards. When tunneling is supposed to occur, it cannot happen; and vice versa.

2.7 Time Domains and Rescattering

The concept of rescattering has been used quite a bit in recent years. It is a concept that is both useful and misunderstood. The term *rescattering* has been applied to both quantum events and to classical events, with no apparent attempt to distinguish

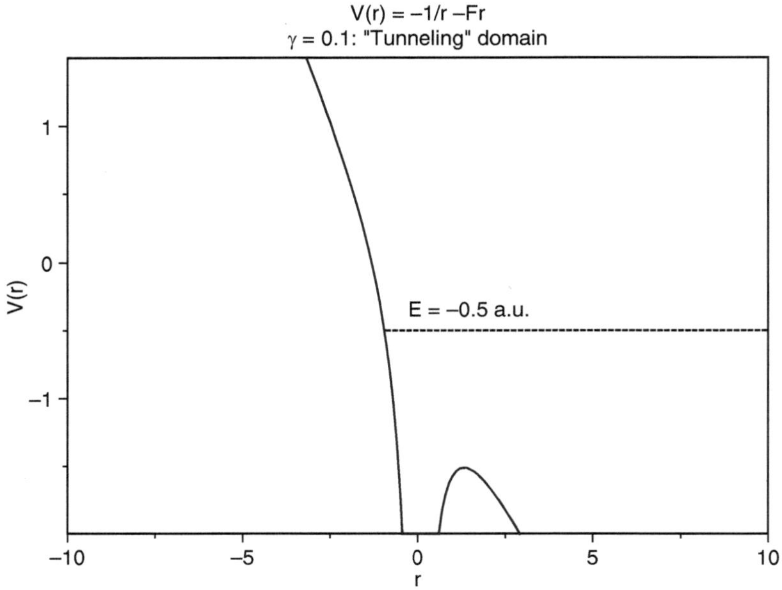

Fig. 2.15 For the extreme "tunneling" example of $\gamma = 0.1$, ionization is very far over the barrier, and certainly the ionization is not by a tunneling process. On the other hand, for $\gamma \gg 1$ the depression of the barrier is not great, the only way for ionization to occur is by barrier penetration, but $\gamma >> 1$ is in the putative multiphoton regime. The entire system contradicts itself

between the two. They are, in fact, quite different, with the criterion for the distinction coming from the very different time scales attached to the two types of rescattering. The atomic unit of time is 2.42×10^{-17} s. One way to derive this is to use the Heisenberg uncertainty principle in the energy–time form rather than the more usual momentum–distance form. When the uncertainty in energy is equal to the binding energy of an electron in the ground state of hydrogen, then the time interval during which it cannot be determined whether the electron is bound or not is the atomic unit of time.

As has just been shown, the atomic unit of time serves to identify the quantum time scale. Next, it is necessary to set the classical time scale. The frequency or wavelength of a laser can be determined by classical optics. The laser operates therefore on a classical time scale. If 800 nm is considered to be a typical laser wavelength, the period of the wave will be of the order of 100 times larger than the quantum time scale.

This means that one can consider an electron leaving the atom by being pushed away by the laser field, and then returning to the atom because the laser field has changed direction and is pushing the electron back in the other direction. The subsequent interaction with the atom is known as classical rescattering. The quantum process of ionization occurs on a much shorter time scale. Since the process of ionization must precede the classical rescattering, all the conservation conditions for the quantum process must be satisfied before the classical process takes place.

It can strongly be questioned whether the term rescattering can properly be used for quantum processes. In the SFA, the final state is approximated by a Volkov solution. A Coulomb correction to this can be applied, and has proven to be useful. There is a plateau in the spectrum from ionization by linearly polarized light that occurs with low probability, and that appears at energies beyond $2U_{\mathrm{p}}$, where the simple SFA falls off very rapidly. This has been explained by including the Coulomb correction to the Volkov solution. This correction has been called rescattering, even though it is a quantum process that is completed long before any classical oscillation can occur. In a perturbation expansion of the complete $\Psi_{\mathrm{f}}(t)$ wave function in powers of the Coulomb interaction, the Volkov solution is the leading term and the first correction term supplies the explanation for the plateau in the spectrum. In a diagrammatic perturbation theory, one can think of correction terms as further interactions with the Coulomb potential, but these do not occur in real time. They are virtual events, all of them connected with the single real time t that occurs in $\Psi_{\mathrm{f}}(t)$. They are unrelated to classical rescattering, which occurs in classical laboratory time. Nevertheless, this Coulomb correction has been called rescattering with no cautions that it is completely unrelated to true rescattering.

Another problem also occurs when there is no account taken of the major time differences between quantum and classical events. In the 1989 Corkum circular polarization experiment [16], approximately 9,000 photons had to be absorbed in order to supply the ponderomotive energy to the ionized electron. In circularly polarized light, the angular momenta of all the photons are aligned. Each photon possesses one quantum unit of angular momentum, independently of the energy of the photon. This means that in this low-frequency ionization process, at least 9,000 quantum units of angular momentum are introduced into the system.

In particular, for the CO_2 laser experiment, the disparity between the classical time and the quantum time is by a factor of more than 10^3 rather than the 10^2 given earlier as a general estimate. Hence, the quantum conservation conditions have to be satisfied before the classical conditions are considered. This means that when a photoelectron comes out of the atom, the electron must have 9,000 units of angular momentum, just corresponding to the most probable kinetic energy for an electron ionized in a circularly polarized field. The energy that must be supplied by the laser to overcome the binding energy in the atom adds even more angular momentum to the total, since every photon absorbed, whether to supply the binding energy or the kinetic energy, contributes one quantum unit of angular momentum. In the CO_2 experiment being used as an example, the total angular momentum required to supply the ionization energy, the ponderomotive energy, and the most probable kinetic energy, is nearly 20,000 quantum units. This is being completely overlooked, leading to incorrect descriptions of what occurs in the subsequent classical motion.

This oversight about the initial angular momentum of the photoelectron creates a huge difference. If the electron comes out with zero angular momentum based on the tunneling concept, it starts with zero angular momentum and not $20{,}000\hbar$. Therefore, the occurrences during the classical phase are completely different.

Nevertheless, there is one qualitative aspect of the two scenarios that is not different, and that has deflected attention away from those aspects that are totally unlike. Both scenarios concur that the electron will not return to the atom after the ionization process.

According to the conventional assumption (Corkum's "simpleman's theory" [40]), an electron emerges with zero energy and zero momentum at the atom. It then responds to the external field and spirals away, never to return. On the other hand, if the electron has to come out with 20,000 units of angular momentum, angular momentum conservation is a strict principle in physics, and the electron must appear outside the atom at the end of the quantum process in a circular orbit around the atom. Considering the parameters of the classical orbit, a classical orbit centered on the atom with the most probable energy from quantum energy conservation, has exactly the right amount of angular momentum to correspond to the nontunneling quantum prescription.

Therefore, the electron does not emerge from the atom with zero velocity and move away with the classical initial conditions of a stationary particle. It comes out with the appropriate angular momentum required by the quantum ionization process. This means that the electron is in an orbit that is cylindrically symmetrical about the direction of propagation of the laser field and it cannot return as long as the field exists. Exactly this kind of motion has been observed recently in experiments by Bergues et al. [41], where photoelectron momentum distribution measurements following from the photodetachment of the negative fluorine ion by circularly polarized light led to a toroidal photoelectron momentum distribution centered on the fluorine atom.

The Corkum model has the electron emerge without regard to quantum conservation conditions, and it acquires its final trajectory parameters as a consequence of classical interactions. This cannot be the case, because of the great difference in quantum and classical time scales. The quantum and classical processes are disjoint, each satisfying its own rules in a quantum-then-classical sequence.

The use of the dipole approximation in AMO physics is important because it is very useful, and normally very well-satisfied. The dipole approximation is such a familiar feature of AMO physics that most people do not pay much attention to it. In atomic physics, the dipole approximation is justified on the basis of the wavelength of laser radiation being considerably larger than the size of an atom. For example, if the wavelength is longer than, say, 10 a.u., the dipole approximation would be justifiable. A field with a wavelength that is ten times larger than the atomic size is actually a very energetic field. It would be completely off the frequency-intensity diagrams shown above, or at the high-frequency end of those diagrams.

What the above justification for the use of the dipole approximation overlooks is that a laser field is a propagating plane-wave field, meaning that there is a magnetic component of the field that is just as large (in Gaussian units) as the electric field. The magnetic field can usually be neglected because the force it exerts on a charged particle (i.e., the electron) is smaller than the force from the electric field by a factor v/c, where v is the velocity of the electron and c is the velocity of light. This velocity ratio is usually so small in AMO physics that neglecting magnetic-field effects is a safe procedure that is so universal that it need not even be mentioned. However, an intense laser field can induce electron velocities that are no longer negligible as compared to the velocity of light.

The effect of the magnetic field on an electron is to change a simple linear oscillation of an electron into a "figure-8" motion, with the long axis of the 8 in the direction of the electric field and the shorter width of the 8 in the direction of propagation of the plane-wave field; that is, in a direction perpendicular to both the electric and magnetic fields (Fig. 2.4). The figure-8 motion is always present in a plane-wave field, but at perturbative intensities the departure from straight-line oscillation is negligible.

As the field intensity increases, the dipole approximation fails because of the figure-8 motion. Once the short axis of the figure-8 becomes large enough that it is a significant fraction of the size of the atom, then the overlap of the electron's wave function with the Coulomb field of the atom becomes significantly altered from that predicted by the dipole approximation, and higher multipole moments must be considered.

The figure-8 shown in Fig. 2.4 is drawn to scale for a relatively high intensity to make its shape clearly visible. Experiments have been carried out in this nondipole domain, but none of these experiments were designed to detect nondipole behavior. For instance, the early experiments [15, 16] with the CO_2 laser were in a frequency-intensity region where the dipole approximation will fail. It was deep within the region labeled "nondipole nonrelativistic" in the frequency-intensity diagram of Fig. 2.1.

The thickness β_0 of the figure-8 motion must be small as compared to the size of the atom, or else this motion in the direction of propagation of the laser field will influence the interaction of the field with the atom. That influence, a direct result of the nonzero magnitude of the magnetic field, will then make the dipole approximation invalid. The absence of dipole-approximation concepts then removes any equivalence between the length gauge and the velocity gauge. The laser field is transverse, and requires a vector potential for its description, making the length gauge meaningless. This figure shows the boundary as occurring between the regions labeled *dipole* and *nondipole* nonrelativistic regions. The true relativistic domain can still be quite distant when tunneling concepts fail completely.

The magnetic field does not matter when v/c is small, because the magnetic term in the Lorentz force equation is negligible for small v/c. As intensity increases to the point where the magnetic field can no longer be neglected, the consequence of this (i.e., nondipole behavior) is called a "v/c effect". A true relativistic effect originates from the relativistic gamma factor (unrelated to the Keldysh γ) that measures relativistic length contraction or relativistic time dilation. This factor is a function of v^2/c^2. An expansion of the relativistic gamma in terms of the electron velocity is thus an expansion in powers of $(v/c)^2$ rather than v/c. True relativistic effects are therefore sometimes referred to as "$(v/c)^2$ effects". Higher intensities are required before $(v/c)^2$ effects become important, and the condition for their onset has a different frequency-intensity dependence than v/c effects.

This is the reason why, in the frequency-intensity diagram of Fig. 2.1, the slope of the line marking the lower limit for nondipole behavior is different from that marking the onset of nonrelativistic behavior.

The line on the frequency-intensity diagram showing the onset of relativistic effects demarcates the region where the ponderomotive energy of the electron in the field, U_p, is one-tenth of the rest mass of the electron: $U_p = mc^2/10$. Since $U_p = I/(2\omega)^2$, the relativistic domain can be entered at relatively low frequencies. From the diagram, it can be seen that it is difficult to achieve relativistic conditions at high frequencies. The new generation of high-frequency FELs (free-electron lasers) will make available a new domain of high-photon-energy laser sources to explore the poorly-known extreme ultraviolet (XUV) and soft X-ray region, but they may produce only limited insight into explicitly strong-field processes. No projected XUV machine can reach relativistic intensity conditions.

To give a specific example, a very intense FEL was operated recently by the RIKEN Institute in Japan in the 60-nm region (photon energy about 20–25 eV). The light beam intensity at the focal point could be as high as 10^{16} or $10^{17}\,\mathrm{V/m^2}$. This value is very high for such a high frequency. This facility is expected to be in the soft X-ray region in the very near future. With regard to the fundamental understanding of the phenomena occurring in such an intense light field in a very short-wavelength region, the above-mentioned experiment is novel. This is because such a photon energy had never previously been available. The situation here is entirely different from the more familiar wavelength region like that of the Ti:sapphire laser, where the photon energy is 1.55 eV. The parameter that measures the applicability of perturbation theory ($z = U_p/\hbar\omega$) has a value of about 0.6. Therefore the RIKEN experimental environment might be nonperturbative. With a combination of such high intensity at a very high frequency, new phenomena might indeed emerge.

2.8 Relativistic Effects

A qualitative discussion of explicitly relativistic phenomena will now be presented. A full treatment of this subject would be a major enterprise that cannot be undertaken in the present context. A separate strong-field community has sprung up that specializes in the relativistic environment. There is one aspect of this work that seems surprising. Relativistic effects are being treated almost entirely by numerical methods. It is surprising because the Dirac equation, the governing quantum-mechanical equation of motion, is immensely demanding of computer capabilities. Almost all work done so far has had to be restricted to one spatial dimension, supplemented by a limited amount of two-dimensional work. The relativistic problem, however, is fundamentally four-dimensional, including all three spatial dimensions. This is because the electric and the magnetic fields are equally important. These fields are perpendicular to each other, and both are perpendicular to the direction of propagation of the laser beam. These properties define a problem that has, irreducibly, three spatial dimensions.

Much effort is also being devoted to second-quantized treatments of relativistic laser problems. The Dirac equation does not require second quantization in an intense electromagnetic field. Second quantization means that the electromagnetic field is treated as a particle field, consisting of a countable number of individual photons, rather than just treating the electromagnetic field as a classically continuous field. It was shown within my strong-field group many years ago that the very large photon numbers in an intense field make the classical-field description equivalent to the second-quantized theory [42, 43]. The additional complication of second quantization is unnecessary. When the Dirac equation is considered by three-spatial-dimensional analytical approximations, instead of by numerical computations in one spatial dimension, results can be obtained that are impossible to achieve in one dimension, and can therefore give far more fundamental information.

There is another curious concept that has taken hold in the strong-field community that leads to a great deal of wasted effort. There is an apparent problem that arises from the fact that both the Dirac equation for spin-$1/2$ particles (like electrons) or the Klein–Gordon equation for spinless particles (bosons) appear to predict negative energy states for the free particle in addition to positive energy states. That is, a free electron can have any energy greater than or equal to its rest energy mc^2, or it can have any energy less than or equal to $-mc^2$. Early in the history of relativistic quantum mechanics, it was thought that the normal state of affairs was one in which all the negative energy states were filled, so that the Pauli principle would forbid any positive energy electron from spontaneously making a transition to a negative energy state. This automatically would imply that any theory must be a many-body theory demanding second quantization of the electron field analogous to the second-quantization of the photon field that was mentioned earlier. This line of reasoning makes no sense because the Klein–Gordon equation also has negative energy states, it describes real particles (like pions), but there is no Pauli principle to prevent spontaneous decay into negative energy states.

The resolution of this problem was given in 1934 [44]. Simply view negative energy states of the particle as positive energy states of the antiparticle. That is, there is no such thing as a negative energy state of a free particle. This has been standard information in the high-energy community for more than 70 years; it is an integral part of the Feynman diagram approach that dates to the 1950s; and yet the strong-field community seems to be unaware of it. There is no need to fill negative energy states, and there is nothing in strong-field physics to force the use of second quantization.

Further evidence comes from the fact that there is a set of Feynman rules derivable from relativistic quantum mechanics [45]. It is identical to the Feynman rules that can be derived from quantum field theory (i.e., the second-quantized theory) [46]. Hence, relativistic perturbation theory, the source of the Feynman diagram technique, is identical to all orders between relativistic quantum mechanics and relativistic quantum field theory. The results obtained are independent of whether second-quantization is used or not. Beyond the radius of convergence of perturbation theory, the work that was done by my group in the 1960s proved the same equivalence for nonperturbatively strong fields. It is inexplicable that so much effort is devoted to second-quantized theories for strong fields.

The avoidance of the simplifying features of relativistic quantum mechanics instead of relativistic quantum field theory, and the insistence on profoundly limited numerical methods in place of analytical approximations, constitute needless complications that impede progress in problems with fields strong enough to produce relativistic effects.

This presentation will conclude with a sampling of effects that arise with relativistic conditions due to strong fields. Four such effects are introduced below. All results come from the SFA, which exists for the Schrödinger, Klein–Gordon, and Dirac problems. The SFA is very useful, since its validity conditions are easily satisfied in the relativistic domain, and the relativistic results reduce identically to the nonrelativistic limit when $z_f \ll 1$. (Recall that $z_f = 2U_p/mc^2$.)

Figure 2.16 compares spectra for photoelectrons as predicted by a fully relativistic SFA Dirac theory [47–50] as compared to the same problem as it would be calculated (inappropriately) for the same conditions using the nonrelativistic SFA. Linear polarization is treated first. The dashed curve is the nonrelativistic spectrum. The atom treated is hydrogen, and the frequency is arbitrarily selected as 1/8 a.u. As mentioned previously, the spectrum peaks at a very low energy. So low, in fact, that the drop-off to zero is not visible on the scale of this graph, which extends out

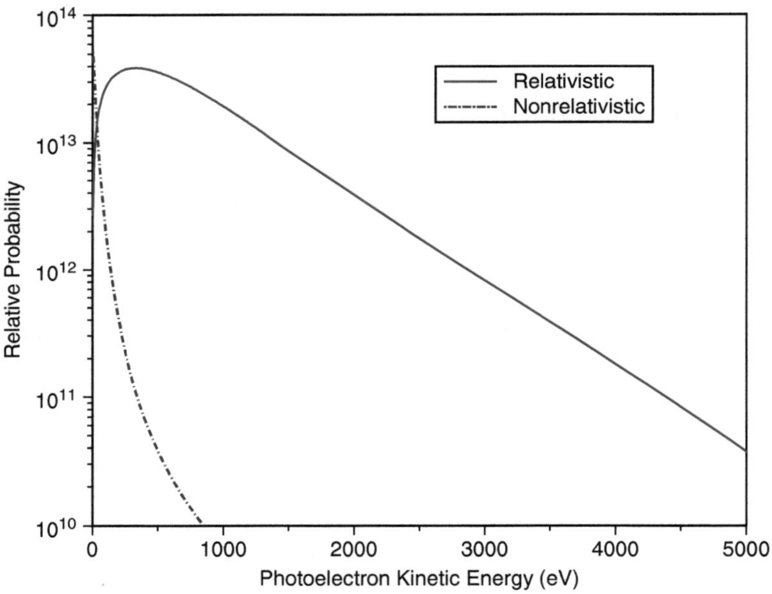

Fig. 2.16 Ionization of a ground-state hydrogen atom with linearly polarized light ($\omega = 1/8$ a.u., $I = 10^{2.5}$ a.u. $= 1.1 \times 10^{19}$ W/cm^2). The energy spectrum of photoelectrons ionized by linearly polarized light under relativistic conditions has a very different appearance in the few-keV region than does a nonrelativistic spectrum

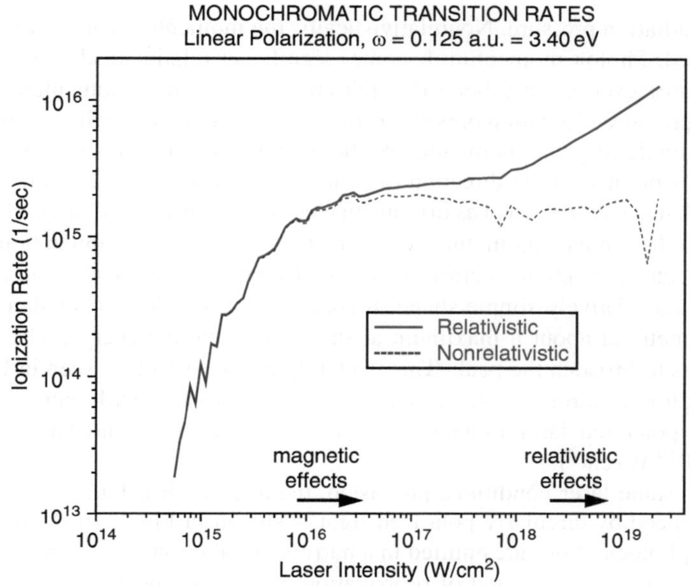

Fig. 2.17 Total rates for photoionization by linearly polarized light show a weak stabilization effect. The relativistic total rate departs from that behavior slightly after magnetic effects become significant, and strongly after fully relativistic conditions obtain

to several keV. The maximum occurs at a few eV. The relativistic spectrum, given by the solid line, starts off being less energetic than the nonrelativistic spectrum, but then continues to rise until about 300 eV. It remains above the nonrelativistic prediction even if the spectrum is extended out to hundreds of keV. At larger energies, both spectra move down further. Eventually, these two plots become parallel with a large vertical displacement between them.

Figure 2.17 illustrates what happens to the stabilization effect when relativistic conditions exist [49]. Stabilization was mentioned earlier. It refers to the fact that many systems exhibit a maximum in transition probability as a function of intensity. Beyond a certain *stabilization intensity*, further increases in field intensity cause a decline in the transition probability. That effect is shown, at least weakly, by the dotted line in the graph, which is a plot of ionization rate as a function of intensity for ground-state hydrogen in a field of frequency 1/8 a.u. This is a "raw" rate, showing neither the smoothing effects arising from averaging over the intensity distribution within the laser focus, nor the effects of a continuous time-dependent change in field intensity. The solid line shows what happens when relativistic effects are included.

Up to about $10^{16}\,\mathrm{W/cm^2}$ in laser intensity, the relativistic and nonrelativistic results are almost identical. Magnetic effects then begin to cause a difference. By about $2 \times 10^{18}\,\mathrm{W/cm^2}$, true relativistic effects appear. These two domains correspond to the v/c and v^2/c^2 regions described earlier. The reason that stabilization does not occur at all in the relativistic calculations can be explained in terms of radiation pressure. Nonrelativistically, the momentum carried by a photon is neglected. Photon momentum becomes significant relativistically, and the subsequent force exerted on a body through absorption of this momentum is called radiation pressure. Radiation pressure is the same as saying that when there is sufficient photon density, the momentum of the photons matters. This momentum adds up and it is pointed in the direction of propagation of the field. It is this radiation pressure that can be regarded as driving an electron that is ionized or nearly ionized away from the remnant atom, thus removing the possibility of stabilization.

The spectrum of photoelectrons produced by intense, circularly polarized light always has a relatively simple shape. It somewhat resembles a Gaussian distribution, symmetrical about a maximum at the ponderomotive energy. The effect of relativity is to broaden the peak. This is plainly a large effect, as seen in Fig. 2.18, showing photoelectron spectra calculated for ground-state hydrogen exposed to circularly polarized laser radiation with $\omega = 1/8$ a.u. at an intensity of 10 a.u. ($3.51 \times 10^{19}\,\mathrm{W/cm^2}$).

For the same laser conditions just listed, the angular distribution of photoelectrons produced by circularly polarized light is shown in Fig. 2.19. Nonrelativistically, the photoelectrons are emitted in a narrow distribution very close to the plane perpendicular to the direction of propagation of the laser beam, and axially symmetric about that propagation direction as the axis. Relativistic effects broaden the distribution somewhat but, most importantly, the entire angular distribution is tilted

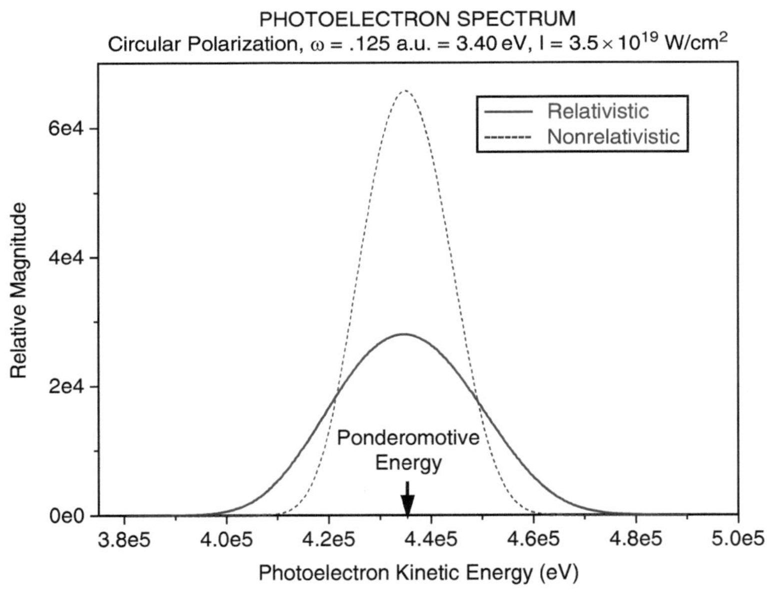

Fig. 2.18 In circularly polarized light, both relativistic and nonrelativistic single-intensity spectra peak at the value of the ponderomotive energy, but the relativistic spectrum is much broader

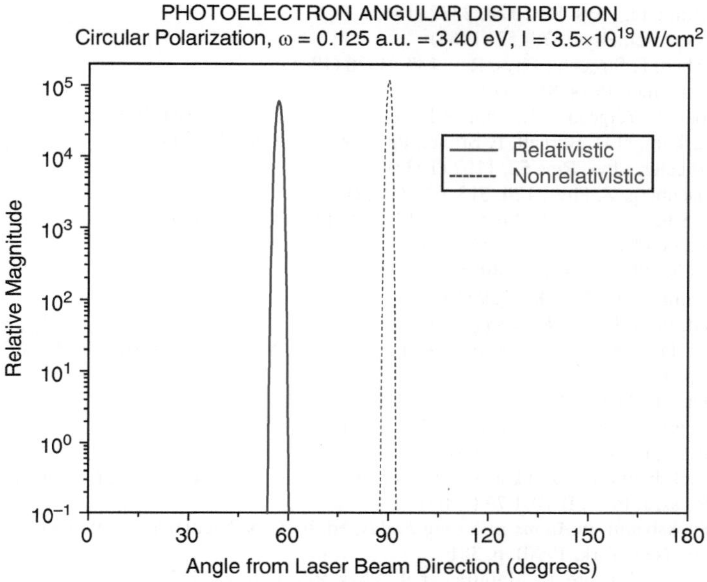

Fig. 2.19 Photoelectron angular distributions resulting from ionization by circularly polarized light show a very narrow distribution for both the relativistic and nonrelativistic cases. Nonrelativistically, this direction is the waist direction, perpendicular to the direction of propagation of the field. The relativistic distribution, by contrast, shows a forward tilt by exactly the amount that can be attributed to the effect of the momentum of the photons necessary to achieve ionization

forward. This is a direct consequence of the radiation pressure mentioned above. The angle by which the distribution is tilted can be accurately predicted by a simple calculation based on the radiation pressure concept. In this example, where the intensity is 10 a.u., the forward tilt is more than 30°.

Much more can be said about relativistic effects, as well as all the other topics discussed today. However, the allotted time is up, and this is a good place to stop.

References

1. W.E. Lamb, R.R. Schlicher, M.O. Scully, Phys. Rev. A **36**, 2763 (1987)
2. H.R. Reiss, Phys. Rev. A **22**, 1786 (1980)
3. P.H. Bashkansky, S. Bucksbaum, Phys. Rev. Lett. **60**, 2458 (1988)
4. L.V. Keldysh, Zh. Eksp. Teor. Fiz. **47**, 1945 (1964) [Sov. Phys. JETP **20**, 1307 (1965)]
5. F.H.M. Faisal, J. Phys. B **6**, L312 (1973)
6. H.R. Reiss, Phys. Rev. A **75**, 031404 (R) (2007)
7. H.R. Reiss, Phys. Rev. A **75**, 013413 (2007)
8. H.R. Reiss, Prog. Quantum Electron. **16**, 1 (1992)
9. S. Mandelstam, Ann. Phys. (N.Y.) **19**, 1 (1962)
10. B.S. DeWitt, Phys. Rev. **125**, 2189 (1962)

11. F.J. Belinfante, Phys. Rev. **128**, 2832 (1962)
12. M. Lévy, Nucl. Phys. **57**, 152 (1964)
13. F. Rohrlich, F. Strocchi, Phys. Rev. **139**, B476 (1965)
14. M. Priou, Nucl. Phys. **81**, 641 (1966)
15. W. Xiong, F. Yergeau, S.L. Chin, P. Lavigne, J. Phys. B **21**, L159 (1988)
16. P.B. Corkum, N.H. Burnett, F. Brunel, Phys. Rev. Lett. **62**, 1259 (1989)
17. J.A. Wheeler, Phys. Rev. **52**, 1107 (1937)
18. W. Heisenberg, Z. Phys. **120**, 513, 673 (1943)
19. E.C.G. Stückelberg, Helv. Phys. Acta **17**, 3 (1943); **18**, 21, 195 (1945)
20. H.R. Reiss, Phys. Rev. A **1**, 803 (1970)
21. D.H. Kobe, Phys. Rev. Lett. **40**, 538 (1978)
22. W.E. Lamb, Phys. Rev. **85**, 259 (1952)
23. Z. Fried, Phys. Rev. A **8**, 2835 (1973)
24. M. Gavrila, in *Atoms in Intense Laser Fields*, ed. by M. Gavrila (Academic Press, New York, 1992), p. 435
25. M. Pont, M. Gavrila, Phys. Rev. Lett. **65**, 2362 (1990)
26. H.R. Reiss, J. Opt. Soc. Am. B **13**, 355 (1996)
27. H.R. Reiss, Phys. Rev. A **42**, 1476 (1990)
28. P.H. Bucksbaum, M. Bashkansky, R.R. Freeman et al., Phys. Rev. Lett. **56**, 2590 (1986)
29. H.R. Reiss, J. Phys. B **20**, L79 (1987)
30. P.H. Bucksbaum, in *Atoms in Strong Fields*, ed. by C.A. Nicolaides, C.W. Clark, M. Nayfeh (Plenum, New York, 1990), p. 381
31. P. Agostini, M. Fabre, G. Mainfray et al., Phys. Rev. Lett. **42**, 1127 (1979)
32. U. Mohideen, M.H. Sher, G.D. Aumiller et al., Phys. Rev. Lett. **71**, 509 (1993)
33. H.R. Reiss, Phys. Rev. A **54**, R1765 (1996)
34. H.R. Reiss, V.P. Krainov, J. Phys. A **36**, 5575 (2003)
35. H.R. Reiss, V.P. Krainov, J. Phys. A **38**, 527 (2005)
36. A.M. Perelomov, V.S. Popov, M.V. Terent'ev, Zh. Eksp. Teor. Fiz. **50**, 1393 (1966) [Sov. Phys. JETP **23**, 924 (1966)]
37. M.V. Ammosov, N.B. Delone, V.P. Krainov, Zh. Eksp. Teor. Fiz. **91**, 12008 (1986) [Sov. Phys. JETP **64**, 1191 (1986)]
38. M. Lewenstein, Ph. Balcou, M.Yu. Ivanov et al., Phys. Rev. A **49**, 2117 (1994)
39. H.R. Reiss, Phys. Rev. Lett. **101**, 043002 (2008)
40. P.B. Corkum, Phys. Rev. Lett. **71**, 1994 (1993)
41. B. Bergues, Y. Ni, H. Helm, I. Yu. Kiyan, Phys. Rev. Lett. **95**, 263002 (2005)
42. Z. Fried, J.H. Eberly, Phys. Rev. **136**, B871 (1964)
43. J.H. Eberly, H.R. Reiss, Phys. Rev. **145** (1035)
44. W. Pauli, V.F. Weisskopf, Helv. Phys. Acta **7**, 709 (1934)
45. J.D. Bjorken, S.D. Drell, *Relativistic Quantum Mechanics* (McGraw-Hill, New York, 1964)
46. J.D. Bjorken, S.D. Drell, *Relativistic Quantum Fields* (McGraw-Hill, New York, 1965)
47. H.R. Reiss, J. Opt. Soc. Am. B **7**, 574 (1990)
48. D.P. Crawford, Ph.D. Dissertation, American University, Washington, DC, 1994
49. D.P. Crawford, H.R. Reiss, Phys. Rev. A **50**, 1844 (1994)
50. D.P. Crawford, H.R. Reiss, Opt. Express **2**, 289 (1998)

Chapter 3
High Intensity Physics Scaled to Mid-Infrared Wavelengths

Pierre Agostini

Abstract I discuss strong laser physics at wavelengths (1–4 microns) longer than typical. Since the ponderomotive energy scales as λ^2, we expect particles with energy that would not be reached at the same intensity and shorter wavelengths. After briefly discussing the Keldysh parameter, I will show strong field ionization experimental photoelectron spectra with two laser sources developed at Ohio State University, compared to TDSE calculations. I will then examine the possibility of reducing the high harmonics group delay dispersion, and discuss the advantages of using long-wavelength laser light in tomographic reconstruction of molecular orbitals.

3.1 Introduction

Recent advances in laser techniques have enabled new sources with mid-infrared wavelength (typically 1–4 μm) thus bridging the gap between the Ti:Sapphire and the CO_2 lasers. It is now possible to generate intense, femtosecond pulses in this range. I will discuss in this section the scaling of atomic ionization in intense mid-infrared pulses. As the wavelength becomes longer at constant intensity, the Keldysh parameter becomes smaller and hence the ionization regime becomes more tunneling.

The concept of quantum tunneling, central to the Keldysh theory can be, and has been [1] criticized. I will therefore first briefly discuss the Keldysh parameter, which is the metric of the ionization regime. I will then compare the experimental energy spectra resulting from mid-infrared interaction with atoms to the two known forms of the Keldysh theory [2, 3] and to the semi-classical model derived from the Simpleman's theory [4]. Some of these results have now appeared in publication [5,6]. I will then discuss briefly another aspect of the interaction, namely the high harmonic generation and more precisely the scaling of the group delay dispersion (GDD) with

---

P. Agostini
The Ohio State University, Columbus, OH, USA

K. Yamanouchi (ed.), *Lectures on Ultrafast Intense Laser Science 1*, Springer Series in Chemical Physics 94, DOI 10.1007/978-3-540-95944-1_3,

the pump wavelength [7] as well as the scaling of the yield. I will end this overview by sketching the recent work on the tomographic reconstruction of the highest occupied molecular orbital (HOMO). This work is still under progress and I will mainly discuss some of the possible advantages of the method first demonstrated by [8] in the mid-infrared.

3.2 Mid-Infrared Sources at OSU

In this section I briefly describe the mid-infrared sources at OSU. For more details we refer the reader to [9] and [6]. The first one is a 2 μm source shown in Fig. 3.1. It is based on a commercial OPA (TOPAS) system. Superfluorescence is generated and the mixing in the nonlinear crystal along with different frequency generation yields a tunable wavelength from 1 to 2 μm. Typically, an output pulse energy of 500–600 μJ is achieved with a pulse duration of 50 fs, corresponding to a seven-cycle pulse at 2 μm. The pulse propagates through a gas cell about 50 cm to 1 m long; a filament is formed that provides self-compression, and we thus generate around three-cycle/pulse of 300 μJ . An image of the filament is shown at the bottom of Fig. 3.1. Figure 3.2 shows an autocorrelation trace.

The 3.6 μm radiation is obtained through Difference Frequency Generation (Fig. 3.3): the nonlinear crystal converts the wavelengths of 0.8 mm (pump) and that at 1.053 mm (idler) into the "signal" at 3.6 mm, subject to the energy and phase-matching constraints. With a KTA (Potassium Titanyl Arsenate, $KTiOAsO_4$) nonlinear crystal and by a suitable choice of the center wavelength of Ti:S light, tunable MIR radiation with centre wavelengths in the range of 3–4 mm can be produced, limited by the transmission dropping.

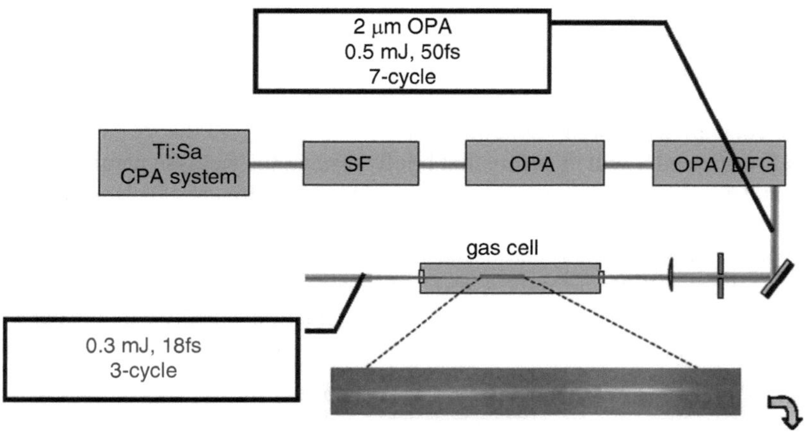

Fig. 3.1 OSU 2-μm source with filament pulse compression (from [10])

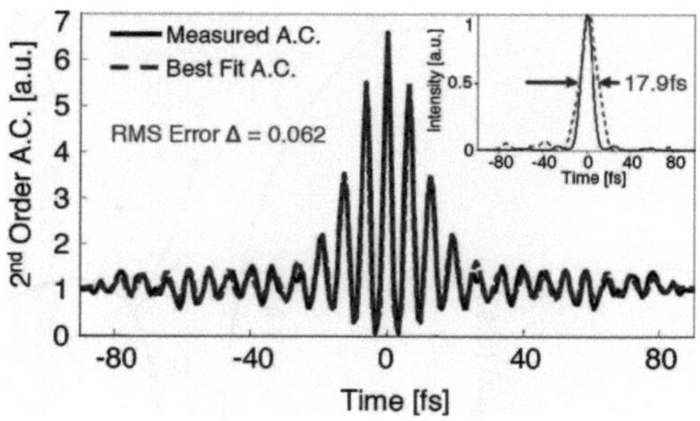

Fig. 3.2 Interferometric autocorrelation of the pulses generated through filamentation. The interferogram shows pulses with approximately 18 fs full width half maximum (two to three cycles) (from [10])

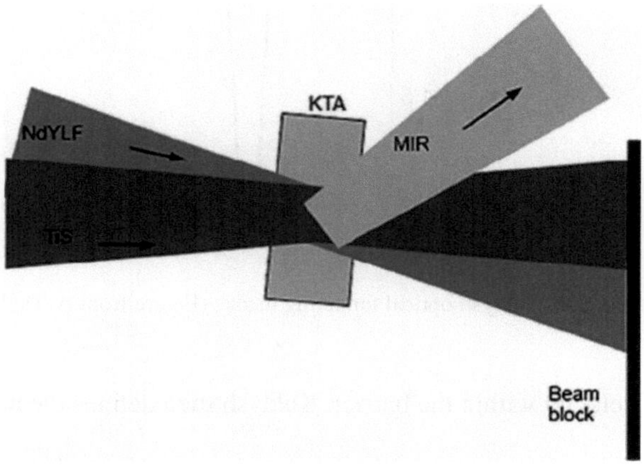

Fig. 3.3 Schematics of the DFG in a KTA crystal generating the 3.6 μm wavelength

3.3 MIR Strong Field Ionization

3.3.1 Keldysh Parameter

I would first like to show that the Keldysh parameter, although defined in the context of tunneling, keeps a definite meaning even in the context of the velocity gauge. Figure 3.4 illustrates the essence of Keldysh tunneling theory. The interaction with the *quasi-static* laser field (the condition of validity for this approximation is $\hbar\omega \ll$ ionization potential) forms a potential barrier with the Coulomb potential and the wave function, solution of the Schrödinger equation has a nonzero amplitude on the classically forbidden side of the barrier. This means that the bound electron may escape through the barrier; if the field is even more intense, over-the-barrier ionization becomes possible. Note that this is a perfectly static image and that the wave function is stationary. Nevertheless, to define his parameter Keldysh considers the barrier length, which is given $l \approx E_B/\varepsilon_0$ and a velocity v within the barrier, $v \approx \sqrt{E_B/2}$ proportional to the square root of the binding energy E_B. Hence a tunneling time defined as $t_{tun} \approx l/v \approx \sqrt{2E_B}/\varepsilon_0$ the barrier length divided by the

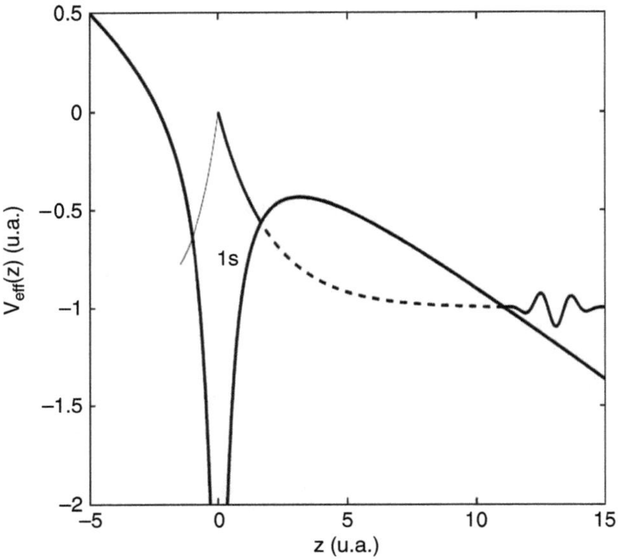

Fig. 3.4 Keldysh optical tunneling theory (Figure from A. De Bohan, PhD Thesis, Paris, 2006.)

velocity within the barrier. Keldysh then defines the adiabaticity parameter γ,

$$\omega t_{\mathrm{tun}} = \gamma = \frac{\omega\sqrt{2E_{\mathrm{B}}}}{\varepsilon_0} \tag{3.1}$$

as the product of ω by the tunneling time.

Despite the fact that the definition of γ is based upon a shaky concept the overall image is very attractive. As γ becomes smaller, either the barrier becomes thinner or the static approximation becomes more valid. In any case, the ionization regime goes deeper into Keldysh' tunneling regime.

Standards expressions of γ include ($U_{\mathrm{p}} = I/4\omega^2$, ponderomotive energy in a field of intensity I and frequency ω, in atomic units):

$$\gamma = \frac{\sqrt{2E_{\mathrm{B}}}}{\varepsilon_0} = \sqrt{\frac{2\omega^2}{\varepsilon_0^2}}\sqrt{E_{\mathrm{B}}} = \sqrt{\frac{E_{\mathrm{B}}}{2U_{\mathrm{P}}}}. \tag{3.2}$$

The first one expresses the dependence on the wavelength.

When this parameter is smaller than one, $\gamma \leq 1$ the ionization rate is proportional to the exponential of minus two times the Keldysh parameter divided by 3ω [11] as

$$\frac{\mathrm{d}w}{\mathrm{d}t} \propto \exp\left[\frac{-2\gamma}{3\omega}\right] \tag{3.3}$$

and is therefore independent of ω, since ω cancels out in the numerator and denominator. Therefore, with Keldysh approximation, the ionization rate is independent of the wavelength as consistent with the quasi static approximation which underlies Keldysh theory.

For a γ-value significantly larger than one, Keldysh' expression for the rate we are in the perturbative multiphoton regime with a rate proportional to the field divided by ω raised to the power $2k$ as

$$\frac{\mathrm{d}w}{\mathrm{d}t} \approx \left(\frac{\varepsilon_0}{\omega}\right)^{2k}. \tag{3.4}$$

Note however [1] that if the intensity is such that $\gamma \ll 1$, the barrier becomes completely suppressed and tunneling becomes meaningless. Since the very definition of γ depends on the choice of the length form of the interaction Hamiltonian, in which it has indeed a very simple interpretation, it is interesting to ask what this interpretation is in the velocity gauge [12]. In this gauge the interaction Hamiltonian writes as the sum of the $\mathbf{p}\cdot\mathbf{A}$ term and the $\mathbf{A}^2$ term where $\mathbf{p}$ is the momentum operator and $\mathbf{A}$ the field vector potential:

$$H_\mathrm{I} = \frac{e}{m}\mathbf{p}\cdot\mathbf{A} + \frac{e^2}{2m}\mathbf{A}^2 = H_{\mathrm{I}_1} + H_{\mathrm{I}_2}. \tag{3.5}$$

The ratio of H_{I_1} to the atomic part of the Hamiltonian, i.e., to p^2 can be easily calculated; we observe that this ratio is inversely proportional to the Keldysh parameter γ:

$$\frac{H_{\mathrm{I}_1}}{H_\mathrm{A}} = \frac{pA}{\frac{p^2}{2}} = \frac{\frac{E}{\omega}}{\frac{E_\mathrm{B} R_\Omega}{2}} = \frac{\sqrt{8}\sqrt{2\,U_\mathrm{P}}}{\sqrt{E_\mathrm{B}}} \approx \frac{\sqrt{8}}{\gamma}. \tag{3.6}$$

Therefore, in the velocity gauge, γ *keeps a meaning that does not involve tunneling*. The meaning is that of the ratio,

$$\gamma \approx \frac{H_\mathrm{A}}{H_{\mathrm{I}_1}}. \tag{3.7}$$

Note that the ratio of H_{I_2} to H_{I_1} is the same as the ratio on the right hand side

$$\frac{H_{\mathrm{I}_2}}{H_{\mathrm{I}_1}} \approx \frac{H_{\mathrm{I}_1}}{H_\mathrm{A}}. \tag{3.8}$$

Keldysh theory was later on reexpressed by [3, 13–15], and many others. Note that Reiss version, which uses the velocity gauge and does not rely on tunneling, depends on two parameters $z = U_\mathrm{p}/\omega$ and $z_1 = 1/\gamma^2$ showing the link between the two approaches.

To conclude this section, although the concept of dc or quasi-static tunneling invoked by Keldysh suffers from many shortcomings, in the limit of not too high intensity and low frequency characteristic of the mid-infrared regime we want to explore, it is a good metric of the nonperturbative character of the ionization process.

Note: the "perturbative" character refers to a perturbative expansion in terms of the field intensity, or H_i/H_A. The Keldysh or Reiss transition amplitudes still require a perturbative expansion in terms of the interaction with the Coulomb potential (see [3] for a full exposition of the S-matrix approach).

3.3.2 Keldysh Scaling

I would like now to introduce the concept of Keldysh scaling which we are frequently using. It is based upon the assumption that if γ is kept constant, the physics remains the same. For instance, let us compare the ionization of two atoms. If I want to keep z and z_1 constant, when the wavelength is varied, then ω'/ω must be equal to the ratio of the binding energies E'/E_B, and also equal to the ratio of the two ponderomotive energies U_p'/U_P in the two situations:

$$\frac{\omega'}{\omega} = \frac{E_B'}{E_B} = \frac{U_P'}{U_P} = \alpha. \tag{3.9}$$

Let's take an example: if I want Xe to behave as He with respect to ionization, since the ratio of the two binding energies is equal to $\alpha \cong 0.5$, the two wavelengths of 800 nm and 1.6 μm match this ratio. This means that Xe at 1.6 μm and 1.25×10^{14} Wcm^{-2} should have the same ionization behavior as He at 800 nm and 1 PW/cm^2. The same reasoning applies to Rb and Ar for which $\alpha \cong 0.27$. Thus Rb at 3 μm and an intensity of 1.8×10^{12} Wcm^{-2} should be similar to Ar at 800 nm at 10^{14} Wcm^{-2}.

3.3.3 Strong Field Ionization Photoelectron Energy Spectra

First of all, in the [2] tunneling theory as well as in the [3] S-matrix theory, the ionization rate includes a delta-function which insures energy conservation:

$$\delta\left(\frac{p^2}{2m} - E_i - n\omega + z\omega\right). \tag{3.10}$$

This function represents a series of peaks at kinetic energies equal to the energy of an integer number of photons $n\omega$ minus the binding energy shifted by the ponderomotive energy $U_p = z\omega$. It is the now well known ATI structure. No attention was paid to this prediction until the 1980s and the physical meaning of the amplitude of the peak of order n given by [3]:

$$\frac{2\pi}{V}\left|\hat{\phi}_i(\vec{p})\right|^2 \omega^2 \sum_n (n-z)^2 \left|J_n\left(z^{1/2}\alpha, -\frac{z}{2}\right)\right|^2. \tag{3.11}$$

remained obscure until the Simpleman's Theory [4]. In (3.11) appears the generalized Bessel function defined as [3]

$$J_n(u,v) = \sum_{k=-\infty}^{\infty} J_{n-2k}(u) J_k(v). \tag{3.12}$$

The basic idea of Simpleman's theory is to assume that the photoelectron energy spectrum is given by the average kinetic energy of the classical motion of a charge (e,m) in the laser field. It is easy to show then that this energy is equal to:

$$\mathrm{KE} = 2\,U_\mathrm{p} \cos^2 \omega t_0 \tag{3.13}$$

and has therefore an upper bound of $2U_\mathrm{p}$.

Figure 3.5 is a typical result that shows the comparison between an experiment (xenon ionization, 3.6 μm, $5\times10^{13}\,\mathrm{Wcm}^{-2}$) with (1) the semiclassical model based on Keldysh tunneling and the simpleman's theory and (2) the S-matrix theory.

In the situation of Fig. 3.5 $\gamma = 0.31$, $z = 174$, $z_1 = 10$. The figure clearly shows the drop of the theoretical rates at 2 (U_p) while the experimental count shows the beginning of the well known rescattering plateau [16]. Note that the experiment is in excellent agreement with the S-matrix prediction between 0.3 and 0.7 (U_p),

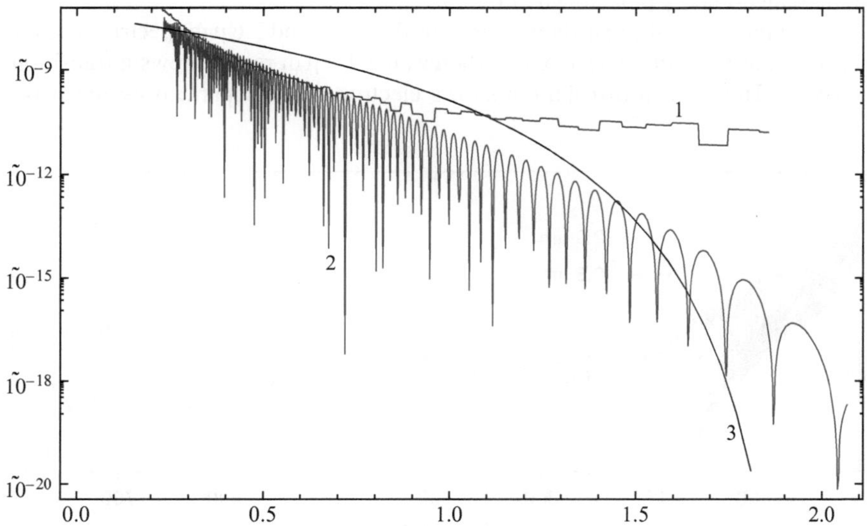

Fig. 3.5 Photoelectron energy spectra (arb. un.) vs. scaled kinetic energy KE/Up: experiment (1), S-matrix (2), semi-classical model (3) (Agostini, unpublished)

in qualitative but poor quantitative agreement with the semi-classical model. To my knowledge this difference has not been analyzed in the literature. In Fig. 3.5, the tunneling rate used is that of Keldysh:

$$W_{\mathrm{K}} = \frac{(6\pi)^{1/2}}{2^{5/4}}\varepsilon_0 \left(\frac{E}{(2\varepsilon_0)^{3/2}}\right)^{1/2} \exp\left(-\frac{2(2\,|\varepsilon_0|)^{3/2}}{3E}\right) \tag{3.14}$$

where E is the field strength and ε_0 the binding energy. Other expressions can be found in [17]. The case of the low energies (0 to $0.3U_{\mathrm{p}}$) will be examined separately hereunder.

3.3.4 Wavelength Scaling of the Photoelectron Spectra

Figure 3.6 shows the result of a numerical solution of the Schrödinger equation [18] which clearly illustrates the scaling of the spectra. One recognizes easily the $2U_{\mathrm{p}}$ and $10U_{\mathrm{p}}$ cutoffs determined by the classical limits [16].

The spectacular increase of the $10U_{\mathrm{p}}$ limit in the virtually exact TDSE calculation agrees with the λ^2 prediction of the simpleman's rescattering model.

In the experimental test, Argon was chosen because is well documented at 0.8 μm, thus providing a benchmark for the longer-wavelength results. Furthermore, the variation in the Keldysh parameter ($\gamma = 1.3 - 0.3$) is sufficient to observe the evolution from the multiphoton behavior at 0.8 μm towards the "nonperturbative" or classical limit at the longest wavelength. A global view of this evolution is illustrated in the scaled-energy plot shown in Fig. 3.7.

A comparison of the distributions reveals that the 2 and 3.6 μm spectra are similar but different from the 0.8 μm case whereas the 1.3 μm result shows a transitional behavior. The 0.8 μm distribution shows electron peaks separated by the photon

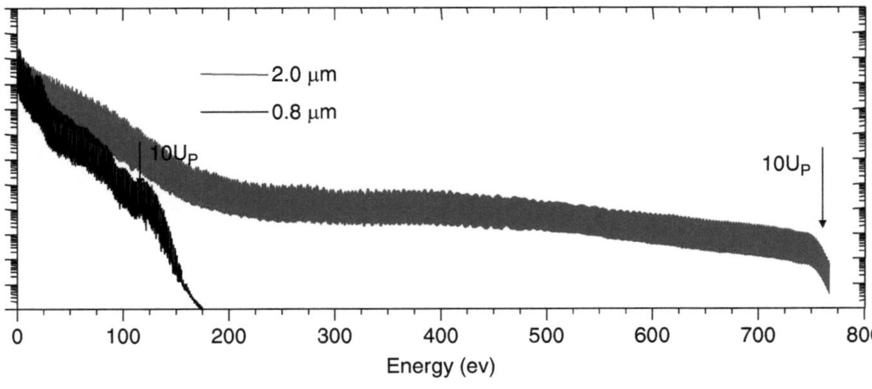

Fig. 3.6 Photoelecton spectra from TDSE. Angle integrated photoelectron spectra for argon exposed to an 8-cycle, flat-top pulse with $I = 2.0 \times 10^{14}\ \mathrm{W/cm^2}$ vs. energy

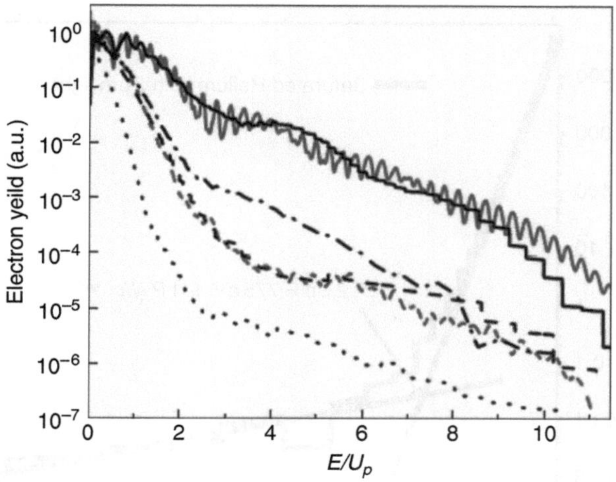

Fig. 3.7 Evolution of the photoelectron distributions in argon for different wavelengths at constant intensity. Excitation is at 0.8 μm (*solid*), 1.3 μm (*dot–dashed*), 2 μm (*dashed*) and 3.6 μm (*dotted*) and intensity is 0.08 PW cm^{-2}. The plots are shown in scaled energy. For 0.08 PW cm^{-2}, U_p is about 5 eV, 13 eV, 30 eV and 100 eV for wavelengths of 0.8 μm, 1.3 μm, 2 μm and 3.6 μm, respectively. Also shown are the TDSE calculations (*gray lines*) at 0.8 μm (*solid line*) and 2 μm (*dashed line*), simulating the experimental conditions (From [6])

energy (ATI), Rydberg structure near zero energy, a broad resonant enhancement near $4.5U_p$ and a slowly modulated distribution decaying to $10U_p$ energy. The 2 and 3.6 μm distributions have a distinctly different structure: no ATI peaks and a rapid decay from zero to near $2U_p$ energy (Simpleman limit), followed by a plateau extending to $10U_p$ (rescattering limit). The broad resonant enhancement near $4.5U_p$, well characterized at 0.8 μm, is either completely absent or broadened over a larger energy range. The 1.3 μm distribution shows both multiphoton (ATI and broad resonance, although diminishing) and tunneling (developing classical distribution) character. It therefore appears that the photoelectron evolves closer to the Simpleman behavior, which is consistent with ionization increasingly approaching the tunneling regime. Note that in this evolution at constant intensity the potential barrier width remains constant and the evolution can be attributed only to the fundamental frequency becoming slow compared to the tunneling frequency, as postulated by Keldysh. This experimental result gives some substance to the tunneling interpretation of γ. Moreover, as the wavelength is increased, the majority of photoelectrons become progressively confined below $2U_p$. The ratio of the number of electrons with energy $>2U_p$ to those with energy $<2U_p$ is found to decrease as λ^{-4}. This scaling is easily interpreted in the rescattering model since the scattering cross-section declines rapidly with energy and impact parameter. The number of returning electrons with a given kinetic energy decreases by λ^{-2} and the wavepacket's transverse spread scales, in area, also as λ^{-2}, thus explaining why the total decrease in the fraction of rescattered electrons scales as λ^{-4}.

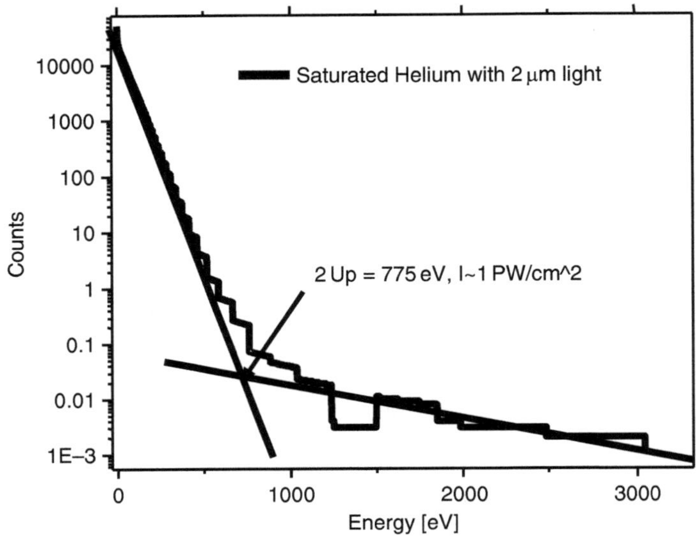

Fig. 3.8 Experimental photoelectron spectrum obtained using He at the light wavelength of 2 μm

The λ^2 scaling of the electron energies leads, in the case of helium to kinetic energies in the kilovolt range for the $10U_\text{p}$ cutoff (Fig. 3.8). This energy corresponds to a net number of absorbed photons of almost 5,000!

3.3.5 Wavelength Scaling of the Ionization Rate: TDSE vs. Tunneling Theory

Let us now look at the scaling of the ionization rate vs. the wavelength of light in Fig. 3.9. These plots show the dependence of the ionization rate on the wavelengths at different intensities based on the TDSE calculation; the rate for Ar at different intensities is found to decrease. A similar result is found for helium. If we remember that the Keldysh rate, in the quasi static regime is independent of the wavelength, it appears that the quasi-static approximation actually fails over this range of wavelength.

3.3.6 Intensity Scaling of the Rescattering Plateau

With regard to the scaling of the rescattering plateau, it is known that in the case of He the height of the plateau scales as $I^{-2.6}$. The TDSE calculation in Ar agrees with this value. We found the scaling of $I^{-2.8}$ as shown in Fig. 3.10.

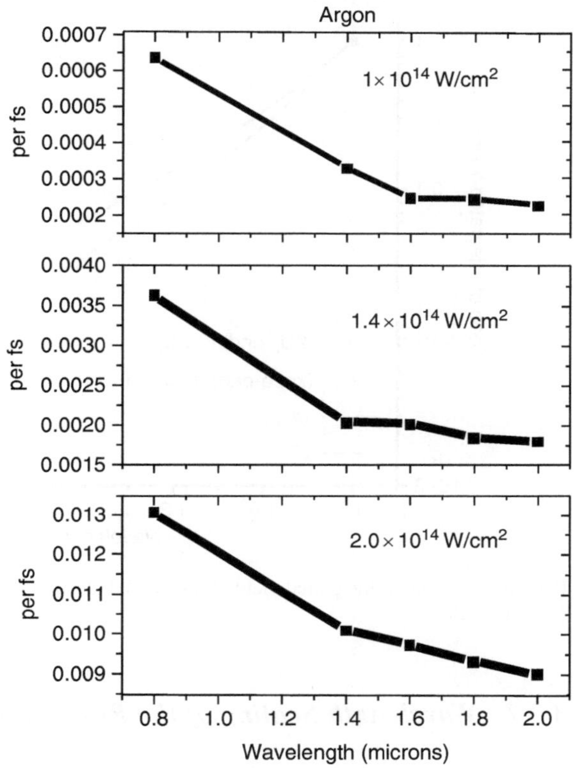

Fig. 3.9 Argon ionization rate as a function of wavelength for three intensities, showing a decrease of about 40% over the range from 0.8 to 2.0 μm

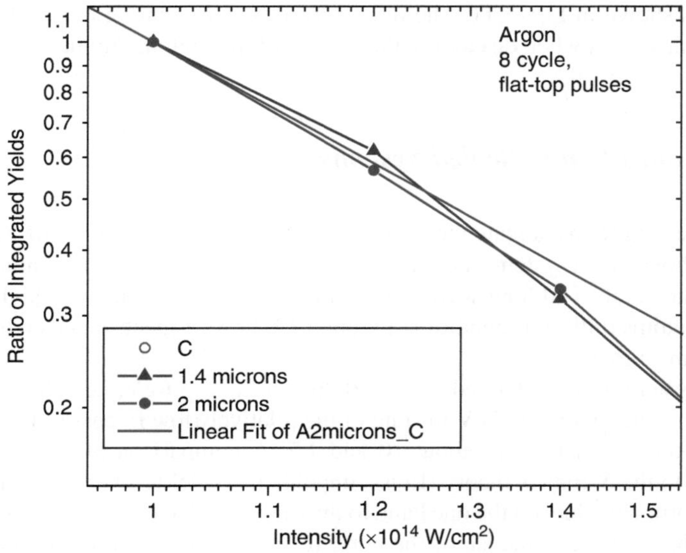

Fig. 3.10 TDSE intensity scaling of the yield of rescattered electrons. The yield scales as $I^{-2.8}$, in good agreement with experiment ([18], unpublished)

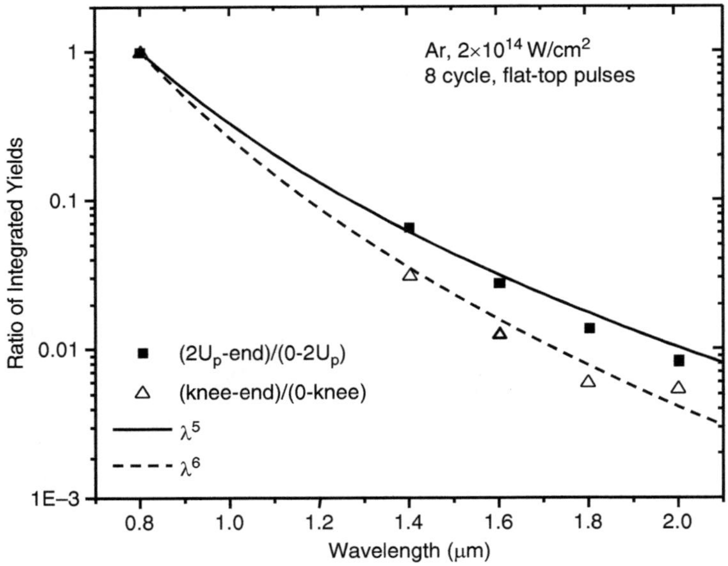

Fig. 3.11 The angle-integrated yield of rescattered electrons as a function of the wavelength, λ, is found to scale as λ^{-5}

3.3.7 *Wavelength Scaling of the Rescattering Plateau*

In the case of Ar, the TDSE calculations show that the height of the plateau scales as λ^{-5} as shown in Fig. 3.11. The reason for this steep scaling is not clear; I will return to this point when discussing the scaling of the high harmonics yield.

3.3.8 *Ionization of Scaled Systems*

It is interesting to investigate the case of scaled systems in the sense of Keldysh as defined above. Such systems are provided by the combination of low binding energy atoms (like alkali) and long wavelength, if the saturation intensity is high enough. Rb or Cs atoms fulfill this condition. Figure 3.12 shows a spectrum taken in Rb with the 3.6 μm source.

This system is characterized by $\gamma = 0.76$ and $U_p = 3.6$ eV, i.e., almost equal to the binding energy of 4.177 eV in spite of the relatively low intensity. The z parameters of Reiss S-matrix are about 10 and 1.7. A comparison of the spectrum of Fig. 3.12 to the S-matrix theory (Reiss, unpublished) yields a very poor agreement which is probably due, on the one hand to the too low value of $z_1 = 1.7$, on the other hand perhaps to the influence of the core. In support of the latter, the low energy

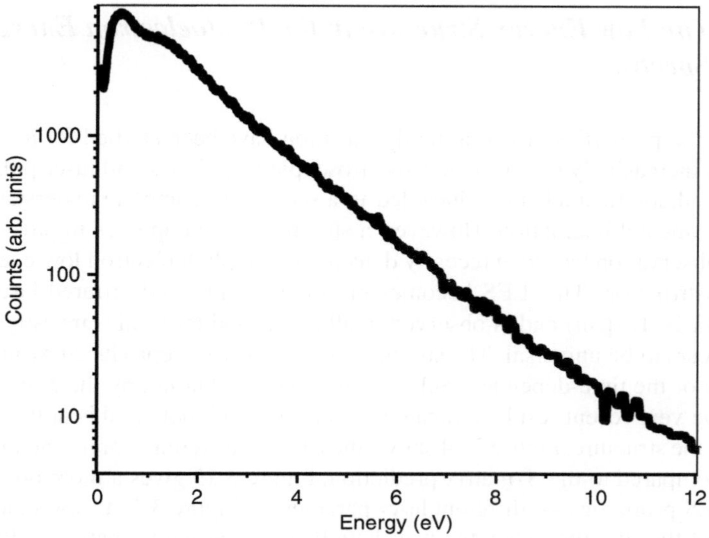

Fig. 3.12 Photoelectron spectrum obtained using Rb at 3.6 μm at an intensity of 3×10^{12} W cm^{-2}

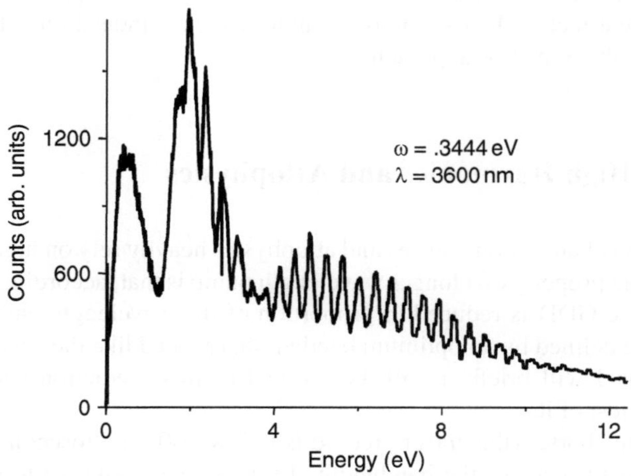

Fig. 3.13 Photoelectron spectrum obtained using Rb at 3,600 nm The fine structure visible in the plot is the photon energy

part of the spectrum shown with high resolution in Fig. 3.13 shows a complicated structure not reproduced by the S-matrix theory.

This makes the link to an interesting and unexpected discovery of a universal structure in the low energy part of the photoelectron spectra when using mid-infrared wavelengths [5, 19].

3.3.9 The Low Energy Structure in the Photoelectron Energy Spectra

Many of the properties of strong field ionization have been studied for a very long time and increasingly sophisticated spectroscopic techniques and laser pulses coupled with theoretical advances have led to a seemingly complete understanding of this fundamental interaction. However, a structure which appears to have hitherto evaded observation has been recently detected in the photoelectron low energy part of the distribution. This LES becomes prominent using mid-infrared laser wavelengths ($\lambda > 1.0\,\mu$m) and is observed in all atoms and molecules investigated and thus appears to be universal. The structure is qualitatively reproduced by numerical solutions of the time-dependent Schrödinger equation but not by the S-matrix theory. Some very recent results indicate that the Coulomb potential is at the physical origin of the structure. Figure 3.14 shows the LES for different atomic and molecular targets compared to the S-matrix prediction. Figure 3.15 gives a more precise idea of the LES properties vs. different laser parameters. Figure 3.15a shows clearly the buildup of the structure when the wavelength increases and probably explains why it was not observed before, while Fig. 3.15c suggest a strong link with the γ parameter and Fig. 3.15d suggest a link with the rescattering since circular polarization suppresses it. An extension of the semi-classical model [20] appears to clearly tie the LES to the effect of the Coulomb potential, which is indeed one of the missing ingredients of the S-matrix approach.

3.4 MIR High Harmonics and Attophysics

The generation of attosecond pulses and attophysics heavily rely on high harmonics. One interesting property of a longer wavelength pump is that, according to the three-step model, the GDD is reduced in proportion of the wavelength and the shortest possible pulse defined by an optimum bandwidth, reduced like the square root of λ. In this section I will briefly recall the rational of this prediction and discuss an experimental test of it.

To generate shorter (the current record is below 100 as) attosecond pulses, one requirement is to increase the bandwidth. To do so, we could try to produce harmonics from ions, but this is not very efficient [21]. Shan and Chang at the Kansas State University [22] showed a method to increase the bandwidth by using longer wave length; they found that because of the increased U_p, there was an increased plateau in the harmonics using long wavelengths. The longest wavelength they used was 1.5 μm. A second condition is to decrease the attochirp or GDD, which appears to be theoretically possible by increasing the wavelength. If the chirp cannot be decreased, it can be compensated for by propagating the pulse in a medium with a suitable group velocity dispersion, which has been successfully done [23]. The shortest pulse currently is actually <100 as.

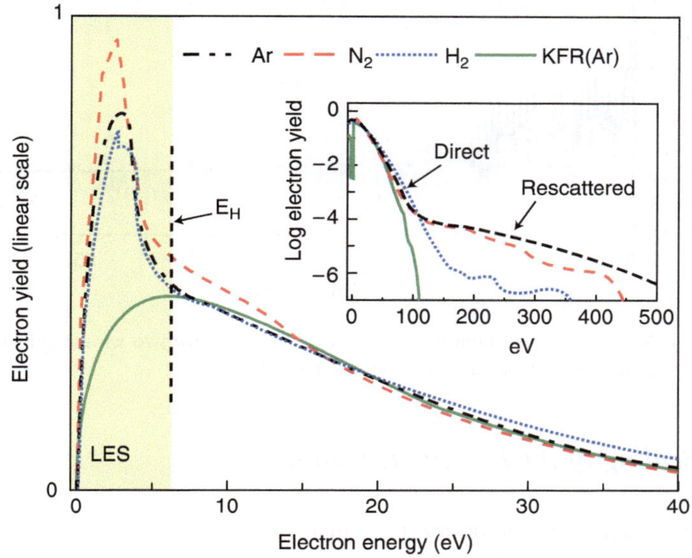

Fig. 3.14 Low-energy part of photoelectron spectra showing the spike-like structure that is not reproduced by the S-matrix calculations (*green solid line*). The insert shows a global view of the spectra in which the $2U_\mathrm{p}$ and $10U_\mathrm{p}$ cutoffs are clearly visible (from [5])

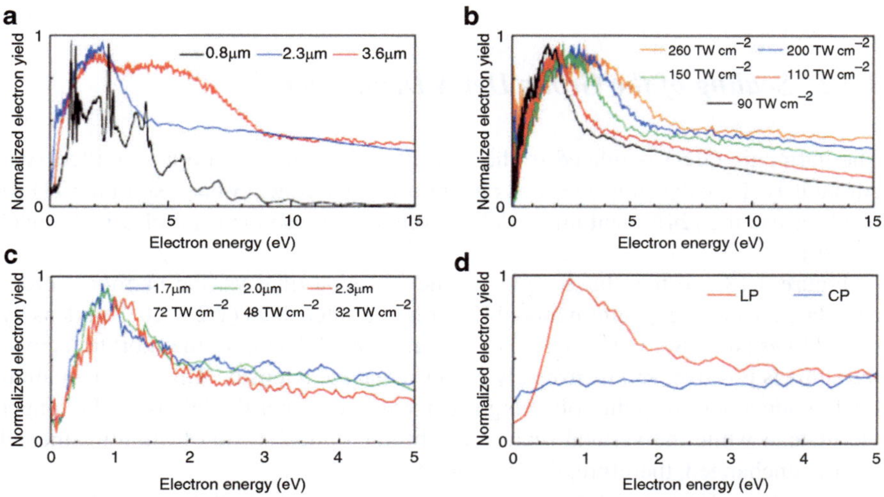

Fig. 3.15 Dependences of the LES on the wavelength at constant intensity (**a**), intensity at constant wavelength (**b**), intensity and wavelength at constant Keldysh parameter (**c**) linear (LP) vs. circular (CP) polarization (**d**) (from [5])

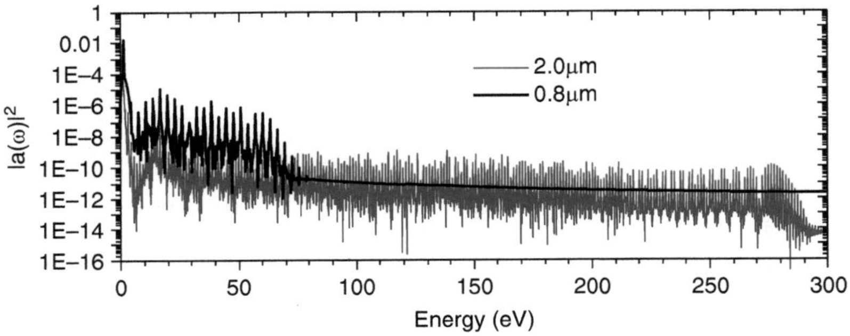

Fig. 3.16 TDSE calculation of high harmonic spectra in argon for two wavelengths at constant intensity clearly showing the increase of the total bandwidth

3.4.1 Scaling of the Harmonic Cutoff

Figure 3.16 is a high harmonic spectrum that is typical for a wavelength of 800 nm in Ar with a cutoff at around 70 eV or larger. For comparison, the gray curve representing the spectrum at 2 μm is shown. This calculation is much easier than the one for the photoelectron spectrum, since the electron state remains close to the core and thus a large box is not required in the calculation. The experiment is shown in Fig. 3.17 [7].

3.4.2 Scaling of the Group Delay Dispersion

The harmonic GDD is related to the fact that all harmonics are not emitted synchronously. To realize this one must either do a time frequency analysis of the TDSE result or use the [24] quantum model or even simply use the semi-classical model [25, 26].

Figure 3.18, we now have a plot of the emission time vs. the harmonic order calculated using the quantum model [27] for two intensities (1.2×10^{14} in densely dotted lines and $3.8 \times 10^{14}\,\mathrm{Wcm}^{-2}$ in sparsely dotted lines); for the short trajectory, the slope is positive, and is inversely proportional the intensity. This can be qualitatively understood from the following argument: the higher the intensity, the longer the plateau while the vertical scale given by the optical cycle of the fundamental laser is unchanged, therefore, the slope tends to be lower.

The effect of the wavelength can be easily deduced from the same calculation. The time on the classical trajectory, which is proportional to the optical period T, scales as λ, and the harmonic cut-off energy, which is proportional to the ponderomotive energy, U_{p}, scales as $I\lambda^2$. Hence, the atto-chirp, i.e., the slope τ/U_{p} scales as $I\lambda^{-1}$. As a result, the pulse duration τ is no longer inversely proportional to the bandwidth B but is given by:

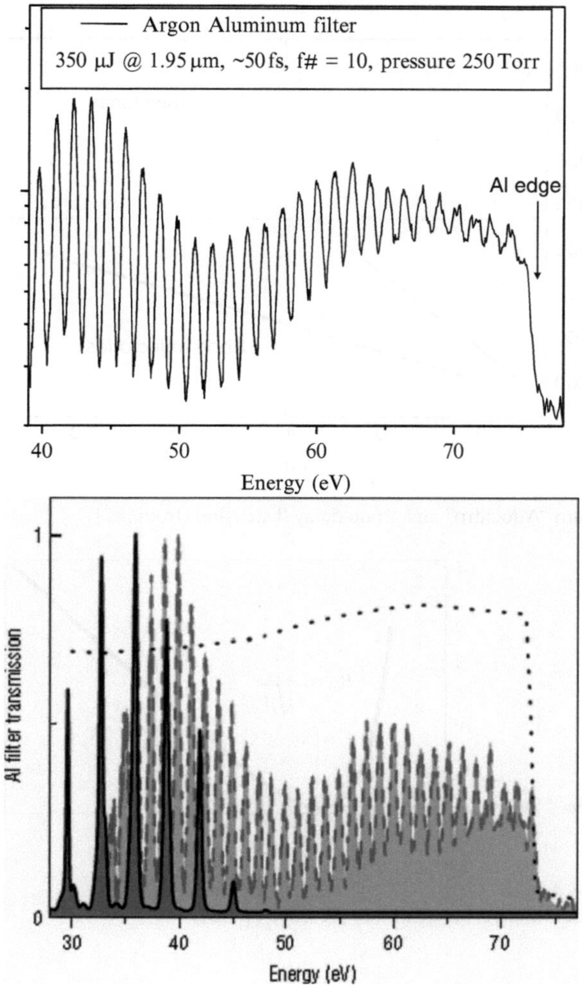

Fig. 3.17 Harmonic spectrum from argon and a 2 mm pump (*left*) compared to a 0.8 mm pump (*right*) The 2 mm spectrum is limited by the transmission of an Al filter. The real cutoff is foend at 225 eV (see [6])

$$\tau \propto \sqrt{\frac{1}{B^2} + \beta^2 B^2} \tag{3.15}$$

and therefore, reaches a minimum (Fig. 3.19). It follows that the minimum pulse duration scales with a square root of 2β, where β is proportional to $1/\lambda$, that is,

$$\beta \propto \frac{1}{\lambda} \tag{3.16}$$

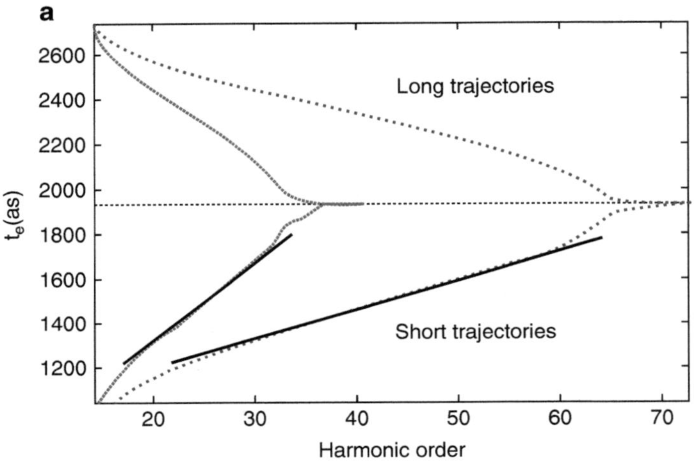

Fig. 3.18 Quantum "Attochirp" and group delay dispersion (from [27])

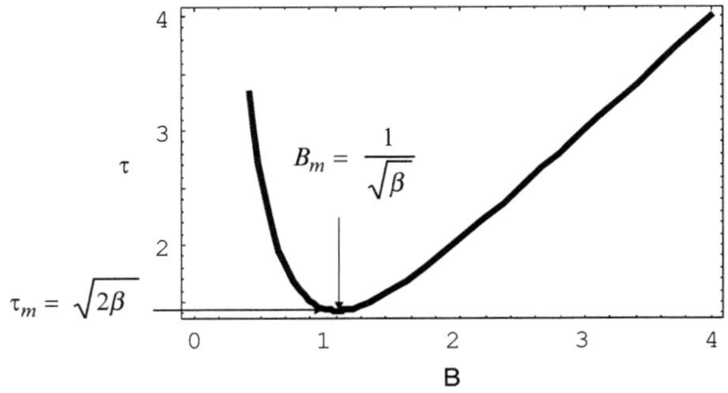

Fig. 3.19 Pulse duration vs. bandwidth in presence of GDD

Consequently,

$$\tau_{\mathrm{m}} \propto \frac{1}{\sqrt{\lambda}} \tag{3.17}$$

Therefore, there should be some advantages in using long wavelength. Experimentally the verification of the theoretical prediction requires the measurement of the GDD at constant intensity. The standard method to characterize the spectral property of the harmonics is the RABBITT method [28].

As shown in Fig. 3.20, the TDSE calculation is in excellent agreement with the classical prediction, namely, $1/\lambda$.

Incidentally, the TDSE calculation also predicts that the attosecond pulse train obtained for a given bandwidth is "cleaner" in the case of 2 μm than in the case of 800 μm using the optimum bandwidths of 24.8 and 37.2 eV, respectively (Fig. 3.21). The 2 peaks per half cycle in the 2 mm plot result from the short and long trajectories

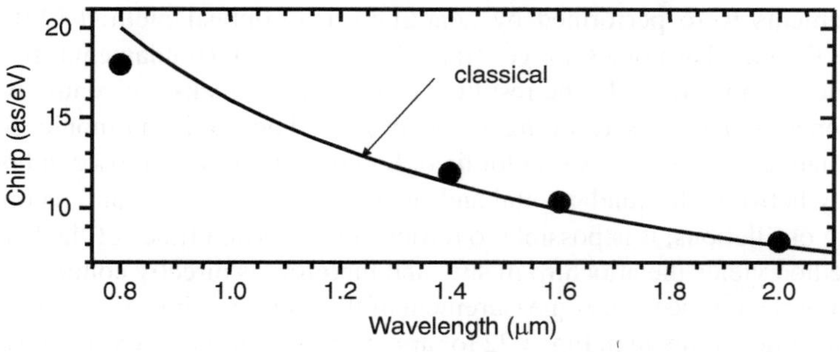

Fig. 3.20 Chirp from TDSE calculation compared to the semi-classical model

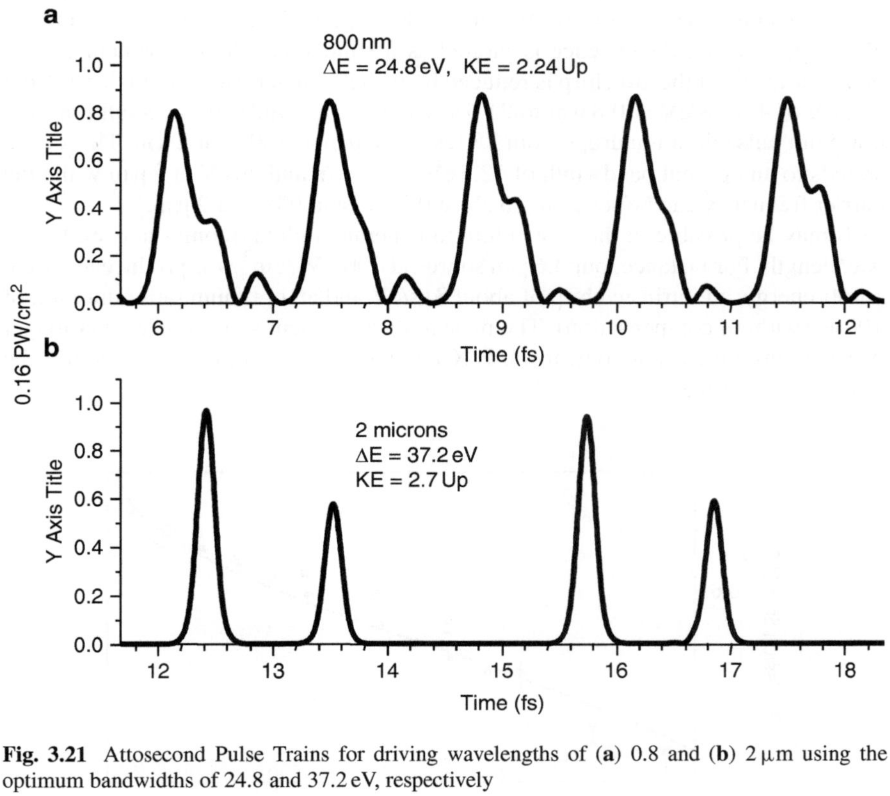

Fig. 3.21 Attosecond Pulse Trains for driving wavelengths of (**a**) 0.8 and (**b**) 2 μm using the optimum bandwidths of 24.8 and 37.2 eV, respectively

contributions (see [18]). In most experiments, the contribution from the long trajectories is eliminated simply because it tends to generate more diverging harmonics than the short trajectories and the second peak would most likely not appear.

Experimentally, the phase of the high harmonic light has been at three different driving wavelengths (0.8, 1.3, and 2 μm) at constant intensity [7]. The measurements were performed by adapting an all-optical method [29]. In this method, the high harmonics are generated by a two-color (fundamental and second harmonic) combined field. The resultant electric field breaks the centro-symmetry of the generation process, resulting in both even- and odd-order harmonics functions of the relative phase of the two-color field. By controlling with attosecond precision the delay between the fundamental and second harmonic pulses and recording the resulting oscillations, it is possible to retrieve the emission times of the harmonics. This method yields the attochirp in situ, and therefore is directly comparable to the theory. For details, see [7]. A measurement of the emission time as a function of the harmonic order is shown in Fig. 3.22 for argon and xenon at approximately the same intensity. Both results agree very well with the universal scaled curve obtained from the semi-classical theory. Figure 3.23 shows the product of the attochirp and the intensity as a function of the fundamental driving wavelength. The figure includes measurements in argon for $80\,\mathrm{TW/cm^2}$, $0.8\,\mu\mathrm{m}$, and $71\,\mathrm{TW/cm^2}$, $2\,\mu\mathrm{m}$ pulses. The expected $1/\lambda$ dependence is plotted as a solid line. The experiment clearly demonstrates that the attochirp is reduced by using a longer wavelength driver from a value of 41.5 as/eV at 0.8 μm to 21.5 as/eV at 2 μm. Under these conditions, the optimum pulse duration drops from 250 as at 0.8 μm to 180 as at 2 μm. This corresponds to an optimal bandwidth of 12.5 eV at 0.8 μm and 20 eV at 2 μm while the carrier frequency can be set up to 25 eV at 0.8 μm and 90 eV at 2 μm.

It may be possible in the near future to generate high harmonics at even longer wavelength. For instance, our 3.6 μm source at $100\,\mathrm{TW/cm^2}$ will produce a 400 eV cutoff energy, an intrinsic chirp of about 8 as/eV and an optimum pulse duration of 100 as (without compensation). The present measurement substantiates that longer wavelengths are a viable route towards X-ray pulses with durations approaching the atomic unit of time.

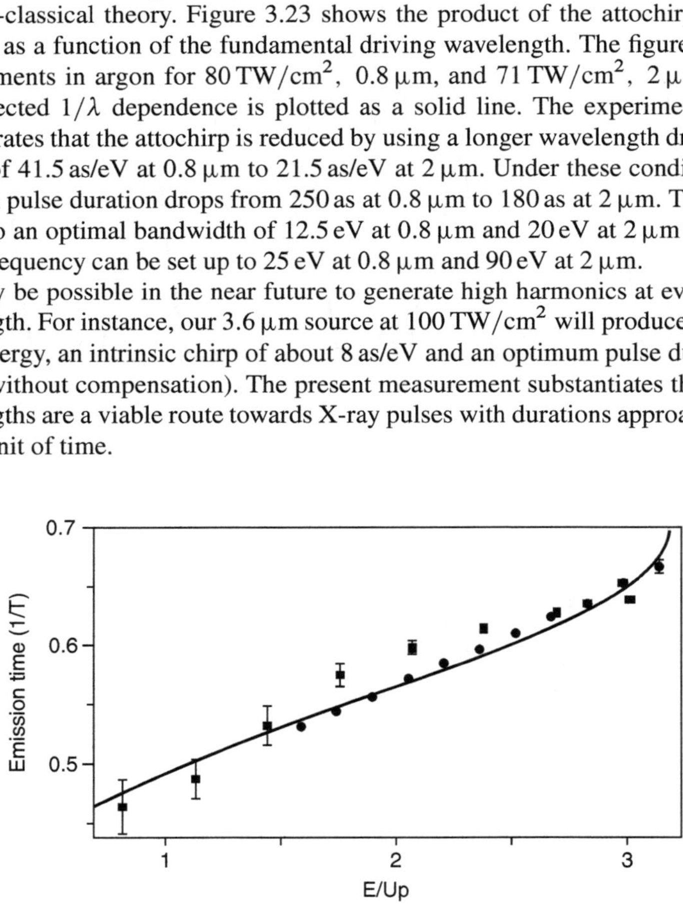

Fig. 3.22 Scaled emission times as a function of the scaled kinetic energy. The data from argon at 1 : 3 μm (*squares*) and xenon at 2 μm (*circles*) are in excellent agreement with the semiclassical calculation (*solid line*) (from [10])

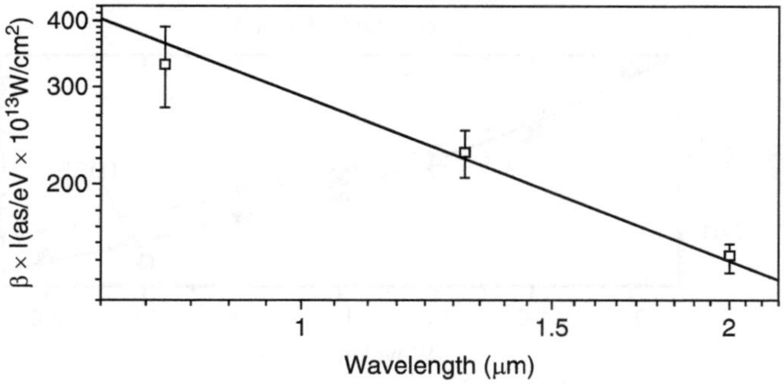

Fig. 3.23 Scaling of the attochirp as a function of wavelength. The product of the attochirp by the intensity is plotted in order to remove the effect of the peak intensity. The straight line corresponds to the predicted scaling $1/\lambda$ (from [10])

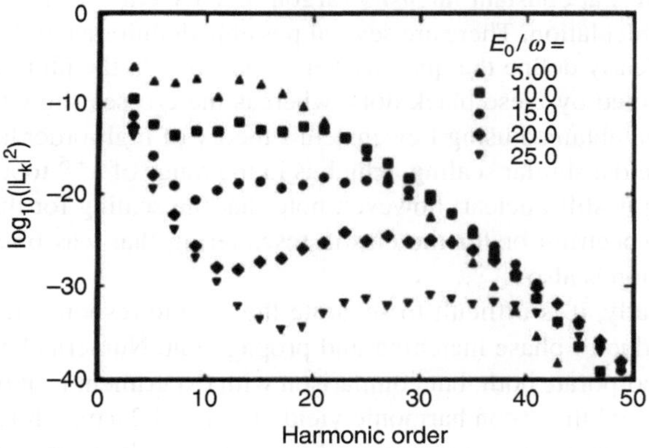

Fig. 3.24 Plot of $|L_k|^2$ in (3.18) at constant intensity and ω vs. the harmonic order $2k + 1$ and different binding energies E_0/ω indicated in the legend. ([30], with permission © APS 1994, http://pra.aps.org/abstract/PRA/v50/i2/p1540_1)

3.4.3 Scaling of the Harmonic Yield

An important quantity is the yield as a function of the pump wavelength at constant intensity. It is interesting to discover if there is a simple scaling for it. This question has been discussed in several papers. Becker et al. [30, 31] using a zero-range potential model, calculated the emission rate of the $(2k + 1)$th harmonic:

$$\frac{dR_{2k+1}}{d\Omega_K} = \omega \frac{r_0}{\pi\lambda}(2k + 1)^3 \varepsilon_x^2 \, |L_k|^2 \tag{3.18}$$

and found the dependence shown in Fig. 3.24. Defining the yield as the height of the plateaus in the graph, it is found that it scales as $\lambda^{-n} (n > 3)$.

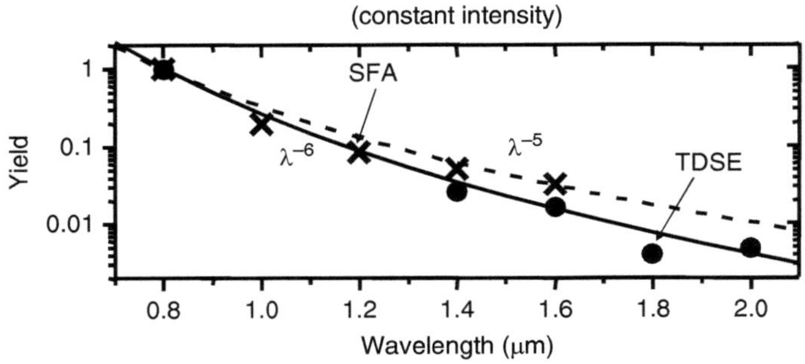

Fig. 3.25 Single atom response harmonic yield vs. pump wavelength at constant intensity TDSE (*dots*), and Lewenstein's model (*crosses*). The lines are the λ^{-5} and λ^{-6} dependences (from [18])

Scaling with λ at constant intensity (argon 0.16 PWcm^{-2}) can also be obtained by the TDSE calculation. There are several possible definitions of the yield and it is important to clearly define the quantity kept constant. On the plot in Fig. 3.25, the results are denoted by these black dots, whereas the crosses show the result of an SFA calculation obtained using Lewenstein's theory of high-order harmonics; both calculations yield a similar scaling, which is in the range of λ^{-5} to λ^{-6}. The reason for this scaling is still unclear; however, note that the scaling for the height of the photoelectron spectrum or for the elastic rescattering that was obtained from the TDSE calculation is also λ^{-5}.

Experimentally, it is difficult to separate the atomic response from the macroscopic effects due to phase matching and propagation. Numerical simulations can in principle incorporate both, but comparison with experiment is at best qualitative. We have measured the argon harmonic yield at 0.8 and 2 μm, while keeping other conditions fixed: intensity, argon density, focusing and collection system. Confining the comparison to a spectral bandwidth (35–50 eV) common to both fundamental wavelengths, the 2 μm harmonics are observed to be 1,000 times weaker than with 0.8 μm and six times less than that predicted by the TDSE calculations. However, independent optimization of only the argon density for maximum harmonic yield at both fundamental wavelengths results in a 2 μm yield that is only 85 times weaker. This suggests that the brightness of a 2 μm harmonic source can be made actually comparable to a 0.8 μm driver in the extreme-ultraviolet region (50 eV) and perhaps larger at higher energies (50–200 eV).

3.5 Tomographic Reconstruction of Molecular Orbitals

Within certain approximations, the dipole moment between the initial HOMO and the continuum, responsible for high harmonic generation, is closely related to the Fourier transform of the orbital. This is the basis of a method proposed by the NRC group to reconstruct the HOMO of a molecule from the high harmonic spectra

Table 3.1 Keldysh parameter for atoms and molecules

Atom	I_p (eV)	E_{OTB} (au)	I_{OTB} (PW/cm^2)	γ @ 0.8 μm	γ @ 2 μm	γ @ 4 μm
Ar	15.8	0.08	0.22	0.8	0.3	0.1
N_2 (X)	15.6	0.08	0.22	0.8	0.3	0.1
NO (X)	9.3	0.03	0.03	1.6	0.6	0.3
NO (A or B)	~3.9	0.005	10^{-3}	5.6	2.3	1.1
Cs	3.9	0.005	10^{-3}	5.6	2.3	1.1
ICN (X)	11.0	0.04	0.06	1.3	0.5	0.2
ICN (A)	~6.0	0.01	0.01	2.3	0.9	0.4

E_{OTB} and I_{OTB} refer to over-the barrier ionization field and intensity respectively

generated by aligned molecules and collected for different Euler angles [8]. During the past 5 years several works, including the NRC group itself, have scrutinized the method and critically examined some of its limitations (see for example [32, 33]). Among those are: the effect of the approximations required to extract the HOMO [33] or, for instance the coupling to lower lying orbital [32].

It is nevertheless interesting to ask whether the method could work better with mid-infrared lasers and if the increase in harmonic bandwidth would provide a corresponding increase in resolution. Indeed the spatial resolution is limited by the recombining electron de Broglie wavelength which is directly related to the harmonic highest frequency through the harmonic cutoff law.

Moreover, the initial paper suggests that the method could be extended to time-resolution in a pump-probe scheme and thus could provide a way to make a "movie" of dissociating molecules.

First, it is clear that one of the most important approximations underlying the method is the plane wave approximation describing the electron wavepacket. This approximation is intimately connected to the tunneling (or high intensity) regime of the ionization process and the spreading of the wavepacket. Since, as we have seen above, it is much easier to generate ionization in the regime of low γ (or high z) at long wavelengths, the conditions required for an efficient tomographic reconstruction should be more easily met in the mid-infrared. Tunneling is also required because of the very fast dependence of its rate on the binding energy which limits the contributions of lower orbitals.

Examples of γ-parameters are shown in Table 3.1. We have included NO, which has a significantly smaller binding energy. Although the binding energy for NO in the ground state is 9.3 eV, it could be as low as 3.9 in the excited state, and therefore, the tunneling regime, and hence the long wavelength light is essential. ICN, which was used in a famous experiment ([34] and references therein) is included too as a possible candidate for a movie of a dissociating molecule.

Acknowledgements Support from USDOE/BES under contracts DE-FG02-04ER15614 and DE-FG02-06ER15833 and from the NSF under contract PHY-0653022 is gratefully acknowledged. I would like to thank all the members of Dr DiMauro's group at OSU and, particularly, Jennifer Tate, Gilles Doumy, Phil Colosimo, Fabrice Catoire, Cosmin Blaga. I would like also to kindly thank Howard Reiss for many illuminating discussions and the communication of results prior to publication.

References

1. H.R. Reiss, Phys. Rev. Lett. **101**, 043002 (2008)
2. L.V. Keldysh, Sov. Phys. JETP **20**, 1307 (1965)
3. H.R. Reiss, Phys. Rev. A **22**, 1786 (1980)
4. H.B. van Linden van den Heuvell, H.G. Muller, in *Multiphoton Processes*, ed. by P.L. Knight, S.J. Smith (Cambridge University Press, Cambridge, 1987), p. 25
5. C.I. Blaga, F. Catoire, P. Colosimo, G.G. Paulus, H.G. Muller, P. Agostini, L.F. DiMauro, Nat. Phys. **5**, 335 (2009)
6. P. Colosimo, G. Doumy, C.I. Blaga, J. Wheeler, C. Hauri, F. Catoire, J. Tate, R. Chirla, A.M. March, G.G. Paulus, H.G. Muller, P. Agostini, L.F. DiMauro, Nat. Phys. **4**, 386 (2008)
7. G. Doumy, J. Wheeler, C. Roedig, R. Chirla, P. Agostini, L.F. DiMauro, Phys. Rev. Lett. **102**, 093002 (2009)
8. J. Itatani, J. Levesque, D. Zeidler, H. Niikura, H. Pépin, J.C. Kieffer, P.B. Corkum, D.M. Villeneuve, Nature **432**, 867 (2004)
9. P. Agostini, L.F. DiMauro, Contemp. Phys. **49**, 179 (2008)
10. C. Hauri, R.B. Lopez-Martens, C.I. Blaga, K.D. Schultz, J. Cryan, R. Chirla, P. Colosimo, G. Doumy, A.M. March, C. Roedig, E. Sistrunk, J. Tate, J. Wheeler, L.F. DiMauro, E.P. Power, Optics Lett., **32**, 868 (2007)
11. M.V. Fedorov, *Atomic and Free Electrons in a Strong Light Field* (World Scientific, Singapore, 1997)
12. V.P. Krainov, H.R. Reiss, M.V. Smirnov, *Radiative Processes in Atomic Physics* (Wiley, New York, 1997)
13. M.V. Amosov, N.B. Delone, V.P. Krainov, Sov. Phys. JETP **64**, 1191 (1986)
14. F.H.M. Faisal, J. Phys. B. At. Mol Opt. Phys. **6**, L89 (1973)
15. A.M. Perelomov, V.S. Popov, M.V. Terent'ev, Sov. Phys. JETP **23**, 924 (1966)
16. G.G. Paulus, W. Nicklich, Xu Huale, P. Lambropoulos, H. Walther, Phys. Rev. Lett. **72**, 2851 (1994)
17. F. Ceccherini, D. Bauer, F. Cornotti, Phys. Rev. A **68**, 053402 (2003)
18. J. Tate, T. Auguste, H.G. Muller, P. Salières, P. Agostini, L.F. DiMauro, Phys. Rev. Lett. **98**, 013901 (2007)
19. F. Catoire, C.I. Blaga, E. Sistrunk, H.G. Muller, P. Agostini, L.F. DiMauro, Laser Phys. **19**, 1574 (2009)
20. W. Quan, Z. Lin, M. Wu, H. Kang, H. Liu, X. Liu, J. Chen, J. Liu, X.T. He, S.G. Chen, H. Xiong, L. Guo, H. Xu, Y. Fu, Y. Cheng, Z. Xu, Phys. Rev. Lett. **103**, 093001 (2009)
21. E.A. Gibson, A. Paul, N. Wagner, R. Tobey, S. Backus, I.P. Christov, M.M. Murnane, H.C. Kapteyn, Phys. Rev. Lett. **92**, 033001 (2004)
22. B. Shan and Z. Chang, Phys. Rev. A **65**, 011804 (2001)
23. R. López-Martens, K. Varjú, P. Johnsson, J. Mauritsson, Y. Mairesse, P. Salières, M.B. Gaarde, K.J. Schafer, A. Persson, S. Svanberg, C.-G. Wahlström, A. L'Huillier, Phys. Rev. Lett. **94**, 033001 (2005)
24. M. Lewenstein, Ph. Balcou, M.Yu. Ivanov, A. L'Huillier, P.B. Corkum, Phys. Rev A **49**, 2117 (1994)
25. P.B. Corkum, Phys. Rev. Lett. **71**, 1994 (1993)
26. K.J. Schafer, B. Yang, L.F. DiMauro, K.C. Kulander, Phys. Rev. Lett. **70**, 1599 (1993)
27. Y. Mairesse, A. De Bohan, L.J. Frasinski, H. Merdji, L.C. Dinu, P. Monchicourt, P. Breger, M. Kovaev, R. Taïeb, B. Carré, H.G. Muller, P. Agostini, P. Salières, Science **302**, 1540 (2003)
28. P.M. Paul, E.S. Toma, P. Breger, G. Mullot, F. Auge, Ph. Balcou, H.G. Muller, P. Agostini, Science **292**, 1689 (2001)
29. N. Dudovich, O. Smirnova, J. Levesque, Y. Mairesse, M.Yu. Ivanov, D.M. Villeneuve, P.B. Corkum, Nat. Phys. **2**, 781 (2006)
30. W. Becker, S. Long, J.K. McIver, Phys. Rev. A **50**, 1540 (1994)
31. W. Becker, S. Long, J.K. McIver, Phys. Rev. A **4**, 4112 (1990)

32. H. Akagi, T. Otobe, A. Staudte, A. Shiner, F. Turner, R. Dörner, D.M. Villeneuve, P.B. Corkum, Science **325**, 1364 (2009)
33. V.H. Le, A.T. Le, R.H. Xie, C.D. Lin, Phys. Rev. A **76**, 013414 (2007)
34. A.H. Zewail, Journ. Phys. Chem **104**, 5660 (2000)

Chapter 4
How Do Molecules Behave in Intense Laser Fields? Theoretical Aspects

Hirohiko Kono

Abstract In this chapter, by referring to our recent theoretical studies, we would like to answer the fundamental question: How do molecules behave in intense laser fields? We discuss first the electronic and vibrational dynamics of $H_2{}^+$ in the near-infrared laser field by solving numerically the time-dependent Schrödinger equation. Then, we describe a bigger molecule, C_{60}, and show how it changes its geometrical structure in an intense laser field. Finally, we introduce the bond dissociation dynamics of ethanol in an intense laser field, and extract the factors that determine the yield ratio of the C–C bond breaking and the C–O bond breaking, which was investigated experimentally by Fumihiko Kannari et al.

4.1 Introduction

In this chapter, by referring to our recent theoretical studies, we would like to answer the fundamental question: How do molecules behave in intense laser fields? We discuss first the electronic and vibrational dynamics of $H_2{}^+$ in the near-infrared laser field by solving numerically the time-dependent Schrödinger equation. Then, we describe a bigger molecule, C_{60}, and show how it changes its geometrical structure in an intense laser field. Finally, we introduce the bond dissociation dynamics of ethanol in an intense laser field, and extract the factors that determine the yield ratio of the C–C bond breaking and the C–O bond breaking, which was investigated experimentally by Fumihiko Kannari et al. (and is presented in Chap. 5).

First, I would like to introduce the characteristic timescales of laser fields and molecules. In most of the following calculations, we use the near-infrared laser field of the 800 nm wavelength. The laser electric field $\varepsilon(t)$ of 800 nm oscillates quickly as shown in Fig. 4.1. The period of oscillation is approximately 2.5 fs. However, the electron can follow even this rapid oscillation if the electronic transition frequency

H. Kono
Tohoku University, Sendai, Japan

K. Yamanouchi (ed.), *Lectures on Ultrafast Intense Laser Science 1*, Springer Series in Chemical Physics 94, DOI 10.1007/978-3-540-95944-1_4,

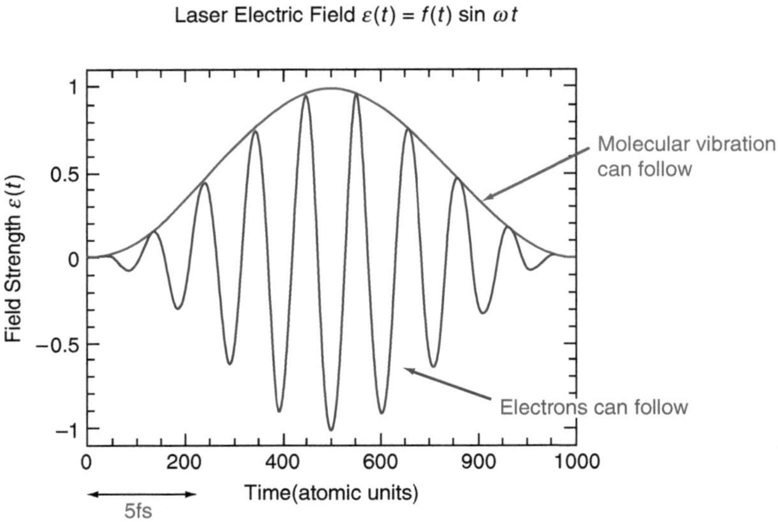

Fig. 4.1 Characteristic timescales of the electronic and nuclear dynamics of molecules in near-infrared intense laser fields ($\lambda = 760$ nm field, $\omega = 1.6$ eV $= 0.06$ in atomic units)

ω_{elec} is sufficiently larger than the optical field frequency ω. This means that the electronic wave function can follow this rapid oscillation adiabatically.

In the case of molecules, there are two different kinds of degrees of freedom – one is for electrons and the other is for nuclei. Therefore, we have two different characteristic timescales. The typical timescale for molecular vibration, $2\pi/\omega_{\mathrm{vib}}$, is 10–100 fs, where ω_{vib} is the vibrational frequency. Therefore, molecular vibrations cannot follow the rapid optical oscillation. Rather, they follow the pulse envelope $f(t)$ itself if $2\pi/\omega_{\mathrm{vib}} < T_{\mathrm{p}}$, where T_{p} is the pulse duration. If this rapidly oscillating electric field is strong, that is, the light intensity I is high, large-amplitude electron motion is induced inside the molecules. As a result, the effective potential for nuclear motion determined by a function of $f(t)$ is highly distorted, which could initiate some kinds of dissociation or reaction processes. Field-induced intramolecular electron transfer also enhances ionization.

4.2 Electronic and Vibrational Dynamics of H_2^+ in a Near-IR Field

We consider the concept of adiabatic and nonadiabatic electronic dynamics against the temporal changes in the laser electric field by taking the simplest case of $H_2{}^+$ as an example. To that end, we must numerically solve the time-dependent Schrödinger equation for $H_2{}^+$. The electronic wave function Φ exhibits a cusp as shown in the left panel of Fig. 4.2. Thus, we invented a new grid point method by introducing

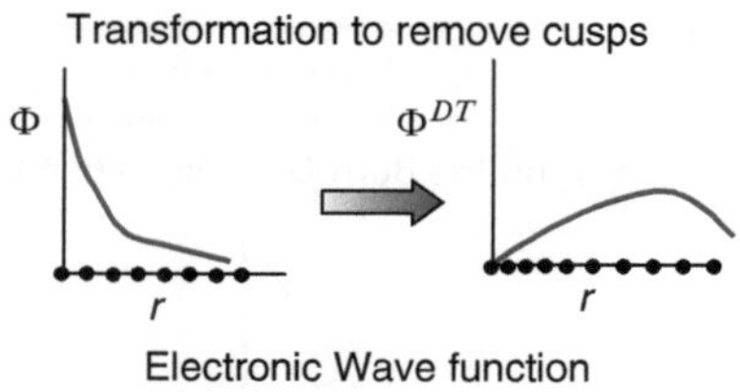

Fig. 4.2 Dual transformation method: Grid point method for wave packet dynamics in Coulomb fields [1, 2]

a consistent transformation of Φ and the electronic Hamiltonian $H_{\rm el}(t)$, the Dual Transformation Method [1–3] shown in Fig. 4.2, to eliminate the cusp of Φ and make it as smooth as $\Phi^{\rm DT}$. This can be done by coordinate transformations $f(\xi)$ and $g(\zeta)$ from the cylindrical coordinates ρ and z to new variables ξ and ζ; $\rho = f(\xi)$ and $z = g(\zeta)$. Φ and $H_{\rm el}(t)$ are transformed to $\Phi^{\rm DT}$ and $H_{\rm el}^{\rm DT}(t)$ as in (4.1) and (4.2):

$$\Phi^{\rm DT}(\zeta, \xi, R) = \sqrt{f(\xi) f'(\xi) g'(\zeta)}\,\Phi(z, \rho, R), \tag{4.1}$$

$$H_{\rm el}^{\rm DT}(t) = \sqrt{f(\xi) f'(\xi) g'(\zeta)}\,H_{\rm el}(t)\frac{1}{\sqrt{f(\xi) f'(\xi) g'(\zeta)}}. \tag{4.2}$$

By using this method, we obtained the simulation results shown in Fig. 4.3. We assumed that the polarization direction of light was parallel to the molecular axis. The coordinate system is drawn in Fig. 4.3a. In this case, two electronic degrees of freedom, ρ and z, are essential. The electronic coordinate z is chosen to be parallel to the polarization direction. In addition to these two electronic coordinates, we also included the internuclear distance R as a quantum mechanical variable. Therefore, we solved the time-dependent Schrödinger equation for the three degrees of freedom.

Figure 4.3b represents a packet in the three-dimensional (3D) space z, ρ, and R. The cloud around $z = 0$ and $R = 2a_0$ is an initial wave packet in the 3D space, where a_0 is the atomic unit of length, i.e., the Bohr radius. Therefore, if the packet goes upward or downward along z (indicated by vertical arrows) ionization is implied; and if the packet goes toward larger internuclear distances, i.e., along the direction of the bold arrow, dissociation is implied.

We applied a laser field as follows: The wavelength is 760 nm and the intensity is 5×10^{14} W/cm^2. The instantaneous potential for the electron is distorted as shown in Fig. 4.3c when a nonzero field is applied. In the initial moment, the packet goes toward longer internuclear distance. This means that the bond is stretching. We have assumed that $H_2{}^+$ is prepared in its lowest electronic state $|1s\sigma_{\rm g}\rangle$ after vertical ionization of H_2. As the internuclear distance approaches $R = 4a_0$, ionizing current can be seen as shown in Fig. 4.3d. The intensity applied is nearly constant from the beginning; this means that the ionization is enhanced at the internuclear distances that are much longer than the equilibrium one. This enhanced ionization is characteristic of molecular ionization induced by an intense field. There are also bound electronic state components. We can see two series of dense clouds around $z = R/2$ and $z = -R/2$ which originate from the lowest two bound electronic states of $H_2{}^+$, namely, $1s\sigma_{\rm g}$ and $1s\sigma_{\rm u}$.

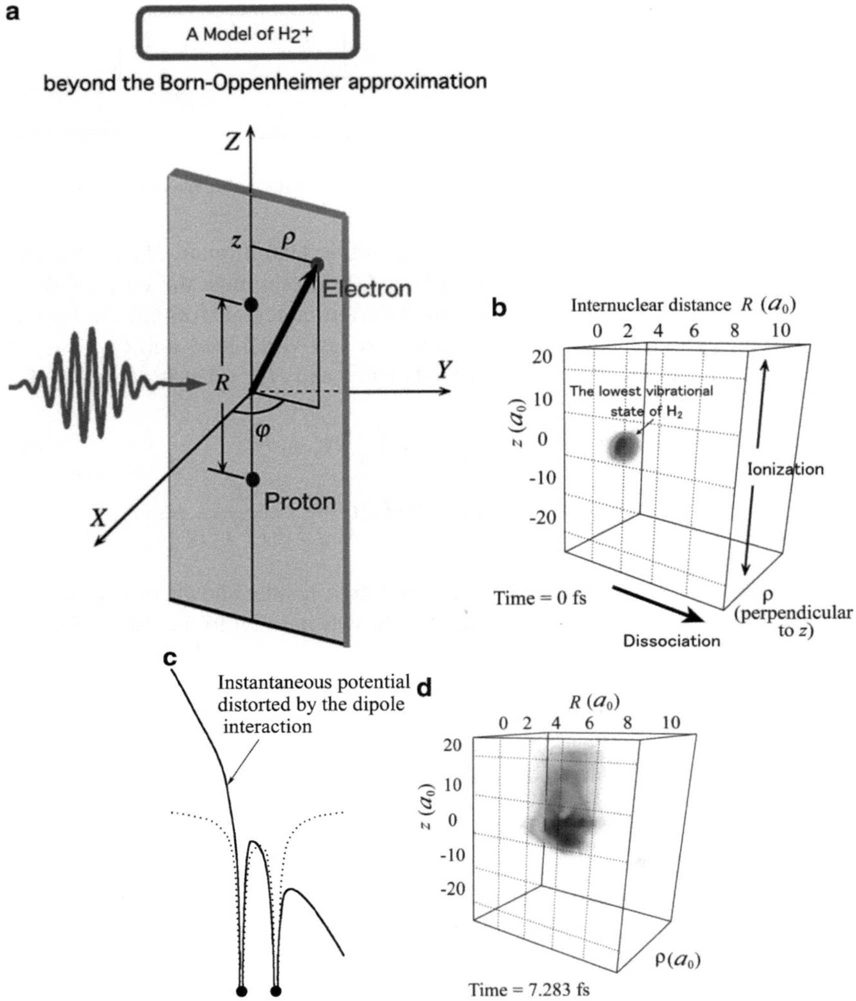

Fig. 4.3 Dynamics of the electronic and nuclear wave packet of $H_2{}^+$ in a nonresonant near-infrared field: (**b**) time = 0; (**d**) time = 7.3 fs. The cylindrical coordinate system for the electron and the internuclear distance R are shown in (**a**). Shown in (**c**) is the instantaneous potential for the electron including the dipole interaction with the field (*solid line*) together with the zero-field pure Coulomb potential (*dotted line*). The potential has two wells corresponding to the two protons denoted by *dots*

4.3 Time-Dependent Adiabatic State Approach and Its Application to Large Amplitude Vibrational Motion of C_{60} Induced by Near-IR Fields

Our next target is larger molecules such as C_{60}. We want to understand how larger molecules behave in an intense laser field. However, application of the exact simulation method presented above is limited to only small systems such as $H_2{}^+$ or H_2. Therefore, we have to invent and develop approaches to deal with molecular dynamics in the intense laser field. To that end, we first analyzed the exact simulation for $H_2{}^+$ in terms of time-dependent adiabatic states.

The time-dependent adiabatic states $|n(R,t)>$ are defined as the eigenfunctions of the instantaneous electronic Hamiltonian $H_{\mathrm{el}}(R,t)$ [3,4]. That is, the electronic Hamiltonian at a molecular geometry, $H_{\mathrm{el}}(R)$, plus the electric dipole interaction with the field, $V_\varepsilon(t)$:

$$\left[\hat{H}_{\mathrm{el}}(R) + V_\varepsilon(t)\right]|n(R,t)\rangle = E_n(R,t)|\, n(R,t)\rangle\,, \tag{4.3}$$

where the explicit form of $V_\varepsilon(t)$ is given by

$$V_\varepsilon(t) = \varepsilon(t) \times (\text{electronic coordinates}). \tag{4.4}$$

In this case, the simple length gauge was chosen. The eigenvalue equation (4.3) can be numerically solved for a general class of molecules. For this, we can use package molecular orbital programs such as GAMESS (General Atomic and Molecular Electronic Structure System) and obtain the eigenvalues of this instantaneous electronic Hamiltonian and also the eigenfunctions.

Then, what does the time-dependent adiabatic state look like? This is a natural extension of the concept of the adiabatic state to the time domain. The instantaneous electronic Hamiltonian of $H_2{}^+$ in atomic units (i.e., the electron mass m_e, Bohr radius $a_0 = 0.53$ Å, elementary electric charge e, hartree $E_\mathrm{h} = 27$ eV, and $\hbar$ are all set to unity) is written as:

$$\begin{aligned} H_{\mathrm{el}}(R) + z\varepsilon(t) = -\frac{1}{2}\frac{\partial}{\partial^2 x} - \frac{1}{2}\frac{\partial}{\partial^2 y} - \frac{1}{2}\frac{\partial}{\partial^2 z} - \\ \frac{1}{\sqrt{x^2+y^2+(z-R/2)^2}} - \frac{1}{\sqrt{x^2+y^2+(z+R/2)^2}} + \frac{1}{R} + z\varepsilon(t). \end{aligned} \tag{4.5}$$

The potential for the electron has two wells corresponding to two nuclei at $z = \pm R/2$. This electronic potential depends on the internuclear distance R as shown in Figs. 4.4a and 4.4b. The solid and dotted horizontal lines indicate the energies of the lowest and second-lowest adiabatic states. The lowest two states in the zero field case of Fig. 4.4a correspond to $1s\sigma_\mathrm{g}$ and $1s\sigma_\mathrm{u}$, respectively. When an electric field is applied, the potential is distorted and the energy levels are also shifted, as shown in Fig. 4.4b. The solid lines in Fig. 4.4c denote the energies of the

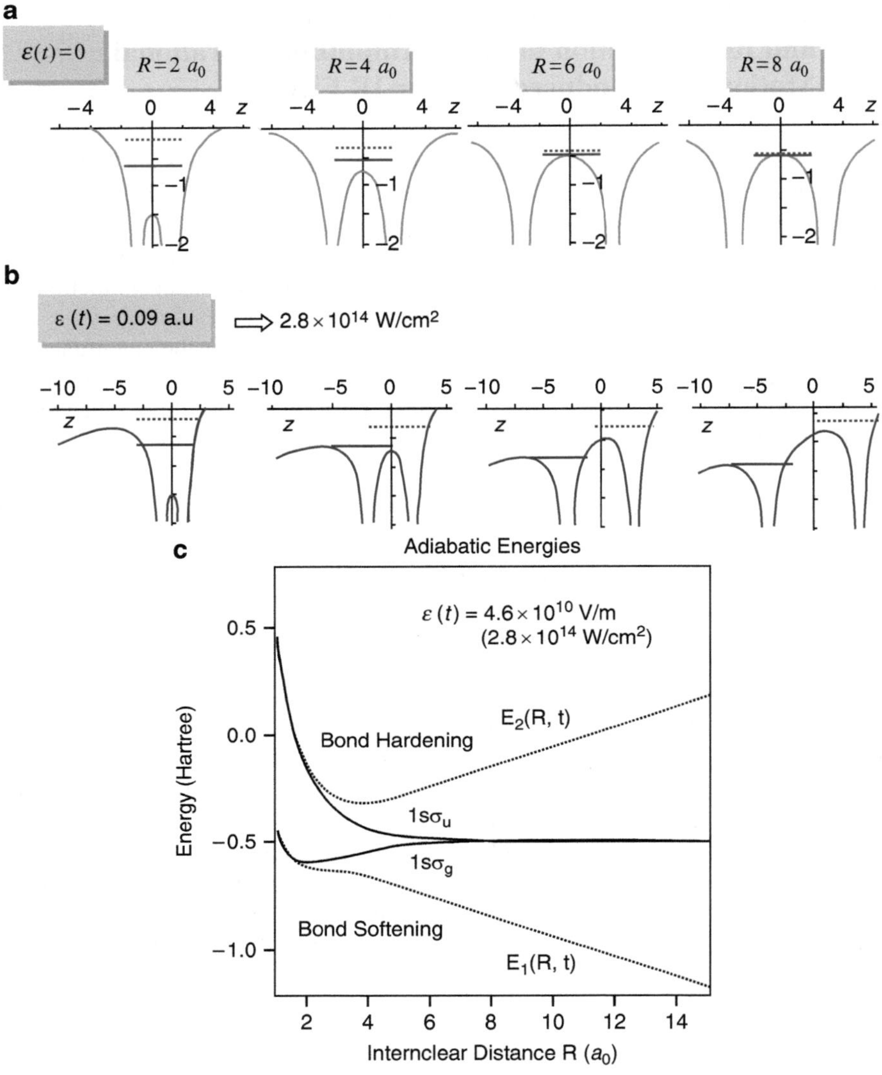

Fig. 4.4 The eigenvalues (time-dependent adiabatic energies) of the instantaneous electronic Hamiltonian of $H_2{}^+$, (4.5). The potential of the instantaneous Hamiltonian at different internuclear distances are shown in (**a**) for $\varepsilon(t) = 0$ and in (**b**) for $\varepsilon(t) = 0.09\ E_h/ea_0 = 4.6 \times 10^{10}\ \mathrm{V/m}$. The lowest and second-lowest adiabatic energies are indicated by *solid* and *dotted horizontal lines*, respectively. Drawn in (**c**) are the lowest two adiabatic potential surfaces, $E_1(R,t)$ and $E_2(R,t)$, which are adiabatically connected to the zero-field surfaces of $1s\sigma_g$ and $1s\sigma_u$

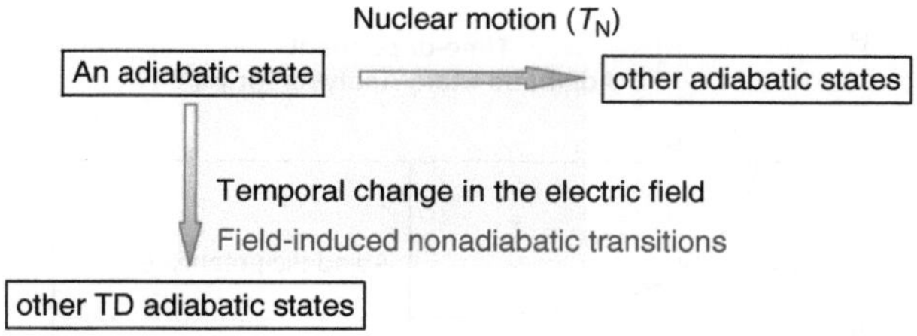

Fig. 4.5 Breakdown of adiabatic properties induced by nuclear motion and by temporal change in the laser electric field

two lowest adiabatic states in the zero field case as a function of R. These potentials against the nuclei vary as the laser electric field changes, i.e., as time passes. For instance, as shown in Fig. 4.4c, the time-dependent potential of the lowest adiabatic state moves down from the solid line to the dotted line as the field strength $\varepsilon(t)$ increases from 0 to 0.09 in atomic units, i.e., 0 to 0.09 E_h/ea_0 (0 to 4.6×10^{10} V/m) and that of the second-lowest state moves up. I would like point out that the electronic wave function of the lowest adiabatic state is localized in the lower potential well (for instance, the left well in Fig. 4.4b) and that of the higher adiabatic state is localized in the upper well.

Thus, we can define time-dependent adiabatic states. If the temporal change in the laser electric field were very slow, we could follow one adiabatic state. However, even in the near-infrared field case, the system has a chance to behave nonadiabatically with respect to the temporal change in the laser electric field unless $\omega_{\text{elect}} \gg \omega$, as denoted by the vertical arrow in Fig. 4.5.

Therefore, we have to analyze the exact wave functions in terms of the time-dependent adiabatic states [5–8]. As noted above, the exact wave functions of $H_2{}^+$ in the optical electric field, $\psi(z, \rho, R; t)$, are available and the time-dependent adiabatic states can be defined for the lowest one and the second-lowest one and so on. Hence, we can map this exact wave functions onto these time-dependent adiabatic states, namely, the lowest two time-dependent adiabatic states $|1(R,t)\rangle$ and $|2(R,t)\rangle$ ($|1(R,t)\rangle = |1s\sigma_g\rangle$ and $|2(R,t)\rangle = |1s\sigma_u\rangle$ for $\varepsilon(t) = 0$) as:

$$|\psi(z,\rho,R,t)\rangle = \chi_1(R,t)\,|1(R,t)\rangle + \chi_2(R,t)\,|2(R,t)\rangle\,, \qquad (4.6)$$

with the coefficients χ_1 and χ_2. These $\chi_1(R,t)$ and $\chi_2(R,t)$ are the vibrational wave functions associated with the time-dependent adiabatic states.

The results obtained by the mapping method are presented in Fig. 4.6. The lowest two time-dependent adiabatic potentials in zero fields, $E_1(R,t)$ and $E_2(R,t)$, are shown in Fig. 4.6a, which will change with time during the interaction with the intense, near-infrared pulse. The initial vibrational component $|\chi_1(R,t=0)|^2$ associated with the state $|1(R,t=0)\rangle = |1s\sigma_g\rangle$ created by vertical ionization of H_2 is also indicated in Fig. 4.6a by the Gaussian-like profile. Initially, the nuclear wave packet propagates on the lowest time-dependent adiabatic state $|1(R,t)\rangle$, and as the intranuclear distance increases, the probabilities of nonadiabatic transitions from the lowest adiabatic state to the higher adiabatic states increase. At $t = 7$ fs, the higher $|\chi_2(R,t)|^2$ is about half of $|\chi_1(R,t)|^2$, as shown in Fig. 4.6b.

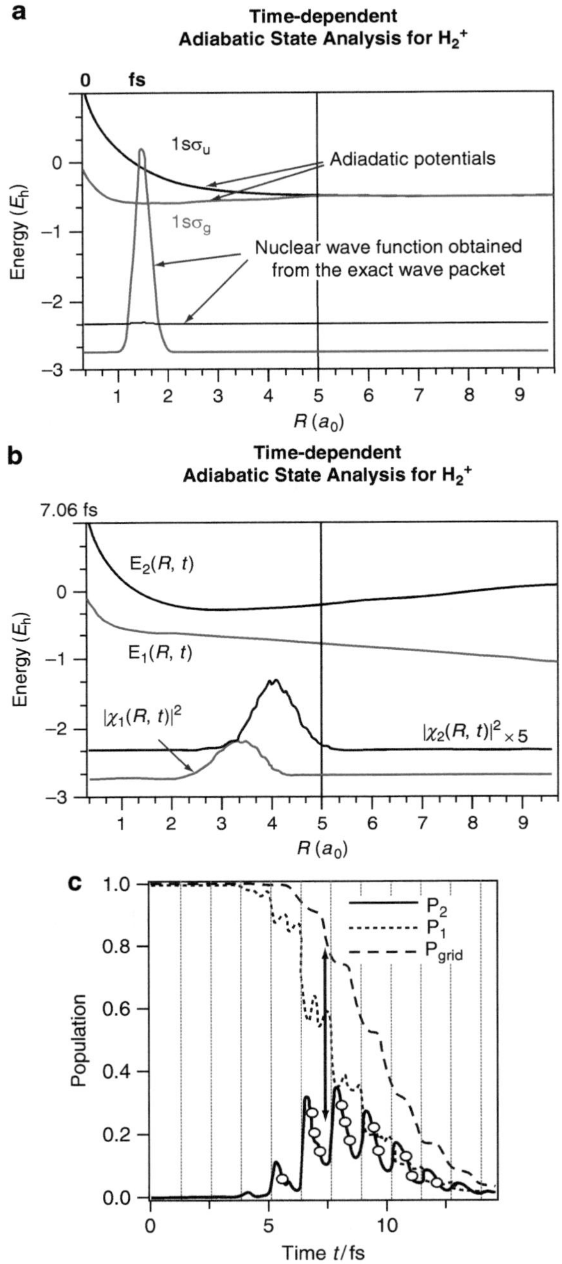

Fig. 4.6 Time-dependent adiabatic state analysis for $H_2{}^+$: (**a**) the vibrational wave functions of the lowest two adiabatic states $|1(R,t)\rangle$ and $|2(R,t)\rangle$ at time $t = 0$; (**b**) the vibrational wave functions at time $= 7.03$ fs; (**c**) the overall populations $P_1(t)$ and $P_2(t)$ of $|1(R,t)\rangle$ and $|2(R,t)\rangle$ (denoted by *dotted* and *solid lines*, respectively). P_{grid} is the population inside the grid space defined for propagating the wave function

Figure 4.6c shows the total populations of the two adiabatic states, namely, $P_1(t) = \int |\chi_1(R,t)|^2 \, \mathrm{d}R$ and $P_2(t) = \int |\chi_2(R,t)|^2 \, \mathrm{d}R$. The dotted line shows the total population of the lowest adiabatic state, $P_1(t)$, and the solid line shows the total population of the higher adiabatic state, $P_2(t)$. The vertical dotted lines indicate the moments at which the field returns to zero. Every time when the field becomes zero, that is, the corresponding two potential surfaces $E_1(R,t)$ and $E_2(R,t)$ come closest to each other, a field-induced nonadiabatic transition from $|1(R,t)\rangle$ to the second-lowest one $|2(R,t)\rangle$ can be found. Soon after that, the population in the second-lowest state decreases. This is due to the ionization from $|2(R,t)\rangle$ or the higher adiabatic states populated via $|2(R,t)\rangle$ [5, 8]. We can say that the second-lowest adiabatic state is a doorway state to the ionization, and the so-called enhanced ionization is considered to occur through this kind of higher lying time-dependent adiabatic states. In short, the higher adiabatic state populated via field-induced nonadiabatic transitions is more easily ionized because the energy of the second-lowest state for $4\,a_0 < R < 11\,a_0$ is located already close to or over the ionization barrier formed in the presence of the field, as shown in Fig. 4.4b.

We can plot the temporal variation of the nuclear wave functions for the lowest two components obtained by the mapping. From the contour maps of $|\chi_1(R,t)|^2$ and $|\chi_2(R,t)|^2$ in Fig. 4.7, we found that at intrernuclear distances longer than $3a_0$, nonadiabatic transitions occur from the lowest state to the higher state when the field $\varepsilon(t)$ returns to zero (denoted by vertical dotted lines in the lower graph of Fig. 4.7).

In the above discussions, we included the internuclear distance as a quantum-mechanical dynamical variable. We also calculated the ionization probabilities at fixed internuclear distances, and plotted them as a function of the internuclear distance as shown in Fig. 4.8. The light intensity of the applied near-infrared field is $10^{14}\,\mathrm{W/cm^2}$ and the wavelength is 1,064 nm.

As shown in Fig. 4.8, the rate of ionization from the initial ground state $|1s\sigma_g\rangle$, Γ, denoted by the solid line, significantly increases as the internuclear distance increases. For example, the ionization rate at the internuclear distance of $9a_0$ is three times as large as that at $6a_0$ and is three orders of magnitude larger than that at the equilibrium internuclear distance of $2a_0$. This enhancement in ionization at the large internuclear distance is ascribable to the increase in the ionization rate from the higher adiabatic state $|2(R,t)\rangle$ nonadiabatically populated from $|1(R,t)\rangle$. This is confirmed by the fact that the ionization rate of the second-lowest adiabatic state at a DC constant field corresponding to the light intensity of $10^{14}\,\mathrm{W/cm^2}$, denoted by the broken line, has two peaks around $R = 6a_0$ and $9a_0$ as Γ. In the range where field-induced nonadiabatic transitions occur, the ionization rates of the ground and second-lowest adiabatic states exhibit a similar tendency as a function of the internuclear distance.

As shown above, the time-dependent adiabatic states can form a good basis set for describing the electronic and nuclear dynamics. Hence, we introduce a simple approach, in which the total wave function is expanded in terms of time-dependent adiabatic states. In this case, we cannot discuss the ionization because we neglect the continuum electronic states. However, we can expect that this approach is sufficient for describing the dynamics of bound electrons within a molecule as well as the vibrational dynamics.

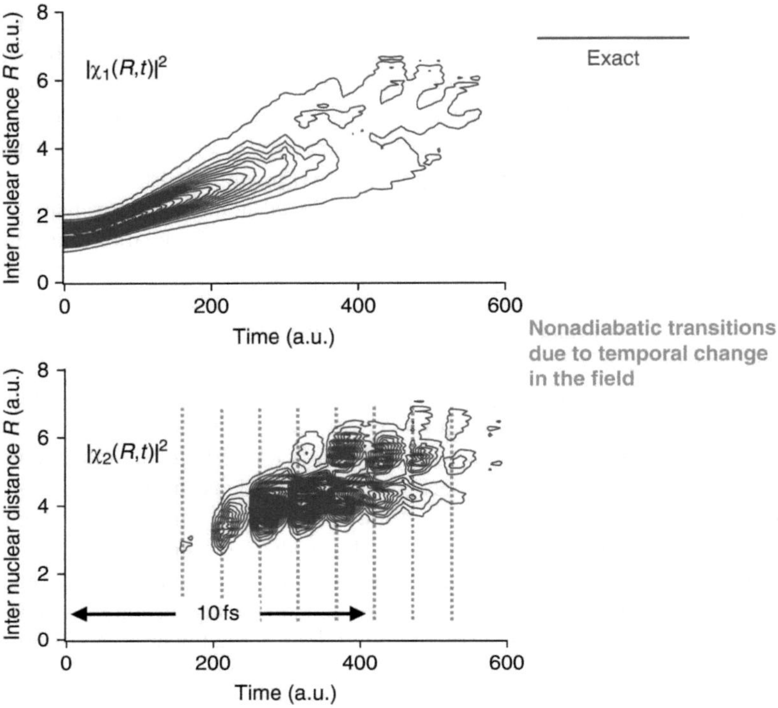

Fig. 4.7 Dynamics of the nuclear wave functions $|\chi_1(R,t)|^2$ and $|\chi_2(R,t)|^2$ of $H_2{}^+$ associated with the time-dependent adiabatic states $|1(R,t)\rangle$ and $|2(R,t)\rangle$ obtained from the exact wave function [5–8]

Thus, we can put, e.g., the two-state expansion:

$$|\psi\rangle \approx \chi_1(R)|\,1(R,t)\rangle + \chi_2(R)|\,2(R,t)\rangle \tag{4.7}$$

into the total Schrödinger equation:

$$\mathrm{i}\hbar\frac{\partial}{\partial t}|\psi\rangle = H_{\text{total}}(t)|\,\psi\rangle \tag{4.8}$$

and derive the time-dependent coupled equations to obtain the vibrational wave functions in the two state model, i.e., approximate vibrational wave functions for $\chi_1(R,t)$ and $\chi_2(R,t)$:

$$\frac{\partial}{\partial t}\chi_1(R) = -\mathrm{i}\left[-\frac{1}{m_\mathrm{p}}\frac{\partial^2}{\partial R^2}+E_1(R,t)\right]\chi_1(R) - \left[\left\langle 1\left|\frac{\partial}{\partial t}\right|2\right\rangle + \cdots\right]\chi_2(R), \tag{4.9}$$

$$\frac{\partial}{\partial t}\chi_2(R) = -\mathrm{i}\left[-\frac{1}{m_\mathrm{p}}\frac{\partial^2}{\partial R^2}+E_2(R,t)\right]\chi_2(R) - \left[\left\langle 2\left|\frac{\partial}{\partial t}\right|1\right\rangle + \cdots\right]\chi_1(R), \tag{4.10}$$

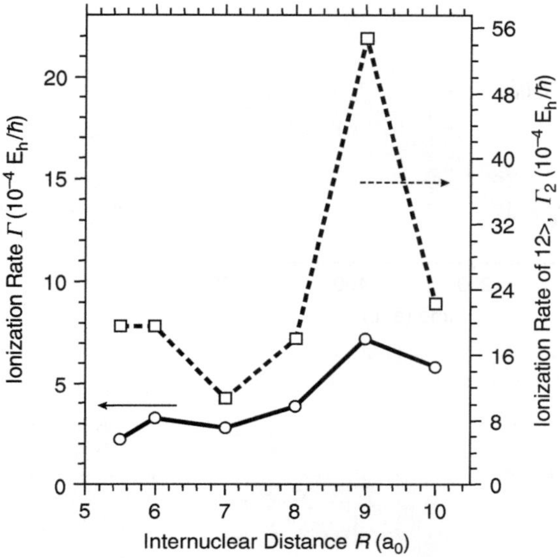

Fig. 4.8 Ionization rates of $H_2{}^+$ as a function of internuclear distance R (in units of $E_h/\hbar = 4.1 \times 10^{16}/s$). The *open squares* denote ionization rates Γ_2 of the upper adiabatic state $|2(R,t)\rangle$ at a DC constant field strength $\varepsilon(t) = 0.0533\, E_h/ea_0$ (which corresponds to $I = 10^{14}\,\mathrm{W/cm^2}$); the *open circles* denote ionization rates Γ in an alternating intense field of $\lambda = 1{,}064$ nm and $I = 10^{14}\,\mathrm{W/cm^2}$ under the condition that the initial state is the ground state 1s σ_g (the field envelope has a five-cycle linear ramp and Γ is defined as the stationary value after the ramp). The scales for Γ_2 and Γ are marked on the *right* and *left* ordinates, respectively. Adapted from [4]

where $E_1(R,t)$ and $E_1(R,t)$ denote the adiabatic energies. To solve these coupled equations, we have to calculate the energies of time-dependent adiabatic states and the field-induced nonadiabatic couplings, which induce population transfer to different time-dependent adiabatic states.

I will first show the exact result of $H_2{}^+$ in Fig. 4.9 (solid lines), and then, superimpose the result based on the two-state model (dotted lines), namely, the expansion using the lowest two states of (4.7) [5–8]. From the comparison between the two results, we found that this approach seems to be quite good for the description of vibrational dynamics. Hence, for general polyatomic molecules, we will expand the total wave function in terms of the time-dependent adiabatic states. We can use package molecular orbital programs to calculate the time-dependent adiabatic energies and field-induced nonadiabatic couplings [6,7].

We take C_{60} as the first example of applications to polyatomic molecules. It is well known that ionized C_{60} molecules undergo fragmentation via the C_2-elimination processes. As reported by Hertel and co-workers [9], when a 5-ps pulse of 800 nm or the shorter wavelength light is used, ionized fragments such as $C_{58}{}^+$, $C_{56}{}^+$, and so on are found, as shown in the upper panel of Fig. 4.10. No odd number large fragments such as $C_{57}{}^+$ are detected. When the pulse length is shorter, highly charged parent cations up to $C_{60}{}^{6+}$ can be produced, but charged

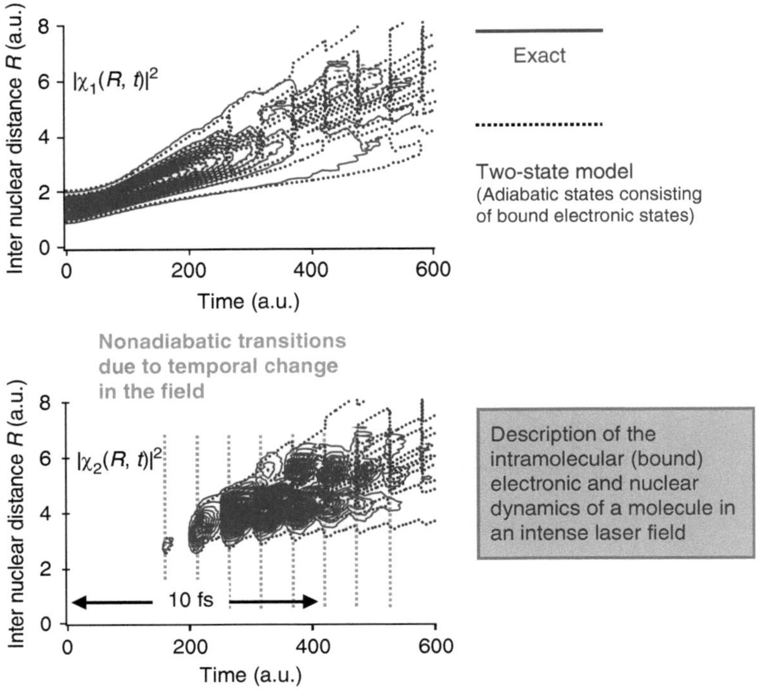

Fig. 4.9 Validity of the two-state model for the description of the nuclear wave functions of $H_2{}^+$ associated with $|1(R,t)\rangle$ (*upper panel*) and $|2(R,t)\rangle$ (*lower panel*): the exact ones (*solid lines*) and those obtained from the two-state model (*dotted lines*) [5–8]

fragments are still found. For example, large dication fragments such as $C_{58}{}^{2+}$ are detected in the case of 25-fs pulses as shown in the lower panel of Fig. 4.10, though appearance of monocations is significantly suppressed.

Recently, Bhardwaj and Corkum carried out an interesting experiment using a longer wavelength of $\lambda = 1{,}800$ nm [10]. The pulse length was 70 fs. They observed only highly charged intact parent cations almost without fragmentation. These highly charged parent cations have lifetimes longer than microseconds. Why is the fragmentation suppressed for the long wavelength excitation? We wanted to answer this question, and we have partially succeeded it as described next.

Highly charged cations of C_{60} must have stable structures for them to survive without fragmentation. We calculated the geometrical structure of C_{60} and its cations and found that cations have a stable structure even up to $C_{60}{}^{14+}$ [11]. The left-hand plot of Fig. 4.11 shows how the equilibrium distances of 90 C–C bonds depend on the charge z of $C_{60}{}^{z+}$. In the neutral case, we have only two typical bond distances; the C–C bond distance in the pentagon, denoted by r_5, and that of the C–C bond shared by the two adjacent hexagons, denoted by r_6. The average bond distances of r_5 and r_6, denoted by $<r_5>$ and $<r_6>$, respectively, increase as the

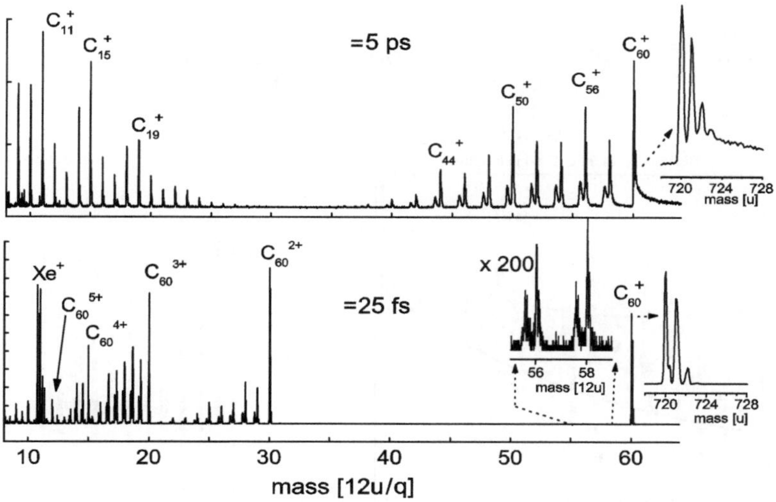

Fig. 4.10 Competition between ionization and fragmentation of C_{60}: Dependence of the mass spectrum on the pulse length [9]. The pulse length used is 5 ps in the case of the *upper panel* and 25 fs in the case of the *lower panel*

charge of C_{60} increases. However, even at $C_{60}{}^{10+}$, the increases in the bond distances are less than 0.03 Å, which are very small. The right-hand plot shows the vibrational frequencies as a function of the charge z. Since all the vibrational frequencies for $C_{60}{}^{z+}(z = 1\text{–}14)$ cations are real, C_{60} cations up to $z = 14$ have a stable structure. There is a tendency that the vibrational frequencies decrease as the change increases, but even at $C_{60}{}^{14+}$ the vibrational frequencies are not so much different from those of neutral C_{60}. The results of Fig. 4.11 suggest that even highly charged C_{60} cations are tough to dissociate.

As shown in Fig. 4.12, we examined the stability of $C_{60}{}^{z+}(z = 1\text{–}14)$ cations, and concluded that parent cations $C_{60}{}^{z+}$ produced by intense 1,800-nm pulses have long lifetimes up to $z = 11$ to 12. We first estimated the energy available for the fission to C_{58} and C_2 cations, i.e., the sum of the excess vibrational energy upon vertical ionization, ΔE, and the thermal energy E_T (~5 eV). The energy required for the fission is the sum of the fission barrier height $D(z)$ [12] and the so-called kinetic shift $S(z)$. The energy available for fission, $\Delta E + E_T$, increases as the charge z increases while $D(z) + S(z)$ decreases. These quantities schematically illustrated in the left graph of Fig. 4.12 are plotted in the right-hand as a function of z. By putting these values into the reaction rate formula of the conventional statistical theory, namely, Rice–Ramsperger–Kassel–Marcus (RRKM) theory [13, 14], we found that the lifetime of $C_{60}{}^{z+}$ dramatically decreased as the charge z increased from +11 to +13 [11]. The lifetime of $C_{60}{}^{11+}$ was on the order of seconds and the lifetime of $C_{60}{}^{12+}$ was on the order of microseconds. This indicates that the parent cations up to $C_{60}{}^{12+}$ can be detected by time-of-flight (TOF) mass spectrometry of microsecond detection time, in agreement with the experimental observation. The lifetime

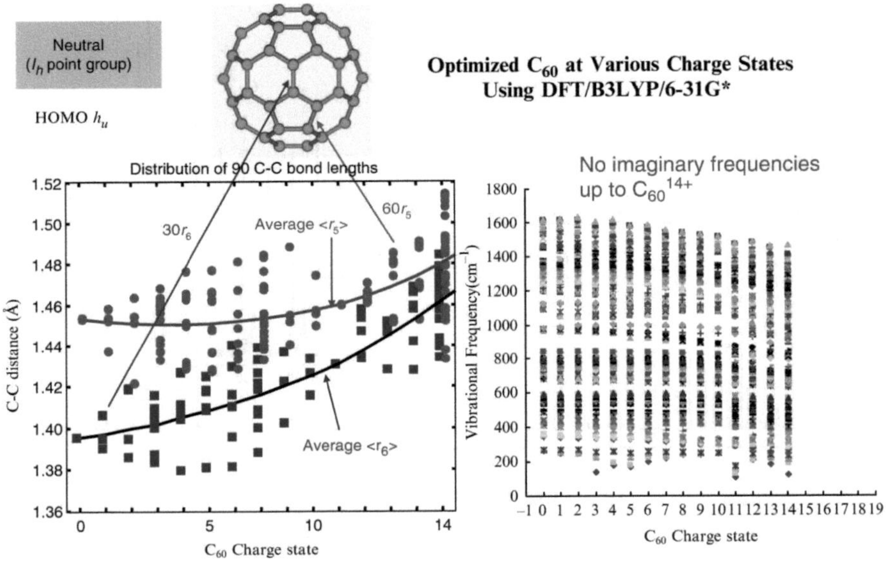

Fig. 4.11 Bond distances (*left*) and frequency distributions (*right*) of C_{60} as a function of the charge z. In the left-hand plot, the distances r_6 of 30 C—C bonds shared by two adjacent hexagons are marked with *squares* and the distances r_5 of 60 C—C bond distances in pentagons are marked with *circles*. In the right-hand plot, the frequencies of 174 normal modes for individual charge states are indicated by *dots*

of $C_{60}{}^{13+}$ was estimated to be less than nanoseconds. The right plot in Fig. 4.12 indicates that the stability condition of $C_{60}{}^{z+}$, $D(z) + S(z) > E(z) + E_T$, holds even for $z \sim 12$.

Figure 4.13 uses results of molecular dynamics simulation to show how tough this molecule is. In this simulation, we assumed that neutral C_{60} molecule was vertically ionized to $z = 14$. Since the equilibrium structure of $C_{60}{}^{+14}$ is larger than C_{60}, the nuclei in the cation start moving after this vertical ionization. The motion is the totally symmetric radial breathing mode (period ~75 fs), in which the bond distances of r_5 and r_6 oscillate in phase and the vibrations continue as shown in Fig. 4.13. Even for $C_{60}{}^{14+}$, we cannot see any fragmentation in the time range of a few picoseconds after vertical ionization.

In the above discussion, we did not include the effects of field-induced vibrational excitation. Next, we wanted to look at how vibration is induced in neutral C_{60} or its cations by an intense laser field. We thought the time-dependent adiabatic state approach would be useful here. Hence, we assumed that the potentials change like shown in Fig. 4.14b according to the change in the field strength, as the points numbered from 1 to 5 in Fig. 4.14a. We can use these potentials for quantum mechanical simulations [5, 8, 15] as well as for classical dynamics simulations [16]. In the simulation introduced above for C_{60}, we treated the motion of all the nuclei

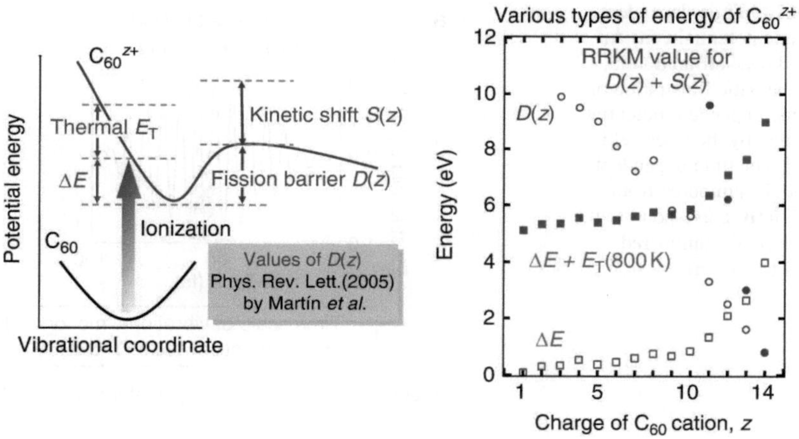

Fig. 4.12 Stability of $C_{60}{}^{z+}$ upon vertical ionization [11]. In the left illustration, $\Delta E(z)$, fission barrier height $D(z)$ for C_2-loss dissociation, total thermal energy $E_T(\sim 5\,eV)$, and kinetic shift $S(z)$ are indicated in the potential surfaces of C_{60} and $C_{60}{}^{z+}$ drawn as a function of a representative vibrational coordinate. $\Delta E(z)$ is the vibrational energy acquired by $C_{60}{}^{z+}$ upon sudden vertical ionization from C_{60}, i.e., the difference between the energy of $C_{60}{}^{z+}$ at the equilibrium structure of C_{60} and the energy of $C_{60}{}^{z+}$ at its equilibrium structure. The sum of $D(z)$ and $S(z)$ is the total vibrational energy required for the cation to fragment within a certain time frame (μs) of the detection experiment. The kinetic shift is obtained by using the RRKM formula. The sum of $\Delta E(z)$ and E_T is available for fissions. Therefore, the stability condition of $C_{60}{}^{z+}$ is that $D(z) + S(z) > \Delta E(z) + E_T$

Fig. 4.13 Dynamics of C–C bonds of $C_{60}{}^{14+}$ after vertical ionization from neutral C_{60}. The distances of 90 C–C bonds are plotted as a function of time (60 r_5-bonds and 30 r_6-bonds). r_6-bonds are indicated by dark lines.

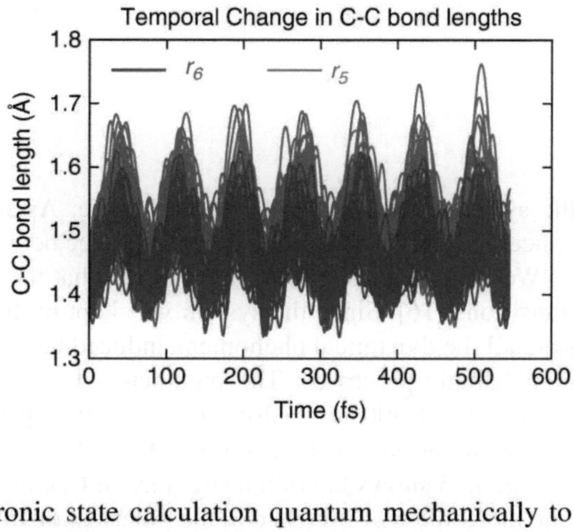

classically, while doing the electronic state calculation quantum mechanically to obtain the time-dependent adiabatic states.

It has been found by numerical simulations that C_{60} and its cations exhibit vibrations of large-amplitude in the presence of an intense near-infrared laser field. In reality, the system is subject to ionization in addition to the large-amplitude vibrational excitation. However, in the following simulation, we assumed that

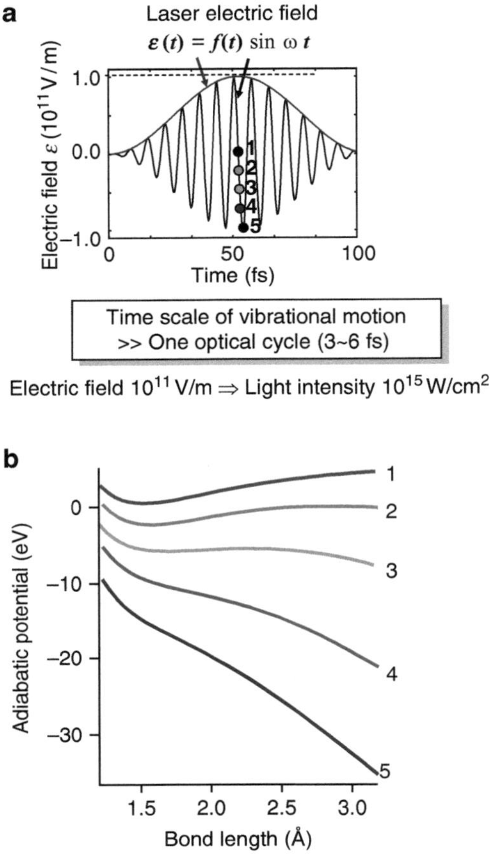

Fig. 4.14 Time-dependent adiabatic potentials: (**a**) An applied near-infrared field; (**b**) Schematic illustration of the time-dependent potential deformed by the laser field [6, 15]. The time-dependent potentials numbered from 1 to 5 in (**b**) correspond to the field strengths numbered from 1 to 5 in (**a**)

the system remained in the neutral stage. As far as the vibrational dynamics is concerned, there is no qualitative difference between the two cases.

We applied a laser field and assumed that there were no nonadiabatic electronic transitions [16]. Since the system was kept in the lowest time-dependent adiabatic state all the dynamical phenomena induced here should be ascribed to vibrationally nonadiabatic processes. The parameters for the optical pulse are chosen as follows: $\lambda = 1{,}800\,\mathrm{nm}$; pulse length $= 70\,\mathrm{fs}$; peak intensity $= 7 \times 10^{14}\,\mathrm{W/cm^2}$. The polarization direction is along the vertical direction in Fig. 4.15. The left panel of Fig. 4.15 shows the initial structure of C_{60} before the interaction with the pulse. The molecule was stretched in the polarization direction as shown in the right panel of Fig. 4.15. Even after the light pulse disappeared completely, the molecule was found to be still moving. This meant that the molecule acquired vibration energy. The vibrational energy increase was found to be 27 eV. Usually, we would expect such a large energy inevitably leads to prompt dissociation of the molecule (fs or ps time domain) because 27 eV is much larger than the typical dissociation energy,

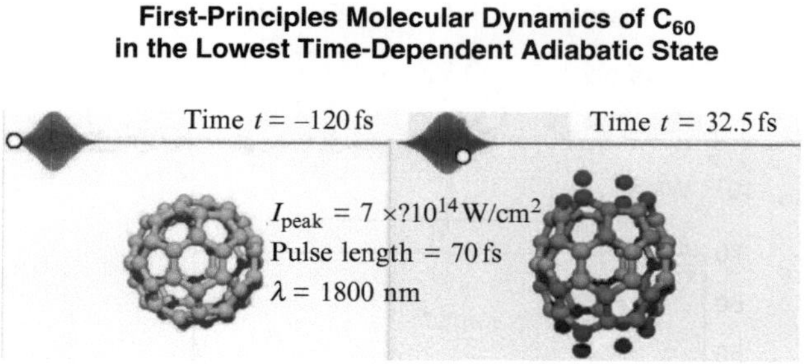

Fig. 4.15 First-principles molecular dynamics of C_{60} in the lowest time-dependent adiabatic state. The length of the applied pulse is 70 fs and the peak of the Gaussian profile is centered at $t = 0$ fs (the peak intensity is $I_{\text{peak}} = 7 \times 10^{14}$ W/cm^2). The *left panel* shows the snap shot of C_{60} nuclei at $t = -120$ fs and the *right panel* shows the snap shot at 32.5 fs. The corresponding field strengths are indicated by *open circles* in the pulse profiles

e.g., the dissociation energy for C_2-elimination that is as low as $\sim$10 eV. However, we found the excess energy was not sufficient for the large C_{60} molecule to decompose in the ps time domain.

We may wonder why this type of large-amplitude vibrational excitation occurs. When we looked at the vibrational motion in the time-dependent adiabatic state, we saw the vibrational mode induced was the so-called $h_g(1)$ vibrational mode. Along with the vibration, C_{60} deforms into the ellipsoid (or discoid) form, and then, into the discoid (or ellipsoid) form as shown in Fig. 4.16. The potential surface is shown in the figure as a function of the displacement corresponding to the shapes of C_{60} above the graph. The potential in the zero optical field and the cycle-averaged potential of the lowest time-dependent adiabatic state at $I = 10^{15}$ W/cm^2 are plotted in the graph. The time-dependent potential averaged over one optical cycle is regarded as the effective potential for vibration because the time-dependent potential changes too fast for the molecular vibration to follow the rapidly oscillating electric field. The potential is most distorted along the Raman active vibrational modes such as a_g and h_g modes. The nuclei move in this suddenly distorted potential, and hence large-amplitude motion is induced in the $h_g(1)$ mode for the present choice of the pulse parameters. The field strength is very strong here, but the potential along the $h_g(1)$ mode is a bound type even in the presence of this very strong field. This is why neutral C_{60} does not exhibit fragmentation in the ps time domain.

The $h_g(1)$ mode is mainly excited in the present case. This allows us to perform a quantum mechanical simulation by using only this mode. We calculated the time-dependent adiabatic states and the simulation results are shown in Fig. 4.17. The broken contour lines in the graph show the effective cycle-averaged potential of the lowest time-dependent adiabatic state when we apply the field shown in the right side of the graph. The maximum distortion of the cycle-averaged potential occurs at the peak of the light intensity. On the other hand, the maximum distortion of the molecule or the vibrational wave packet occurs with a delay of around 30 fs. Therefore, we can clearly say that this excitation is induced by a vibrationally

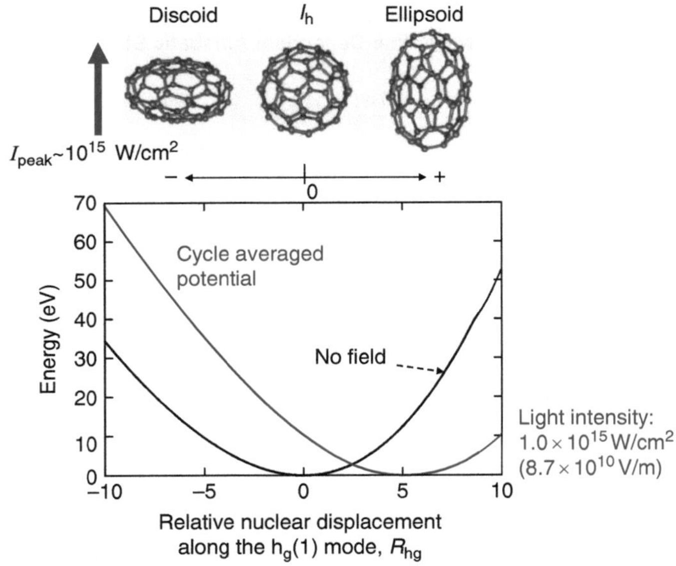

Fig. 4.16 Zero-field potential and optical cycle average of the time-dependent adiabatic potential for the $h_g(1)$ normal mode of C_{60}. The intensity of the applied pulse is $I = 7 \times 10^{14}$ W/cm^2

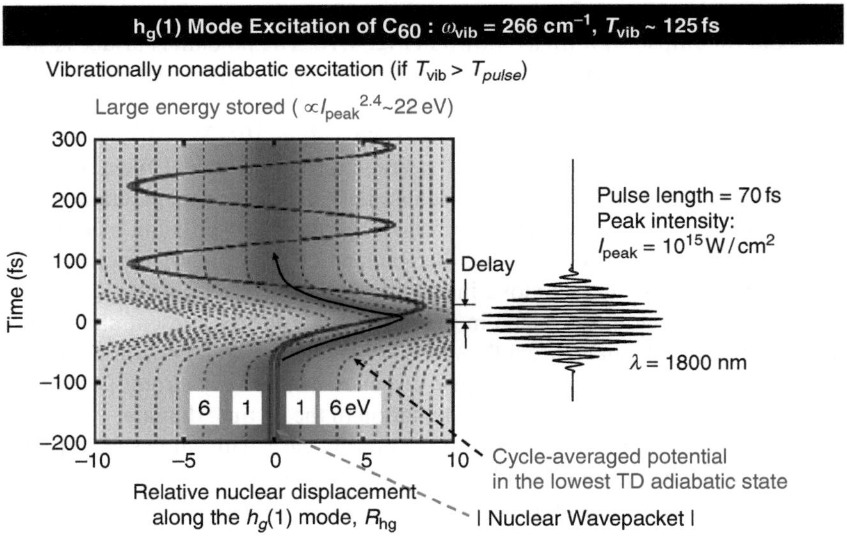

Fig. 4.17 Impulsive Raman excitation by a near-infrared pulse: Wave packet propagation along the normal coordinate of the $h_g(1)$ mode. The cycle-averaged potential is denoted by broken contour lines at intervals of 5 eV. The potential minimum, which is set to zero irrespective of time, follows the curve with an *arrow* in the graph. The delay in time between the maximum distortions of the molecular structure and the cycle-averaged potential is ~30 fs

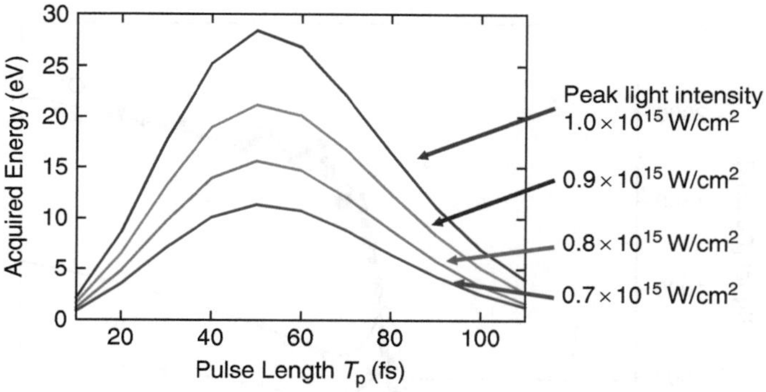

Fig. 4.18 Energy acquired in C_{60} by impulsive Raman excitation at four different light intensities as a function of the pulse length T_p. The acquired energy is maximized around $T_p = 50$–60 fs and is nearly proportional to ${I_{peak}}^{2.4}$ at $T_p = 50$ fs

nonadiabatic process. This might also be regarded as a type of impulsive Raman excitation. Since this pulse length is shorter than the vibrational period of the $h_g(1)$ mode, the vibrational wave packet cannot follow the sudden change in the cycle-averaged potential. This is the reason why we have a large vibrational excitation. The center of the wave packet vibrates even after full decay of the applied pulse. The net energy gain was 22 eV.

We have examined the vibrational energy gained by the intense laser field as a function of the pulse length for the four different peak light intensities. We determined that the maximum peak appeared around the pulse length of 50–60 fs, as shown in Fig. 4.18. For efficient vibrational excitation, the pulse length should not be too short or too long; the pulse length should be around half of the vibrational period. The pulse length of $\sim$ 60 fs is close to half of the vibrational period of the $h_g(1)$ mode ($\sim$125 fs).

It is valuable to look at one more experiment for C_{60} which was recently carried out by Professor Hertel and his group [17, 18]. As shown earlier above, the major fragments, particularly the large ion fragments, are generated via C_2-elimination, that is, ${C_{58}}^+$ or ${C_{56}}^+$ and so on, are generated. If this type of C_2-elimination occurs after the statistical randomization of the vibrational energies given from the external light field, it should take nanoseconds or microseconds for the cations to dissociate. This is characteristic of statistical fragmentation. However, their experimental results shown in Fig. 4.19 clearly demonstrated that within 1 ps there were other channels than the C_2-elimination. They used a 400 nm pump and 800 nm probe pulses and also 800 nm pump and probe pulses; similar results were obtained for both cases. No ion signal appeared when only a probe pulse was introduced. Therefore, they concluded that C_{60} was decomposed by the pump pulse into neutral fragments (in addition to cation fragments) and then the electronically excited neutral species were ionized by the probe pulse. The ionized fragments were detected by time-of-flight mass spectrometry. In this case, even within 1 ps, C_3 fragments

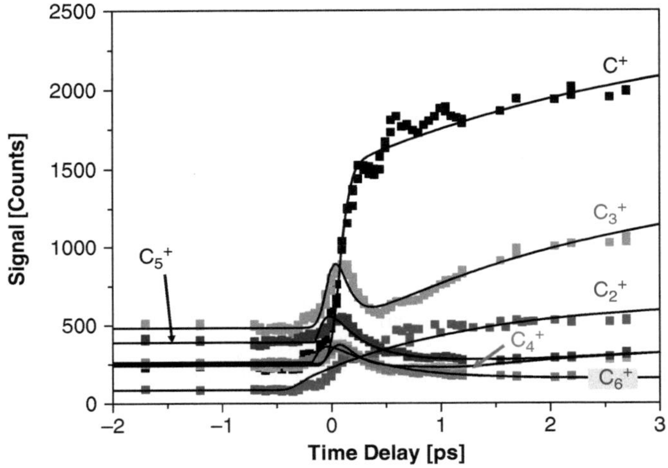

Fig. 4.19 Nonstatistical ultrafast fragmentation of C_{60} detected by a pump (400 nm) – probe (800 nm) experiment [18]. The signals of small fragments in the TOF are plotted as a function of the delay between the pump and probe pulse. C^+ and $C_3{}^+$ fragments appear in the femtosecond domain as well as $C_2{}^+$

appeared. If this is a simple process in which C_3 is ejected from C_{60}, the appearance of C_{57} fragments could be expected. However, as usual, the $C_{57}{}^+$ signal was not detected. The decomposition within a very short timescale was probably triggered by some nonstatistical effects such as field-induced electronically nonadiabatic processes [6, 7] in addition to field-induced potential distortion. Therefore, in order to estimate the exact branching ratios in different fragments, we have to include electronically nonadiabatic processes as well as vibrationally nonadiabatic processes.

4.4 Bond Dissociation Dynamics of Ethanol: Branching Ratio of C–C and C–O Dissociation

In the final example chosen for discussion, we show how important the nonadiabatic electronic transitions are by presenting the simulation results of bond breaking processes of ethanol in intense laser fields. Ethanol CH_3–CH_2–OH has two skeletal bond types: the C–C and the C–O bonds. Professors Kannari and Yamanouchi carried out a very interesting experiment in which they showed that the major fragmentation process was dissociation of the C–C bond if the pulse length was shorter than 30 fs [19, 20]. When they stretched the pulse length, the yield ratio of the C–O dissociation increased. In order to understand this, we proposed the following simple model presented in Fig. 4.20. We first assumed that molecules are randomly oriented. Further, we selected two cases. One case was when the C–O bond is

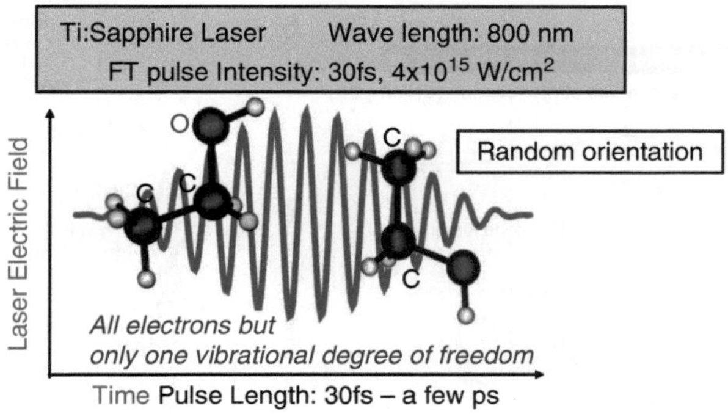

Fig. 4.20 One-dimensional models for bond dissociation of ethanol by near-infrared pulses: (*left molecule*) the C—O bond is parallel to the laser electric field; (*right molecule*) the C—C bond is parallel to the laser electric field [19]

parallel to the laser electric field, and the other one was when the C—C bond is parallel to the laser electric field. These two configurations are expected to be the best configurations for the dissociation of the C—O and C—C bonds, respectively.

First, we calculated the time-dependent adiabatic potentials as a function of the C—C bond length and the laser field was applied parallel to the C—C axis [6, 7]. We know that ethanol is ionized from the neutral state to a monocation without significant structural deformation. Therefore, for the monocation state, we calculated the time-dependent potential surfaces as a function of the bond length as shown in Fig. 4.21. The electronic structure calculation for obtaining the potential surfaces is performed by using a molecular orbital method, namely, the first-order configuration interaction (CI) method with the 6–311G(d,p) basis set. The upper graph of Fig. 4.21a shows potential surfaces in zero fields. We saw a field-induced avoided crossing between the ground and repulsive time-dependent adiabatic states that occurred near $R_{\mathrm{C-C}} = 2$ Å when the electric field of 0.1 E_{h}/ea_0 was pointed from the middle carbon to the end carbon, as shown in Fig. 4.21b. As the intensity increased, nonadiabatic transitions occurred from the lowest adiabatic potential to the higher ones near the equilibrium bond distance.

We also calculated the propagation of the nuclear wave packet as shown in Fig. 4.22. The parameters of the applied pulse are as follows: The wavelength is 800 nm, the intensity is $5 \times 10^{14}\,\mathrm{W/cm^2}$, and the pulse length is 200 fs. The nuclear wave packets in the lowest, second lowest, and third lowest time-dependent adiabatic states (denoted by $|1\rangle$, $|2\rangle$ and $|3\rangle$, respectively) are shown in the left, center, and right graphs, respectively. The initial state is the lowest vibrational state of the field-free ground electronic state of $C_2H_5OH^+$. Initially, the C—C distance increases in the lowest time-dependent adiabatic potential, and when the packet is approaching the field-induced avoided crossing point, the nonadiabatic transition to the higher adiabatic states occurred and the dissociation began. Therefore, we ascribed this dissociation to the field-induced potential crossing. The dissociation probability of the C—C bond was 0.39.

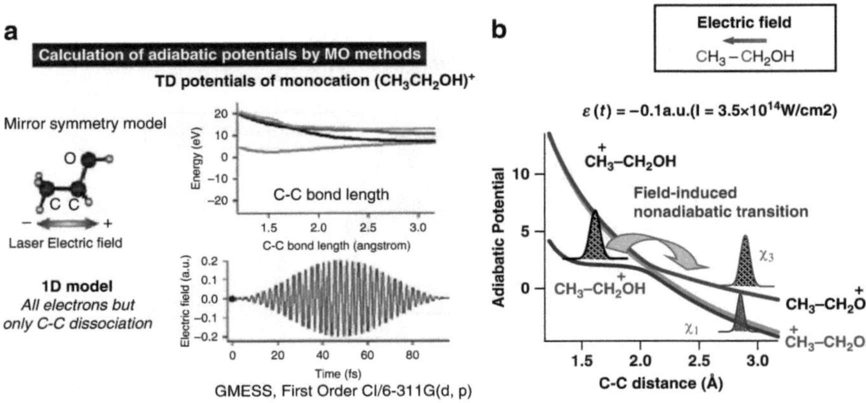

Fig. 4.21 Molecular orbital calculation of adiabatic potentials of ethanol monocation $(CH_3CH_2OH)^+$ as a function of the C—C bond length: (**a**) potential surfaces in zero fields; (**b**) potential crossing in a nonzero field

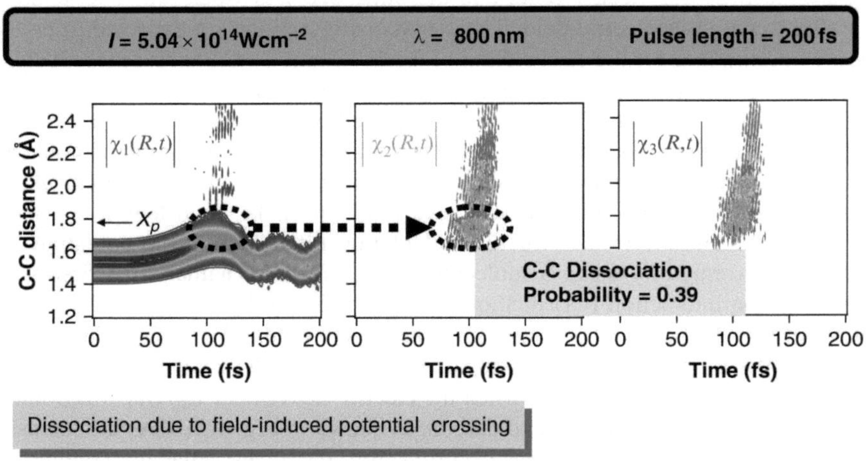

Fig. 4.22 Nuclear wave packet dynamics along the C—C axis in the lowest three time-dependent adiabatic states $|1\rangle$, $|2\rangle$ and $|3\rangle$. The nuclear wave packet of $|1\rangle$, $\chi_1(R,t)$, is ramified around the C—C distance $R = 1.8$ Å by field-induced nonadiabatic transitions to the upper adiabatic states. Once transferred to the upper adiabatic states, dissociation is promoted because the cycle-averaged potentials of the upper adiabatic states are dissociative. Field-induced nonadiabatic transitions at $R = 1.8$ Å occur only when the electric field is pointed from the middle carbon to the end carbon (defined as a negative field as shown in Fig. 4.21b)

We also performed the simulation for the C—O dissociation case as shown in Fig. 4.23, and found a crossing regime, similar to the one in the C—C dissociation case. However, the distance between the equilibrium C—O distance and this avoided crossing point was found to be larger than the corresponding distance for the C—C dissociation case. Hence, it took longer time for the nuclear wave packet to reach

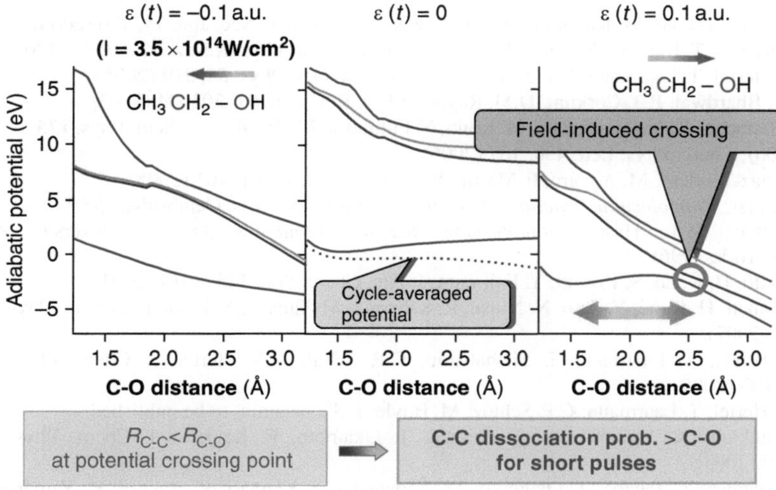

Fig. 4.23 Adiabatic potentials of $(CH_3CH_2OH)^+$ along the C—O axis at different field strengths: (*left*) $-0.1E_h/ea_0$, (*middle*) 0, and (*right*) 0.1 E_h/ea_0. The sign of the electric field is chosen to be positive when the field vector points from the middle carbon atom to the oxygen atom. In comparison with the case where the C—C axis is parallel to the field, field-induced avoided crossings occur in a range far away from the equilibrium C—O distance (= 1.5 Å) when the field is positive

the crossing region in the case of the C—O dissociation. This meant that a longer pulse was required for steering the wave packet to this crossing region.

We were able to explain the ratio of the dissociation probabilities of the C—O and C—C dissociation processes. This will also be explained by Professor Kannari later from an experimental point of view. The electronically nonadiabatic transition induced by an intense laser field is a very important factor to determine the branching ratios of different fragmentation processes and reaction processes of molecules in an intense laser field.

References

1. H. Kono, A. Kita, Y. Ohtsuki, Y. Fujimura, J. Comput. Phys. **130**, 148 (1997)
2. I. Kawata, H. Kono, J. Chem. Phys. **111**, 9498 (1999)
3. I. Kawata, H. Kono, Y. Fujimura, Chem. Phys. Lett. **289**, 546 (1998); J. Chem. Phys. **110**, 11152 (1999)
4. A.D. Bandrauk, H. Kono, in *Molecules in Intense Laser Fields: Nonlinear Multi-Photon Spectroscopy and Near-Femtosecond to Sub-Femtosecond (Attosecond) Dynamics, Advances in Multi-Photon Processes and Spectroscopy*, ed. by S.H. Lin A. Villaeys, Y. Fujimura (World Scientific, Singapore, 2003), vol. 15, p. 147
5. H. Kono, Y. Sato, Y. Fujimura, I. Kawata, Laser Phys. **13**, 883 (2003)
6. H. Kono, Y. Sato, N. Tanaka, T. Kato, K. Nakai, S. Koseki, Y. Fujimura, Chem. Phys. **303**, 203 (2004)

7. H. Kono, Y. Sato, M. Kanno, K. Nakai, T. Kato, Bull. Chem. Soc. Jpn. **79**, 196 (2006)
8. M. Kanno, T. Kato, H. Kono, Y. Fujimura, F.H.M. Faisal, Phys. Rev. A **72**, 033418 (2005)
9. I.V. Hertel, T. Laarmann, C.P. Schulz, Adv. At. Mol. Opt. Phys. **50**, 219 (2005)
10. V.R. Bhardwaj, P.B. Corkum, D.M. Rayner, Phys. Rev. Lett. 91, 203004 (2003)
11. R. Sahnoun, K. Nakai, Y. Sato, H. Kono, Y. Fujimura, M. Tanaka, J. Chem. Phys. **125**, 184306 (2006); Chem. Phys. Lett. **430**, 167 (2006)
12. S. Díaz-Tendero, M. Alcamí, F. Martín, Phys. Rev. Lett. **95**, 013401 (2005)
13. W. Forst, *Unimolecular Reactions* (Cambridge University Press, Cambridge, 2003)
14. T. Baer, W.L. Hase, *Unimolecular Reaction Dynamics* (Oxford University Press, New York, 1996)
15. Y. Sato, H. Kono, S. Koseki, Y. Fujimura, J. Am. Chem. Soc. **125**, 8019 (2003)
16. K. Nakai, H. Kono, Y. Sato, N. Niitsu, R. Sahnoun, M. Tanaka, Y. Fujimura, Chem. Phys. **338**, 127 (2007)
17. M. Boyle, T. Laarmann, I. Shchatsinin, C.P. Schulz, I.V. Hertel, J. Chem. Phys. **122**, 181103 (2005)
18. I.V. Hertel, T. Laarmann, C.P. Schulz, M. Boyle, I. Shchatsinin, to be published
19. R. Itakura, K. Yamanouchi, T. Tanabe, T. Okamoto, F. Kannari, J. Chem. Phys. **119**, 4179 (2003)
20. H. Yazawa, T. Tanabe, T. Okamoto, M. Yamanaka, F. Kannari, R. Itakura, K. Yamanouchi, J. Chem. Phys. **124**, 204314 (2006)

Chapter 5
Pulse Shaping of Femtosecond Laser Pulses and Its Application of Molecule Control

Fumihiko Kannari

Abstract Advanced sophisticated techniques of manipulating and controlling pulse shapes of ultrashort laser pulses are reviewed by referring to our recent studies on the selective bond breaking processes of ethanol in intense laser field. Some remaining issues in the pulse shaping technologies such as spatiotemporal coupling and replica pulse formation are discussed.

5.1 Introduction

It is useful to know the extent to which we can manipulate the ultrashort laser pulse, what is the limitation and what types of errors are caused in this manipulation technology.

When the term "manipulation" is used in this text, it applies to manipulation of only three parameters: the amplitude of the envelope of the laser pulse; the phase of the pulse or in other words, the instantaneous frequency inside the laser pulse; and the polarization.

$$E(t) = \vec{A}(t)\exp\left[\mathrm{i}\omega t + \mathrm{i}\phi(t)\right]. \tag{5.1}$$

Ideally, this is similar to the function generation of the fs laser pulse (Fig. 5.1).

How can we manipulate the temporal shape of the laser pulse? This is a very fundamental definition in linear signal filtering. Linear filtering in the time domain is shown by

$$e_{\mathrm{out}}(t) = e_{\mathrm{in}}(t) \times h(t), \tag{5.2}$$

$$E_{\mathrm{out}}(\mathrm{w}) = E_{\mathrm{in}}(\mathrm{w})H(\mathrm{w}), \tag{5.3}$$

where h is the filter function [1]. In a sampling oscilloscope, $h(t)$ is the delta function. If we have to manipulate the amplitude or phase in this scheme, we have to use

F. Kannari
Keio University, Yokohama, Japan
e-mail: kannari@elec.keio.ac.jp

K. Yamanouchi (ed.), *Lectures on Ultrafast Intense Laser Science 1*, Springer Series in Chemical Physics 94, DOI 10.1007/978-3-540-95944-1_5,

Fig. 5.1 A concept of femtosecond laser pulse function generator

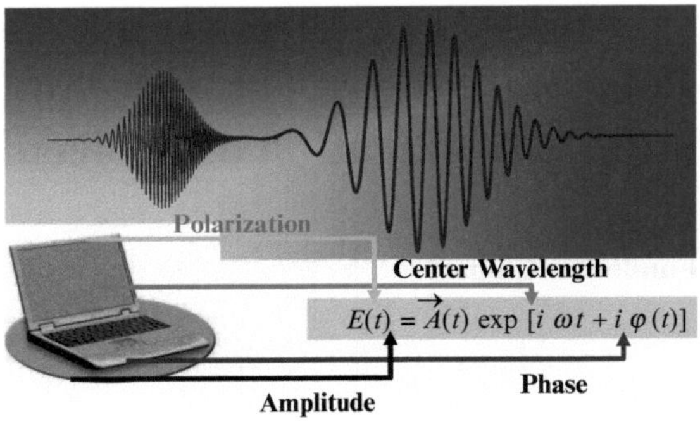

a device or material, which has a very fast response in the timescale of femtosecond. However, that is almost impossible to do; thus, when we carry out a Fourier transform of this equation (5.2), we can find that it is a product of the spectrum with a mask function in the spectral domain. If we consider this in the spectral domain, the filtering will be very straightforward. This idea was presented in the 1970s [1]. In the early 1990s, Weiner of Bellcore demonstrated this kind of pulse shaping for the femtosecond laser using a computer controlled spatial light modulator (SLM) [2].

5.2 Femtosecond Laser Pulse Shaping with a 4f Pulse Shaper

We consider the 4f-type pulse shaper that Weiner demonstrated using a SLM [2,3]. Part of the modulator is the same as a spectrometer, where the dispersed frequency components will arrive at the Fourier plane. The other part is for the inverse Fourier transform. When we add some phase modulation or amplitude modulation at the Fourier plane, this reflects to the pulse shape in the time domain.

The dispersion at this Fourier plane is defined by the focal length and grating, which is typically one or two nanometers per millimeter:

$$\frac{\mathrm{d}\lambda}{\mathrm{d}x} \sim \frac{d \cos \theta}{f} \sim 1 - 2 \text{ nm/mm}. \tag{5.4}$$

Hence, if we use a computer-controlled SLM as in Fig. 5.2, we can manipulate the pulse shape. This is the idea of 4f-type pulse shaping technology.

The pulse shaper in our laboratory uses a commercially available computer controlled liquid-crystal SLM (Fig. 5.3). The modulator consists of arrays of liquid crystal pixel. An electrical field manipulates the retardation for each segment and thus each segment can control each frequency mode of the ultrashort laser pulse. This kind of commercial SLM is readily available. There are 128 segments, which is typical; recently, modulators are also available with 640 segments.

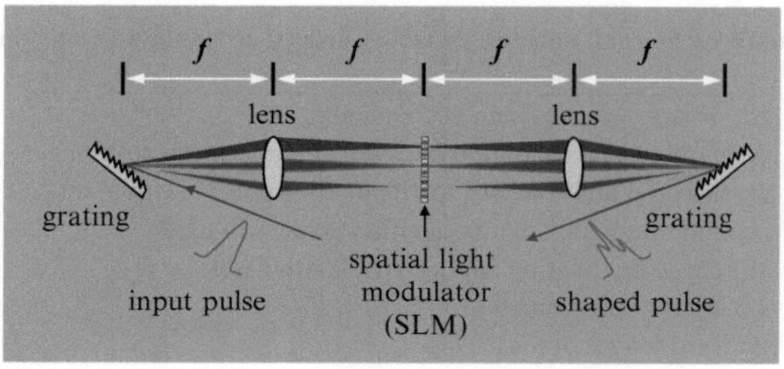

Fig. 5.2 4f-type pulse shaper with a computer controlled spatial light modulator (SLM)

a

b

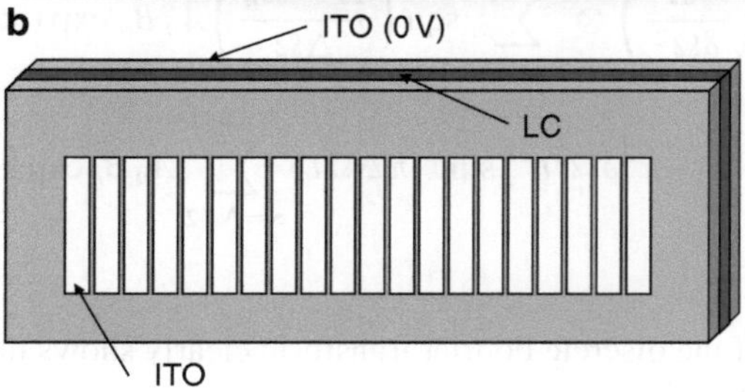

Fig. 5.3 Picture of a 4f-type pulse shaper (**a**) and a schematic view of an arrayed liquid crystal SLM (**b**)

Table 5.1 Summary of the notation for variables needed to explain the mathematics of pulse shaping [4]

$\Omega_n = \omega_n - \omega_0$, ω_0: center frequency
$\Delta\Omega$: frequency separation between neighboring pixel center
B_n: spectral amplitude of the input laser pulse of nth pixel
A_n: amplitude modulation applied by the nth pixel
ϕ_n: phase modulation applied by the nth pixel
D: input beam diameter
f: focal length of lens
$\delta x = 4\pi f/\lambda D$: the spot size at the Fourier plane
N: total pixel number
$\delta\Omega = \delta x \Delta\Omega/\Delta x$: spectral resolution
$\mathrm{squ}\,(x) = 1$ for $\lvert x\rvert \leq 1/2$
$\quad 0$ for $\lvert x\rvert > 1/2$

The equations necessary to understand the pulse shaping are as follows (Table 5.1) [4]:

$$M\,(x) = S\,(x) \otimes \sum_{n=-N/2}^{N/2-1} \mathrm{squ}\left(\frac{x - x_n}{\Delta x}\right) A_n \exp\left(\mathrm{i}\phi_n\right) \tag{5.5}$$

$$\Rightarrow M\,(\Omega) = \exp\left(\frac{-\Omega^2}{\delta\Omega^2}\right) \otimes \sum_{n=-N/2}^{N/2-1} \mathrm{squ}\left(\frac{\Omega - \Omega_n}{\Delta\Omega}\right) A_n \exp\left(\mathrm{i}\phi_n\right), \tag{5.6}$$

$$E_{\mathrm{in}}\,(\Omega) = \sum_{n=N/2}^{N/2-1} \mathrm{squ}\left(\frac{\Omega - \Omega_n}{\Delta\Omega}\right) B_n, \tag{5.7}$$

$$E_{\mathrm{out}}\,(\Omega) = M\,(\Omega)\, E_{\mathrm{in}}\,(\Omega) \tag{5.8}$$

$$= \exp\left(\frac{-\Omega^2}{\delta\Omega^2}\right) \otimes \sum_{n=N/2}^{N/2-1} \mathrm{squ}\left(\frac{\Omega - \Omega_n}{\Delta\Omega}\right) A_n B_n \exp\left(\mathrm{i}\phi_n\right)$$

$$\underset{\mathrm{IFT}}{\Leftrightarrow} e_{\mathrm{out}}\,(t) \propto \exp\left(-\pi^2\delta\Omega^2 t^2\right) \mathrm{sinc}\,(\pi\Delta\Omega t) \sum_{n=N/2}^{N/2-1} A_n B_n \exp\left[\mathrm{i}\,(2\pi\nu_n t + \phi_n)\right]. \tag{5.9}$$

This description of the discrete Fourier transform clearly shows that we can Fourier synthesize the laser pulse.

The output of the pulse shaping apparatus is summarized as follows [4]:

$$e_{\mathrm{out}}\,(t) \propto \exp\left(-\pi^2\delta\Omega^2 t^2\right) \;\sin c\,(\pi\Delta\Omega t) \sum_{n=N/2}^{N/2-1} A_n B_n \exp\left[\mathrm{i}\,(2\pi\nu_n t + \phi_n)\right]. \tag{5.10}$$

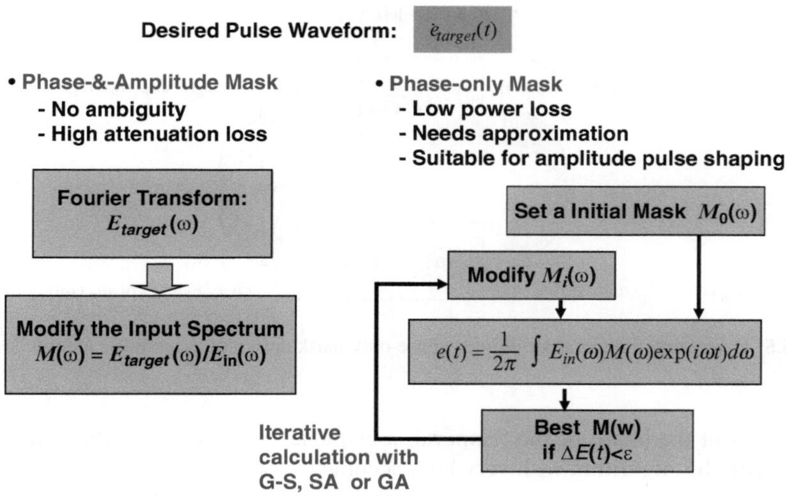

Fig. 5.4 Two approaches for pulse shaping mask design

The last term corresponds to the spectral modulation by the shaping mask, amplitude, and phase modulation at the Fourier transformed plane. The sinc function corresponds to the Fourier transform of the square shape of each pixel. The first term corresponds to the spectral resolution, which is defined by the grating and lens. The product of the first term and the sinc function correspond to the time window that can be achieved with this pulse shaping technique.

How can we design the shaping mask? The answer to this is described in Fig. 5.4. When we can manipulate both spectral phase and amplitude with a SLM, there is no ambiguity. The target pulse has to be Fourier transformed; then the input spectrum that will be generated directly from the laser system is obtained. So, we can obtain the complex function of the shaping mask and there is no ambiguity in this design. However, because we cannot amplify the laser pulse at the pulse shaper, the manipulation of the amplitude implies that it always attenuates the power.

Therefore, in many cases, in order not to lose the laser pulse energy, phase-only masks are used to manipulate only the spectral phase to produce the desired output laser pulse. However, design of this phase-only mask is not straightforward, and some approximations are needed.

In general, optimization is performed iteratively to obtain the best phase mask. The initial mask is assumed and the output pulse is calculated. Then, the difference is measured between the desired pulse and the shaped output pulse, and the parameters are varied so that the difference becomes smaller. These processes are repeated so that the desired pulse can finally be obtained. There are several such optimization algorithms.

One of them is the Gerchberg–Saxton algorithm shown in Fig. 5.5 [5]. It is very powerful tool to shape only the amplitude of the laser pulse. Thus, if we are not

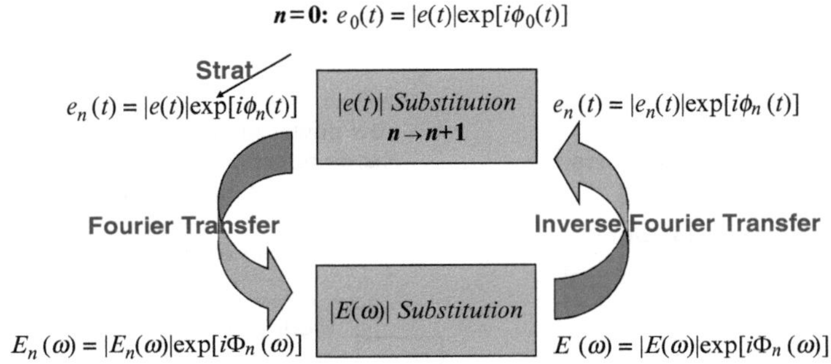

Fig. 5.5 Gerchberg–Saxton algorithm for phase-only mask design to shape pulse amplitude

interested in the phase of the temporal laser pulse, we can design the phase-only mask with this algorithm with very high accuracy.

Two physical parameters are fixed in the Gerchberg–Saxton algorithm. One is the temporal pulse shape that we intend to shape. The other is the spectral amplitude shape, which is the resource of our actual laser pulse. First we assume a certain spectral phase. After the inverse-Fourier transform, the calculated temporal shape may not match the targeted laser pulse shape. So only the temporal amplitude is replaced and the spectrum is calculated via Fourier transform. Next, the calculated spectral amplitude is replaced with the actual spectrum and these processes are repeated. After approximately 100 iterations, the algorithm provides the optimum design of the phase mask.

If we have to design both the temporal amplitude and the phase, we can do that by the simulated annealing algorithm, shown in Fig. 5.6, which is another type of optimization algorithm. Initially, we can set any type of spectral phase, apply that phase on our laser spectrum and calculate the temporal shape. At this stage, we calculate the cost function. We can set any function for the cost function; it simply depends on the type of the optimization we require. In the general pulse shaping, this cost function may correspond to the difference between the shaped pulse and the desired laser pulse. The important point is that we have to set one single value for the cost function. This algorithm reduces this cost function to the minimum; in other words, the optimization algorithm implies that we have to find the global optimization point. Then the question becomes how do we reach this global optimization point?

We accept a new phase mask only if the cost function is lower than the last one. However, by using only this algorithm, we cannot get out of this local minimum. Therefore, sometimes even if the cost function is larger than the last one, we have to accept the change to the new mask. The probability is controlled by an exponential function in the algorithm using the inverse of the temperature T during the annealing processes. Hence, we have to control T during the optimization. The concept is the same as in a genetic algorithm, but our group prefers this simulated annealing algorithm (Fig. 5.6).

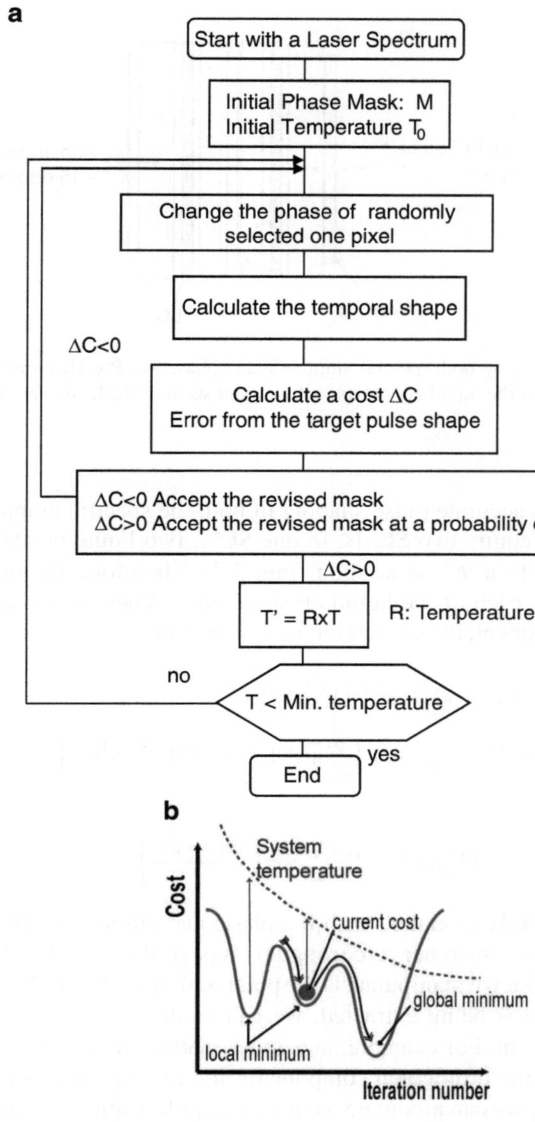

Fig. 5.6 (**a**) Design of phase-only shaping masks by simulated annealing algorithms. (**b**) Schematic view of iterative optimization process with Simulated Annealing. Even if the cost is trapped at a local minimum, this algorithm can release it from the local minimum and guide toward the global minimum

Both these types of optimization algorithms allow us to design the phase-only mask for generating the desired amplitude pulse or the approximated amplitude and phase in the pulse. The design of the target laser pulse shape depends on the user, i.e., the preknowledge of the user, and on the use, i.e., the kind of light and the matter interaction.

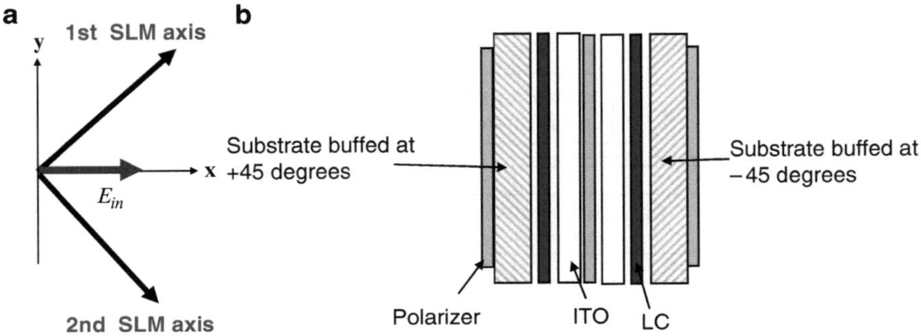

Fig. 5.7 Hybrid SLM for shaping both spectral amplitude and phase. (**a**) Polarization of the incident laser pulse E_{in} relative to the liquid crystal axis of first and second SLM. (**b**) The structure of a hybrid SLM

Next we consider the accurate pulse shaping in both the spectral amplitude and phase. In this case, we require two SLMs. In one SLM, two liquid crystal devices are attached together with a 90° separation (Fig. 5.7). Therefore, the orthogonal component is shaped by each of the liquid crystal plates. When we are extracting the x-polarization component, the expression is as shown in

$$\begin{aligned} E_{\mathrm{out}} &= \tfrac{|E_{\mathrm{in}}|}{2}\left\{\mathrm{e}^{\mathrm{j}\phi_1}\,(x+y) + \mathrm{e}^{\mathrm{j}\phi_2}\,(x-y)\right\} \\ &= |E_{\mathrm{in}}|\,\mathrm{e}^{(\mathrm{j}\phi_1+\mathrm{j}\phi_2/2)}\left\{x\,\cos\left(\tfrac{\phi_1-\phi_2}{2}\right) + \mathrm{j}y\,\sin\left(\tfrac{\phi_1-\phi_2}{2}\right)\right\} \end{aligned} \tag{5.11}$$

and

$$E_{\mathrm{out}_x} = |E_{\mathrm{in}}|\,\mathrm{e}^{\mathrm{j}(\phi_1+\phi_2)/2}\cos\left(\frac{\phi_1-\phi_2}{2}\right). \tag{5.12}$$

Thus, we can independently modulate both the phase and amplitude. This is very handy and recently, many researchers are using this type of the hybrid SLM.

The third parameter that we manipulate is the polarization. Unless only one directional linear polarization is being extracted, we can modulate the polarization of each quadrature component. For example, in a Jones matrix, the first liquid crystal plane modulates one of the orthogonal components and the second one modulates the other one. Therefore, we can modulate the temporal polarization shape using the hybrid SLM as in

$$\begin{aligned} &\begin{bmatrix} \cos\dfrac{\pi}{2} & -\sin\dfrac{\pi}{2} \\ \sin\dfrac{\pi}{2} & \cos\dfrac{\pi}{2} \end{bmatrix} \begin{bmatrix} \mathrm{e}^{\mathrm{i}\phi_2} & 0 \\ 0 & 1 \end{bmatrix} \begin{bmatrix} \cos\dfrac{\pi}{2} & \sin\dfrac{\pi}{2} \\ -\sin\dfrac{\pi}{2} & \cos\dfrac{\pi}{2} \end{bmatrix} \\ &= \frac{1}{2}\begin{bmatrix} \mathrm{e}^{\mathrm{i}\phi_2}+1 & \mathrm{e}^{\mathrm{i}\phi_2}-1 \\ \mathrm{e}^{\mathrm{i}\phi_2}-1 & \mathrm{e}^{\mathrm{i}\phi_2}+1 \end{bmatrix} \end{aligned} \tag{5.13}$$

and

$$\begin{bmatrix} \cos\frac{\pi}{2} & \sin\frac{\pi}{2} \\ -\sin\frac{\pi}{2} & \cos\frac{\pi}{2} \end{bmatrix} \begin{bmatrix} e^{i\phi_3} & 0 \\ 0 & 1 \end{bmatrix} \begin{bmatrix} \cos\frac{\pi}{2} & -\sin\frac{\pi}{2} \\ \sin\frac{\pi}{2} & \cos\frac{\pi}{2} \end{bmatrix} = \frac{1}{2}\begin{bmatrix} e^{i\phi_3}+1 & -e^{i\phi_3}+1 \\ -e^{i\phi_3}+1 & e^{i\phi_3}+1 \end{bmatrix}. \tag{5.14}$$

However, the available polarization states are very limited. Moreover, the spectral polarization is distorted by a grating or dielectric mirrors. These two factors cannot be easily compensated for by this modulator.

Therefore, recently we proposed the four cascaded manipulations of the polarization which is explained in the following equations:

#1 SLM

$$\begin{bmatrix} 1 & 0 \\ 0 & 0 \end{bmatrix} \begin{bmatrix} \cos\frac{\pi}{2} & -\sin\frac{\pi}{2} \\ \sin\frac{\pi}{2} & \cos\frac{\pi}{2} \end{bmatrix} \begin{bmatrix} e^{i\phi_1} & 0 \\ 0 & 1 \end{bmatrix} \begin{bmatrix} \cos\frac{\pi}{2} & \sin\frac{\pi}{2} \\ -\sin\frac{\pi}{2} & \cos\frac{\pi}{2} \end{bmatrix} = \frac{1}{2}\begin{bmatrix} e^{i\phi_1}+1 & e^{i\phi_1}-1 \\ 0 & 0 \end{bmatrix}, \tag{5.15}$$

#2 SLM

$$\begin{bmatrix} \cos\frac{\pi}{2} & \sin\frac{\pi}{2} \\ -\sin\frac{\pi}{2} & \cos\frac{\pi}{2} \end{bmatrix} \begin{bmatrix} e^{i\phi_3} & 0 \\ 0 & 1 \end{bmatrix} \begin{bmatrix} \cos\frac{\pi}{2} & -\sin\frac{\pi}{2} \\ \sin\frac{\pi}{2} & \cos\frac{\pi}{2} \end{bmatrix} = \frac{1}{2}\begin{bmatrix} e^{i\phi_3}+1 & -e^{i\phi_3}+1 \\ -e^{i\phi_3}+1 & e^{i\phi_3}+1 \end{bmatrix}, \tag{5.16}$$

#3 SLM

$$\begin{bmatrix} \cos\frac{\pi}{2} & -\sin\frac{\pi}{2} \\ \sin\frac{\pi}{2} & \cos\frac{\pi}{2} \end{bmatrix} \begin{bmatrix} e^{i\phi_2} & 0 \\ 0 & 1 \end{bmatrix} \begin{bmatrix} \cos\frac{\pi}{2} & \sin\frac{\pi}{2} \\ -\sin\frac{\pi}{2} & \cos\frac{\pi}{2} \end{bmatrix} = \frac{1}{2}\begin{bmatrix} e^{i\phi_2}+1 & e^{i\phi_2}-1 \\ e^{i\phi_2}-1 & e^{i\phi_2}+1 \end{bmatrix}, \tag{5.17}$$

#4 SLM

$$\begin{bmatrix} e^{i\phi_4} & 0 \\ 0 & 1 \end{bmatrix}. \tag{5.18}$$

If #2, #3, and #4 SLMs are combined, we can obtain the Jones matrix expression given in (5.19) [6]. When we use all the specialized modulators from #1 to #4, the polarization manipulation is more flexible. Hence, we can precompensate for the distortion in the polarization waveform caused by the grating or dielectric mirrors.

$$\#2-\#3-\#4\begin{bmatrix} e^{i\phi_4} & 0 \\ 0 & 1 \end{bmatrix} \times \frac{1}{2}\begin{bmatrix} e^{i\phi_3}+1 & -e^{i\phi_3}+1 \\ -e^{i\phi_3}+1 & e^{i\phi_3}+1 \end{bmatrix} \times \frac{1}{2}\begin{bmatrix} e^{i\phi_2}+1 & e^{i\phi_2}-1 \\ e^{i\phi_2}-1 & e^{i\phi_2}+1 \end{bmatrix}$$

$$= \frac{1}{2}\begin{bmatrix} e^{i\phi_4}\left(e^{i\phi_2}+e^{i\phi_3}\right) & e^{i\phi_4}\left(e^{i\phi_2}-e^{i\phi_3}\right) \\ e^{i\phi_2}-e^{i\phi_3} & e^{i\phi_2}+e^{i\phi_3} \end{bmatrix}, \tag{5.19}$$

$$\begin{aligned} &\#1-\#2-\#3-\#4\begin{bmatrix} e^{i\phi_4} & 0 \\ 0 & 1 \end{bmatrix} \times \frac{1}{2}\begin{bmatrix} e^{i\phi_3}+1 & -e^{i\phi_3}+1 \\ -e^{i\phi_3}+1 & e^{i\phi_3}+1 \end{bmatrix} \\ &\times \frac{1}{2}\begin{bmatrix} e^{i\phi_2}+1 & e^{i\phi_2}-1 \\ e^{i\phi_2}-1 & e^{i\phi_2}+1 \end{bmatrix} \times \frac{1}{2}\begin{bmatrix} e^{i\phi_1}+1 & e^{i\phi_1}-1 \\ 0 & 0 \end{bmatrix} \\ &= \frac{1}{4}\begin{bmatrix} e^{i\phi_4}\left(e^{i\phi_2}+e^{i\phi_3}\right)\left(e^{i\phi_1}+1\right) & e^{i\phi_4}\left(e^{i\phi_2}-e^{i\phi_3}\right)\left(e^{i\phi_1}-1\right) \\ \left(e^{i\phi_2}-e^{i\phi_3}\right)\left(e^{i\phi_1}+1\right) & \left(e^{i\phi_2}+e^{i\phi_3}\right)\left(e^{i\phi_1}-1\right) \end{bmatrix}. \end{aligned} \tag{5.20}$$

There is another useful device for pulse shaping, a two-dimensional SLM. The advantage of the two-dimensional SLM is that in addition to using the x-direction for the spectral manipulation, another direction can be used as well. Feurer and his coworkers have produced a spatially dependent ultra-short laser pulse and further, they have generated acoustic phonons in SiO_2 and designed the pattern of the acoustic wave inside the material [7]. This kind of spatially dependent ultrashort pulse and matter interaction can be manipulated by the two-dimensional SLM (Fig. 5.8).

There are two serious issues that must be dealt with regarding the basic technologies of pulse shaping that we have been discussing. The first issue is spatiotemporal coupling in the $4f$-type pulse shaper and the second issue is the replica pulse formation.

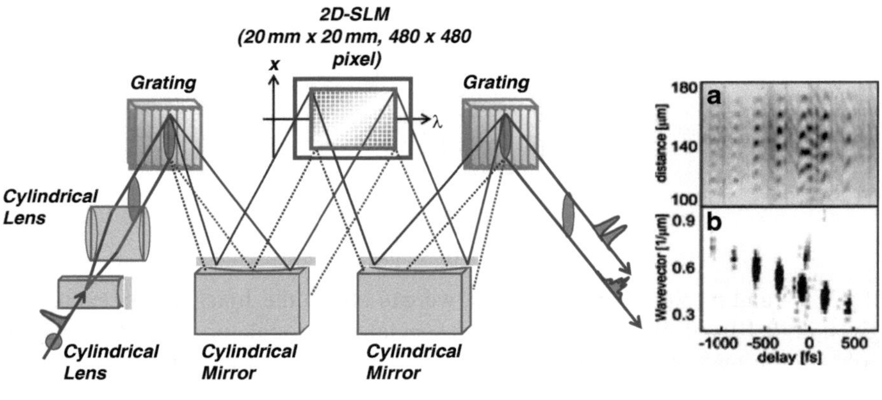

Fig. 5.8 Spatiotemporal shaping with a 2-D SLM [7]. The *inset* shows an example of shaped laser pulses

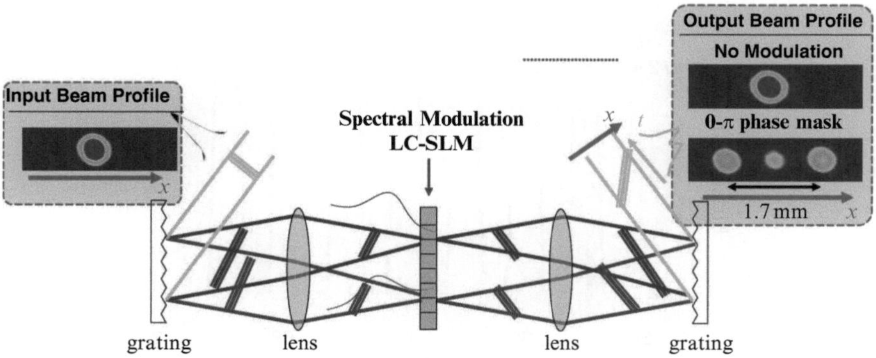

Fig. 5.9 Spatiotemporal coupling at a $4f$-type laser pulse shaper

5.3 Spatiotemporal Coupling at $4f$ Pulse Shapers

Figure 5.9 shows another expression of this pulse shaping. This calculation includes the diffraction of the grating and the Fourier transform with the Fourier lens. If we exactly calculate the pulse propagation, there is a definite coupling between the time and the space, and thus we cannot separately consider them in this $4f$ optical layout [3]. Some spatial diffraction caused by this SLM cannot be compensated for using the inverse Fourier transform. This diffraction somehow reflects the pulse of the shaped laser pulse [3].

$$E_{\text{out}}(k,\omega) = \tilde{E}_{\text{in}}(-k,\Omega)m\left(-\frac{\lambda_0 f}{2\pi}(\gamma\omega+\beta k)\right). \tag{5.21}$$

A typical result for the shaped laser pulse with the $0-\pi$ type phase mask is reproduced in Fig. 5.10. The alternating 0 and π phase pattern is applied on the SLM. Here, γ/β is defined by optical wavelength, grating and the incident angle to the grating.

$$\frac{\gamma}{\beta} = \frac{\lambda}{cd\cos\theta_{\text{i}}}. \tag{5.22}$$

The lower right image in Fig. 5.10 is the measured spectrogram [9]. The x-axis is the delay time, i.e., the time domain, and the y-axis is the space in the direction of the diffraction at the grating. There is a shift of the spatial laser pattern; thus, if we measure the pulse shape at 0 mm point, we obtain the double pulse. We obtained good coincidence between the calculations and measurements. The slope corresponds to the parameter, γ/β. If γ/β is larger, this spatiotemporal shift will be much steeper.

The cause of this spatiotemporal coupling is the diffraction at the SLM. Therefore, this is the worst case because the neighboring phase difference is the highest in this $0-\pi$ phase mask [9].

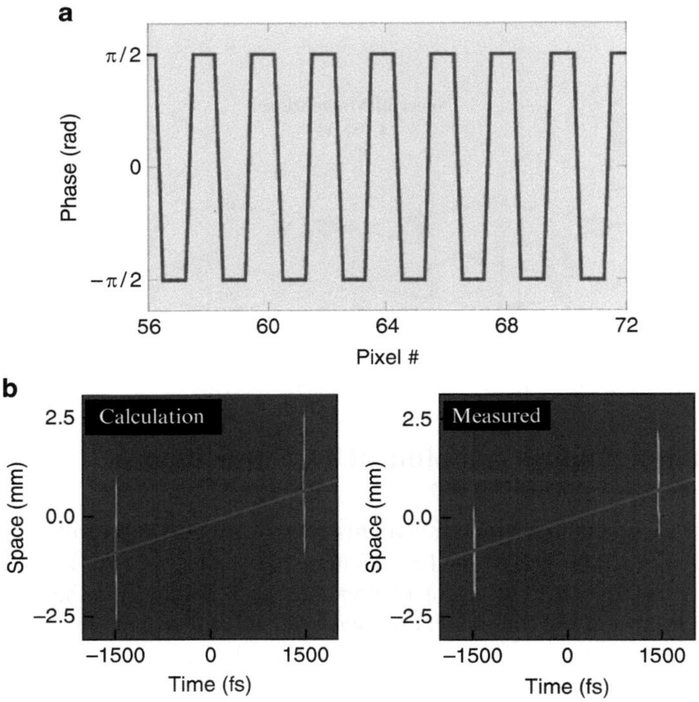

Fig. 5.10 Spatiotemporal pattern of the ultrashort laser pulse shaped by a $0 - \pi$ phase mask [9]. (**a**) a $0 - \pi$ phase mask applied on the SLM. (**b**) Spatially varying shaped laser pulses obtained by calculation and experiment. The slope corresponds to γ/β

In order to observe the actual features of the spatiotemporal coupling, I will show one more calculation which is the Wigner function representation of the shaped laser pulse. Figure 5.11 shows the near field after the pulse shaper. We can clearly see the spatial separation and the temporal and spatial coupling. With the Wigner function along the temporal frequency or the spatial frequency [10], there is no transformation and only a single peak is found. This means that in the far field we can obtain only an ideal double pulse.

Figure 5.12 is a typical setup for the $4f$-type pulse shaper [8]. In order to examine the spatiotemporal coupling, we observed the ablation pattern created by this shaped laser pulse at the focus of the lens.

With this setup, when we used a $0 - \pi$ type shaping mask, we obtained two ablation patterns reproduced in Fig. 5.13 [8]. Hence, there are two spatial peaks as predicted by our spatiotemporal coupling. This is the case without any modulation of the pulse shaper (a null mask). This represents the worst case when using the $0-\pi$ type phase mask; we always have to consider the temporal spatial coupling when we use the shaped laser pulse for any kind of light and matter interaction experiments.

We have calculated the laser field distribution near the focus of the lens.

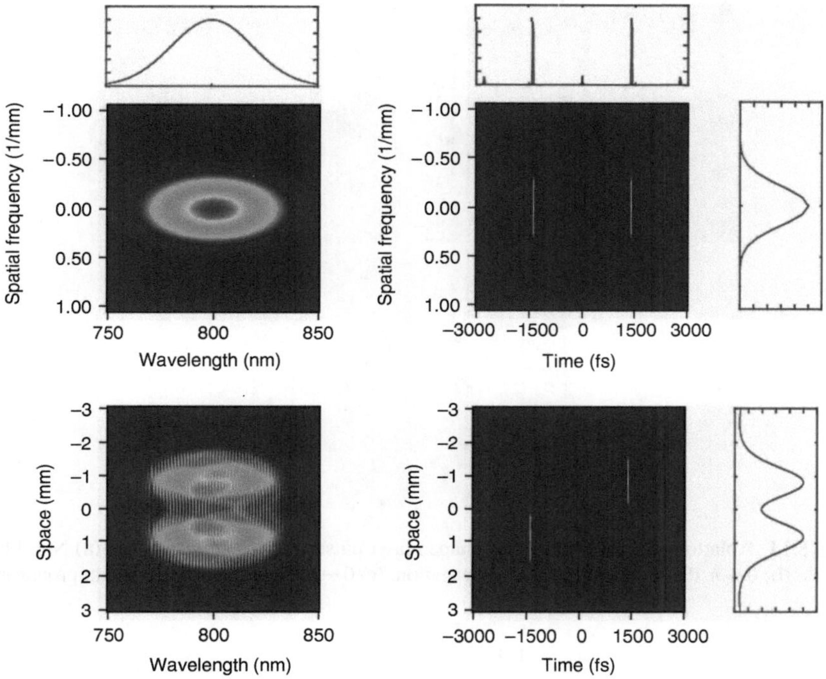

Fig. 5.11 Wigner function representation of the output laser pulse shaped by a $0 - \pi$ phase mask [10]

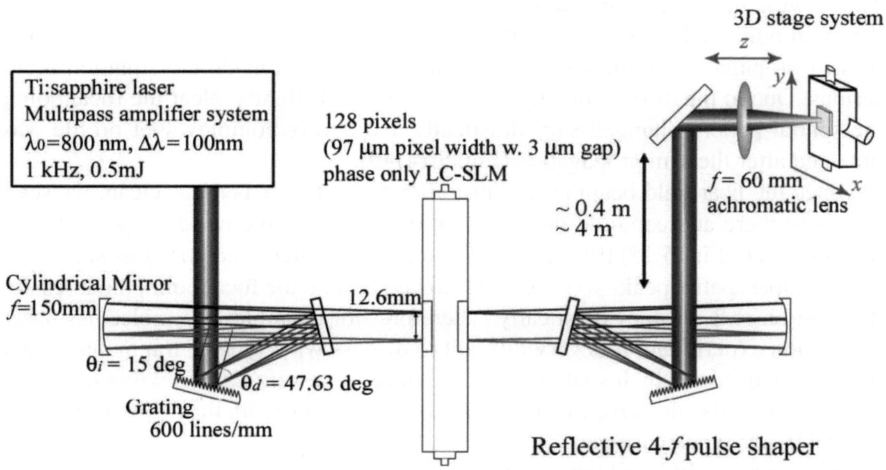

Fig. 5.12 Experimental setup to observe the spatiotemporal coupling through a laser ablation pattern [8]

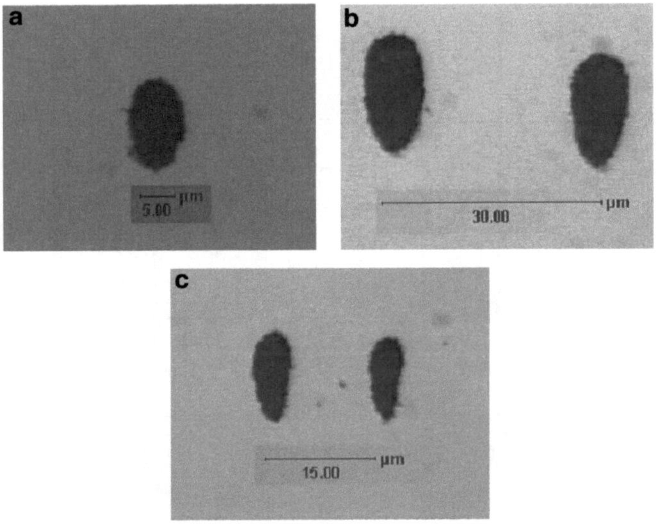

Fig. 5.13 Ablation patterns obtained by shaped laser pulses on a Cr film samples. (**a**) Null Phase Mask. (**b**) $0-\pi$ Phase mask after 4-m propagation. (**c**) $0-\pi$ Phase mask after 40-cm propagation

$$E_{\text{focus}}(x,\omega) = \left(\frac{2\pi}{\lambda_0 f}\right)^{1/2} \tilde{E}_{\text{out}}\left(\frac{x}{\lambda_0 f},\omega\right)\exp\left(-\mathrm{i}\frac{\pi L}{\lambda_0 f^2}x^2\right), \tag{5.23}$$

$$\tilde{E}(k,z,\omega) = \tilde{E}_{\text{focus}}(k,\omega)\exp(-\mathrm{i}\pi\lambda_0 z k^2). \tag{5.24}$$

This is without any shaping modulation; hence, it exhibits a very clean spot site near the focal point. When we applied $0-\pi$ type phase mask, we included diffraction of 4-m propagation in the calculation. This propagation made the situation more serious. Due to the diffraction, the focal point shifted slightly. Near the focal point, the spatial pattern changed very drastically and a very complex spot profile was obtained after the 4-m propagation (Fig. 5.14) [9].

Since the near field beam profile just after the pulse shaper was clean, we set a telescope there and captured the image of the output of the pulse shaper onto the target surface (Fig. 5.15) [9]. In this case, the pulse pattern becomes just single; no more double spatial peaks were seen. However, around the focal point, the intensity distribution still changes drastically. Therefore, for example in a molecules–laser interaction experiment, it is very difficult to discuss what type of fragments would be generated from which spatial point. We have to be very careful while using the shaped laser pulse; however, I would like to remind you again this is the worst case using a $0-\pi$ type phase mask.

When we use different types of phase mask, for example, we can produce triple pulses with a relatively shorter interval (Fig. 5.16) [9]. In this case, the neighboring phase shift in the phase mask is smaller than π; thus, there is some distortion near the focal frame, but not much.

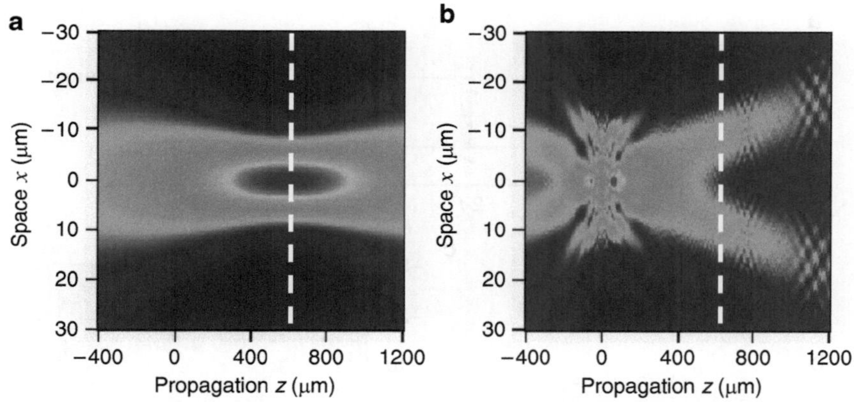

Fig. 5.14 Theoretical analysis of the focusing profile after beam propagation [8]. (**a**) After 4-m propagation from a pulse shaper: Null Mask. (**b**) After 4-m propagation from a pulse shaper: $0-\pi$ phase mask

If we increased the time separation for these triple pulses, the phase shift between the neighboring masks became larger (Fig. 5.17) [8]. Again, we found a larger distortion at the focal point from which we concluded this effect depended on the phase mask.

5.4 Replica Pulse Formation with a Pixelated SLM Pulse Shaper

The second issue is the replica pulse formation. We explained the pulse shaping using the discrete Fourier transform as shown in (5.25); however, the transform itself implies that the series members repeat themselves with a period given by the reciprocal of the frequency increment.

$$e_{\text{out}}(t) \propto \sum_{n=-N/2}^{N/2-1} A_n B_n \exp\left[\mathrm{i}\left(2\pi\nu_n t + \phi_n\right)\right]. \quad (5.25)$$

Therefore, this discretion naturally produces the pulse train of the time domain. However, as was shown before, the sinc function and a Gaussian function are multiplied by this mask function, and the longer and shorter delay parts are suppressed by the sinc function and Gaussian function.

We can understand this mathematics in a different manner. In order to temporarily shift the laser pulse, we have to apply a linear phase shift of the spectrum.

We wrap the spectral phase, based on the mathematical equivalence of phase values that differ by integer multiples of 2π and eventually we will obtain the

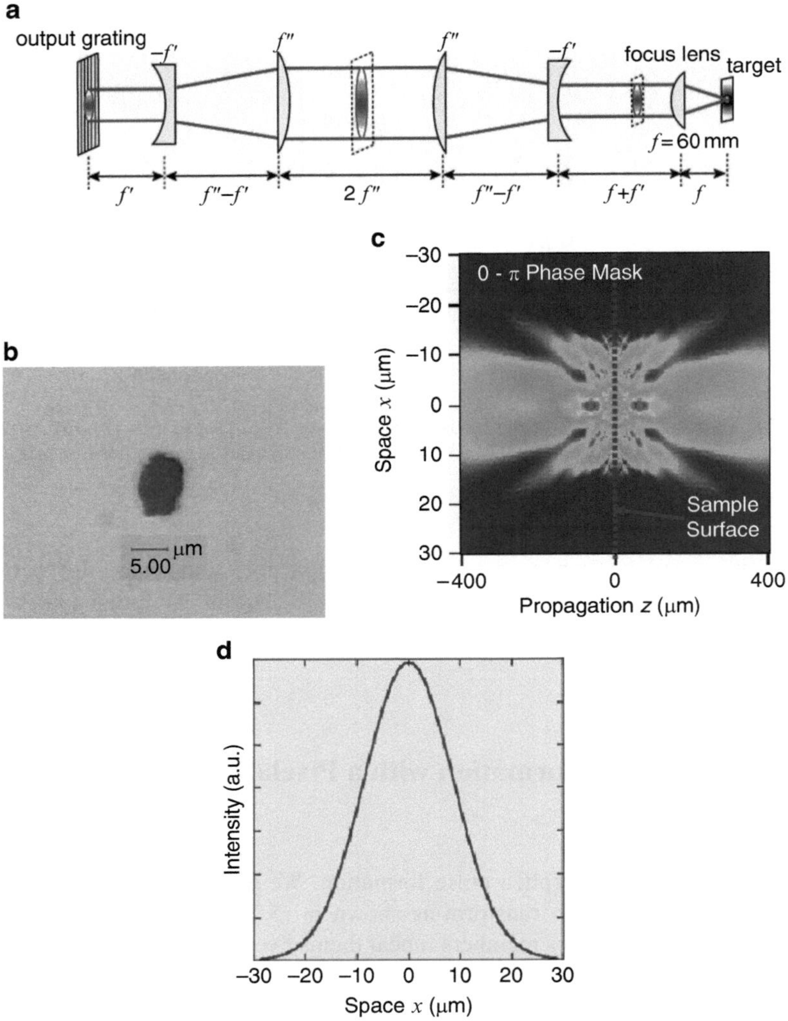

Fig. 5.15 Theoretical analysis of the focusing profile after beam propagation [8]. Diffraction is compensated by a telescope. (**a**) Schematic diagram of an image relay lens installed after the pulse shaper. (**b**) Ablation pattern obtained with the telescope. (**c**) Beam Focusing Profile at the spot when an alternative o-p phase mask is applied while diffraction is compensated for by image relay kenses. (**d**) The energy density distribution at the focal spot.

desired phase shown by the shallow positive slope of Fig. 5.18. The stair-type discrete phase is the actual phase applied to the SLM. The sharp positive and negative slopes correspond to the sampling replica phase of the shallow positve slope. The sharp-positive-slope replica phase generates the delayed replica pulse, whereas the sharp-negative-slope replica phase generates the preceding replica pulse.

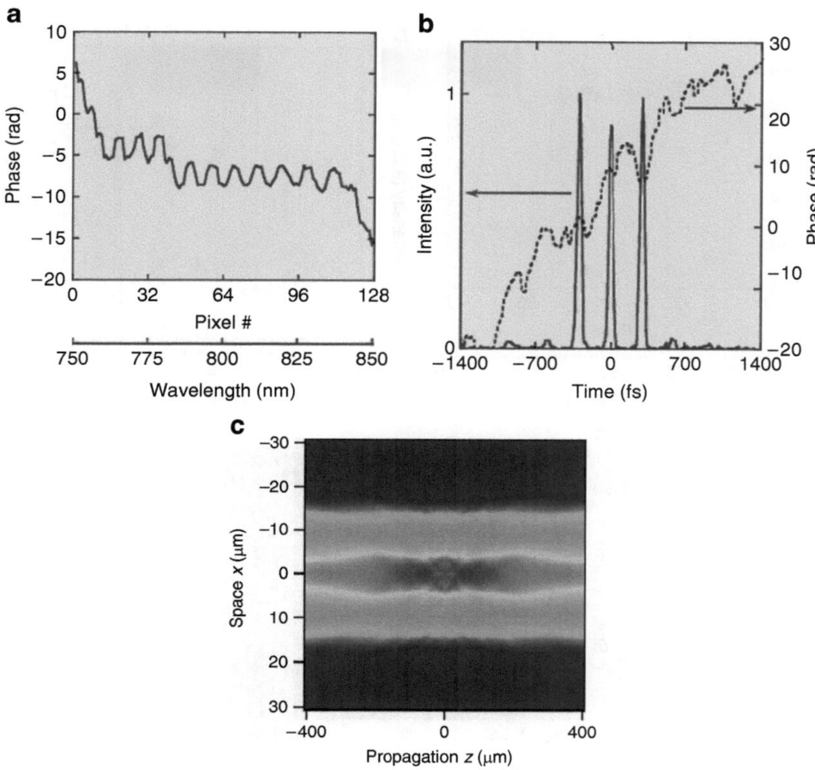

Fig. 5.16 Dependence on the spatial profile near the focus for the pulse shaped by a phase mask with less phase variation between the neighboring pixels [8]. (**a**) Phase modulation applied to the SLM to create triples pulses shown in (**b**). (**b**) Shaped triple pulses with a relatively short interval. (**c**) Spatial energy density distribution at the beam spot.

At the Fourier transform plane, the frequency separation is not equal. The dispersion created by the grating and lens is described by the formula of (5.26); therefore, frequency is not linearly dispersed at the Fourier plane.

$$x(\omega) = f \tan\left[\arcsin\left(\frac{2\pi c}{\mathrm{d}\omega} - \sin\theta_{\mathrm{i}}\right) - \theta_{\mathrm{d}}(\omega_0)\right]. \tag{5.26}$$

The x-axis in the left graph in Fig. 5.19 represents a nonlinear dispersion of the Fourier Transform plane. The first order sampling replica phase and the minus first order sampling replica phase are these slightly curved phases [4]. In Fig. 5.19a the sharp positive slope and the negative slope correspond to the chirped replica pulses. If we measure the output pulse using a FROG (frequency-resolved optical grating) [11] or an auto-correlator, we will lose this type of replica. But by using a cross-correlation technique, we can find these replica pulses on both sides of the shaped targeted pulse.

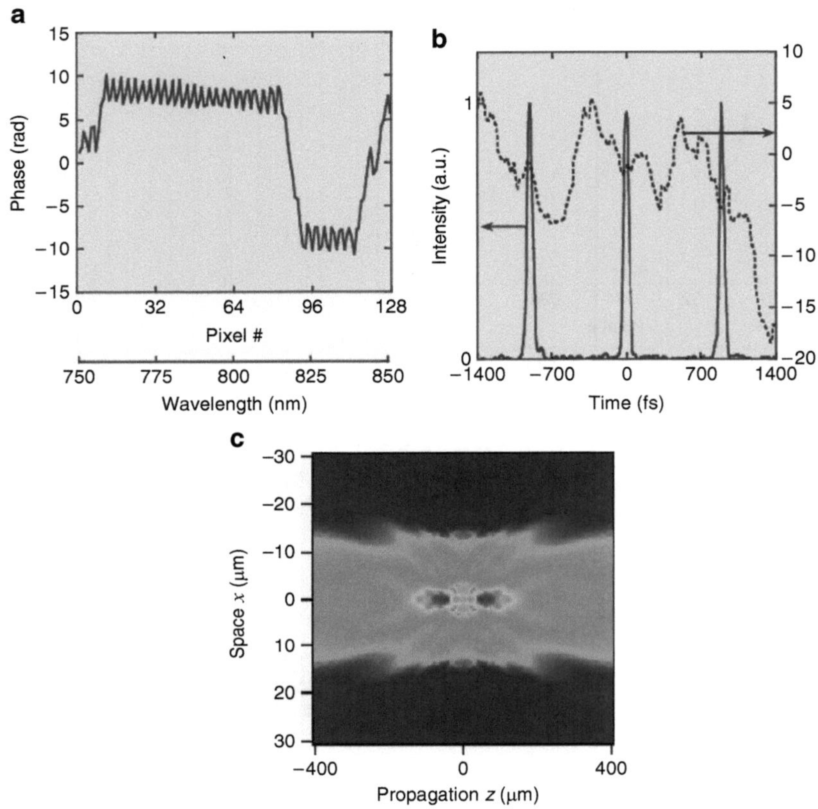

Fig. 5.17 Dependence on the spatial profile near the focus for the pulse shaped by a phase mask with large phase variation between the neighboring pixels [8]. (**a**) Phase modulation applied to the SLM to create triples pulses shown in (**b**). (**b**) Shaped Triple pulses with a relatively long interval. (**c**) Spatial energy density distribution at the beam spot.

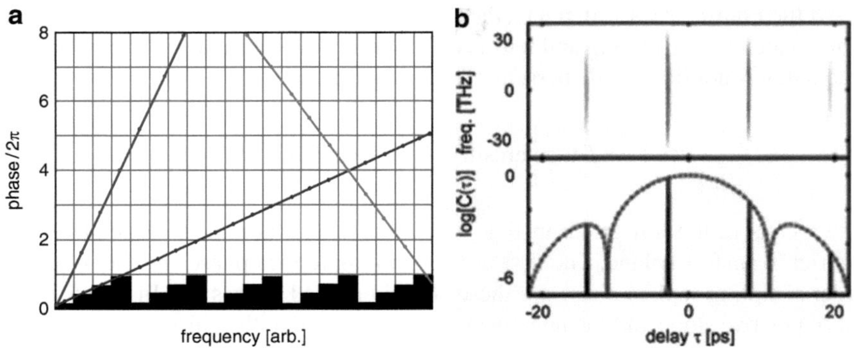

Fig. 5.18 Replica pulse formation by the discrete sampling of a SLM [4]. (**a**) In the case of linear spectral dispersion at the Fourier plain. (**b**) Simulated XFROG and cross-correlation measurements of the corresponding output waveforms

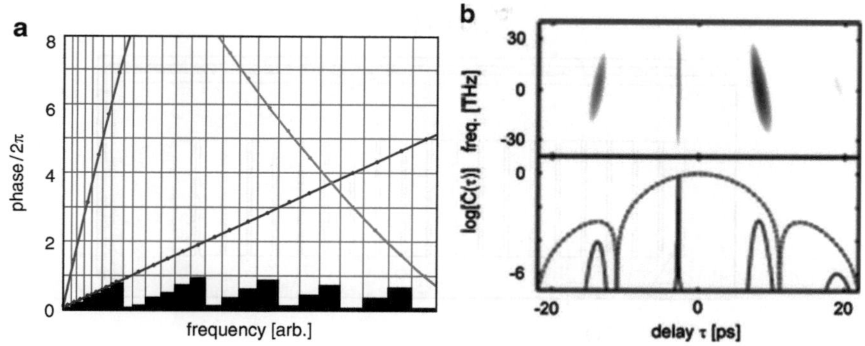

Fig. 5.19 Additional replica pulse features caused by nonlinear spectral dispersion of a diffractive grating [4]. (*Left*) In the case of linear spectral dispersion at the Fourier plain. (*Right*) Simulated XFROG and cross-correlation measurements of the corresponding output waveforms

Another replica pulse formation is caused by the gap between the pixels in the SLM (Fig. 5.20) [4]. For example, in a commercial modulator, each pixel is typically 100 μm wide and there is a gap of about 3 μm. This gap creates the sequential output of replica pulses. Moreover, for example, when we intend to generate a linear delay of 0.4 ps then we actually apply this linear phase in the SLM by wrapping the phase. After shaping, we can measure the actual spectral phase. When we did that, there was some slight error of the phase at each gap point; we can always find this kind of a slight phase jump discontinuity at the spectral light modulator gap. This error always generates a very complicated pulse strain. This replica pulse feature caused by the pixel gap depends on the SLM. Furthermore, this replica pulse effect appears for any kind of shape.

5.5 Femtosecond Laser Pulse Shaping with an AOPDF

Recently, instead of the $4f$-type pulse shaper, a new device based on an acoustic-optic (AO) crystal was invented and it is now commercially available; it is called the AOPDF [12–14].

The AOPDF is a single crystal and it has an ordinary axis and extraordinary axis (Fig. 5.21). If the original signal pulse is incident along the ordinary axis, the pulse interacts with the acoustic wave inside the crystal. Since this acoustic wave is designed to generate a certain pulse shape, diffraction occurs at some distance when the phase matching is satisfied between the $\boldsymbol{k}$ vector of the light wave and the $\boldsymbol{k}$ vector of the acoustic wave. Then, the laser axis changes to the extraordinary axis. Since the group velocities are different, the total delay time depends on the position at which the spectral component diffracts inside the crystal.

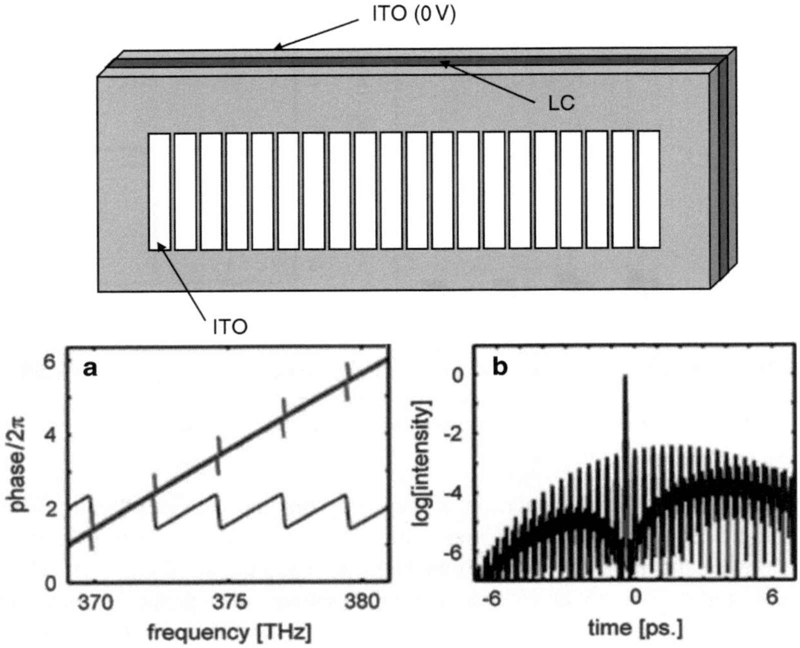

Fig. 5.20 Replica pulse formation by pixel gaps and nonsharp pixel edges [4]. *Upper*: Schematic of a pixelated SLM. (**a**): Simulated spectral phases with slope $2\pi \times 0.4$ ps. Unwrapped phase exhibit phase jumps at the pixel gaps. (**b**): simulated output pulse at -0.4 ps with weak modulator replica pulses separated by 0.4 ps.

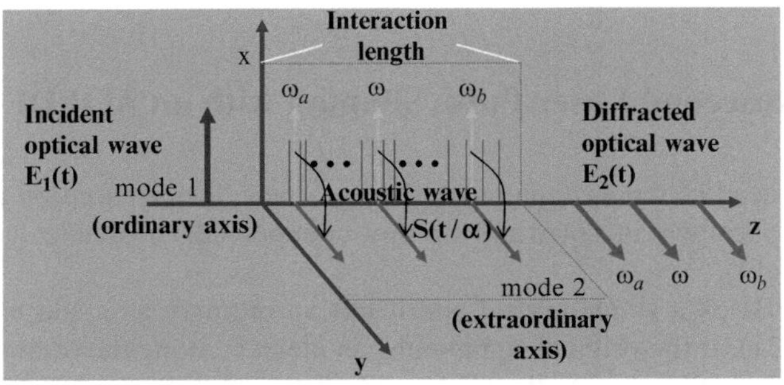

Fig. 5.21 Pulse shaping with an AOPDF (acousto-optic programmable dispersive filters) [12–14]

The mathematics behind the pulse shaping by the AOPDF is very simple. It is a combination of the input pulse and acoustic wave. However, we have to consider the interaction in the time domain, whereas spectral phase was designed in the frequency domain in the $4f$-type pulse shaper. We have to scale the time over the acoustic wave using following equation.

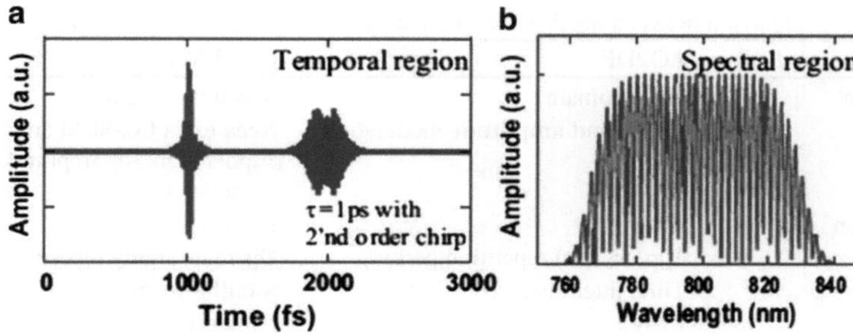

Fig. 5.22 Design of acoustic signal to generate arbitrarily shaped laser pulses [13]

$$E_{\text{out}}(t) = E_{\text{in}}(t) \otimes S\left(\frac{t}{\alpha}\right) \Rightarrow \quad S\left(\frac{t}{\alpha}\right)$$
$$\alpha = \Delta n \frac{V_a}{c}, \quad \Delta n = |n_e - n_o| \,. \tag{5.27}$$

Δn is the difference between the refractive index of the two axes, V_a is the acoustic wave, and c is the light wave velocity. Therefore, by scaling α, we can calculate the shaped out pulse with this AOPDF. The left graph in Fig. 5.22 shows a designed acoustic wave; in other words, it is a RF excitation wave. First, the pulse is transformed to a limited pulse and then the second pulse is designed, they are linearly chirped and the separation is 1 ps. The Fourier transform of this temporal pulse corresponds to the spectrum in Fig. 5.22b. From the viewpoint of Fourier synthesis, this is a Fourier mask; however, the actual physics is a direct interaction between this acoustic wave and light inside the crystal.

The design is very simple; however, it may not be straightforward to image the pulse shaping (Table 5.2).

The AOPDF alignment is very easy to do compared with the $4f$-type pulse shaper and its dynamic range is very high, being more than 50 dB. Another advantage of the AOPDF is that we can manipulate both the phase and amplitude by using this single crystal. One disadvantage is because the repetition rate of the acoustic wave is low, the AOPDF cannot be applied to an 80-MHz type oscillator.

5.6 How to Generate the Desired Ultrashort Laser Pulse in an Actual Laser System: Case 1: We Know the Desired Pulse Shape

Theoreticians may calculate the optimal pulse shape using the Optimization Control Theory model; in such a case, laser engineers create the desired laser pulse with higher accuracy using the pulse shaper and an ultrashort laser source (Fig. 5.23). Then a light-matter interaction experiment can be straightforwardly carried out. If the interaction result is reported to the theoreticians, they can modify their model. If

Table 5.2 Comparison between the AOPDF and the LC-SLM

	AOPDF	LC-SLM
Shaping in	Time domain	Spectral domain
	Phase and amplitude modulation	Need extra LC-SLM and polarizers for amplitude modulation
Modulation	Active	Passive
	Typ. 58 (μs) repetition period	Operates at **any** repetition rate
	Time jitter observed	**Small** time jitter
Alignment	**Easy**	Difficult
Programming	Complicate	**Easy**
	Typ. 4096 × 16 (bits)	Typ. 128 × 16 (bits)
Dynamic range	**High**	**Low**
	Typ. > 50 (dB)	Typ. < 20 (dB)
Time window	Limited by crystal length	Determined by spectral resolution (lateral dispersion) on Fourier plane
	Typ. ∼ 3 (ps)	Typ. ∼ 3 (ps)
Time resolution	Determined by bandwidth of rf signal	Determined by spectral bandwidth
	Typ. ∼ 6.7 (fs)	Typ. ∼ 21 (fs) with $\Delta\lambda = 100$ (nm)
Other characteristics	**Small** spatial distortions	Large spatiotemporal coupling

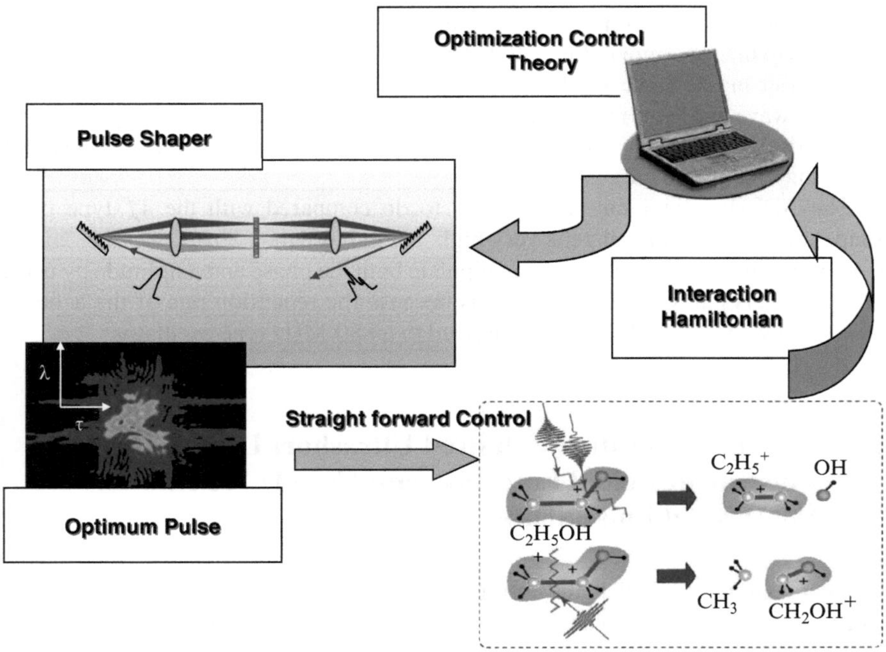

Fig. 5.23 Straightforward pulse shaping when a desired pulse shape is already known

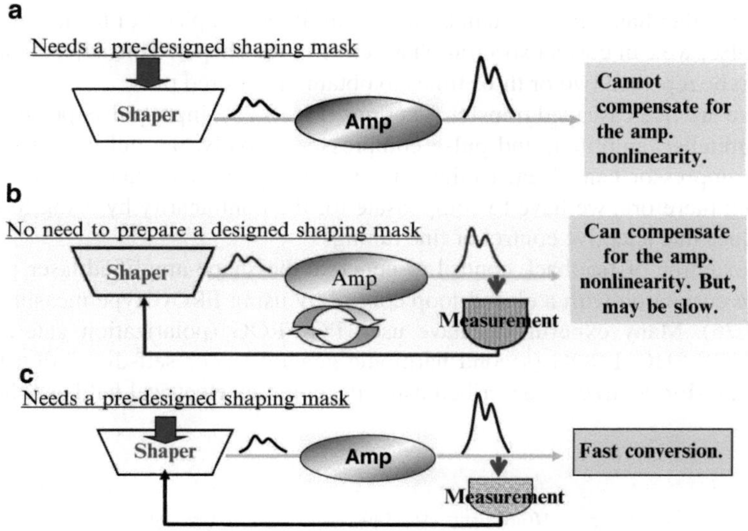

Fig. 5.24 Pulse shaping of amplified femtosecond laser. (**a**) Direct amplification of shaped pulses. (**b**) Close-loop adaptive shaping. (**c**) Direct amplification and some corrections

we have this kind of preknowledge about the optimal pulse shape, we can produce the desired pulse. Therefore, our task is to accurately create the desired laser pulse.

Since we are interested in high field physics, we need an arbitrarily shaped amplified laser pulse with intensity greater than $10^{13}\,\mathrm{W/cm^2}$ (Fig. 5.24). We can place the pulse shaper after the amplifier to get this pulse. However, there are two disadvantages. One disadvantage is that the damage threshold of the pulse shaper is not high. After the regenerative amplifier, we may be able to manage that damage, for example, by using a cylindrical lens instead of a spherical lens for Fourier transform. However, after the second power amplifier, we cannot use the liquid crystal type pulse shaper any more. Instead of the liquid crystal type pulse shaper, we must use an adaptive mirror, although we cannot manipulate the amplitude shape with that mirror. Therefore, in our experimental set-up, the pulse shaper is located after the oscillator and before the pulse stretcher. Thus, we amplify the shaped laser pulse.

If the amplifier is described by the linear transfer function, we can generate linearly amplified shaped laser pulses. However, the amplifier is no more a linear device; it exhibits self phase modulations and also gain-narrowing. Hence, there are some distortions between the input and the output pulse shapes. We have to compensate for this nonlinear transfer function. One scheme is to use closed-loop adaptive shaping. The output is measured, then the feedback correction is given to the pulse shaper and this closed-loop shaping is repeated until the amplified laser pulse becomes the desired target shape.

On the other hand, if we obtain both the amplitude and phase of the target output laser pulse, we can correct spectral phase error in the shaper. This adjustment only needs to be repeated two or three times to obtain the desired pulse.

These are five cascaded transfer functions (Fig. 5.25): input pulse, pulse shaper, pulse stretcher, amplifier, and pulse compressor. Usually, the pulse stretcher and pulse compressor cancel each other out. However, the amplifier is not a linear function; therefore, we have to compensate for the nonlinearity by using feedback techniques like adaptive control or fine tuning.

An example of feedback control to generate the shape-amplified laser pulse is iterative correction with a closed-loop control by using FROG-type measurements (Fig. 5.26). Many experiments have used PG-FROG (polarization gate-FROG) (Fig. 5.27), SHG–FROG (second harmonic generation) or self-diffraction FROG [11]. A PG-FROG trace is described using the complex electrical field as follows:

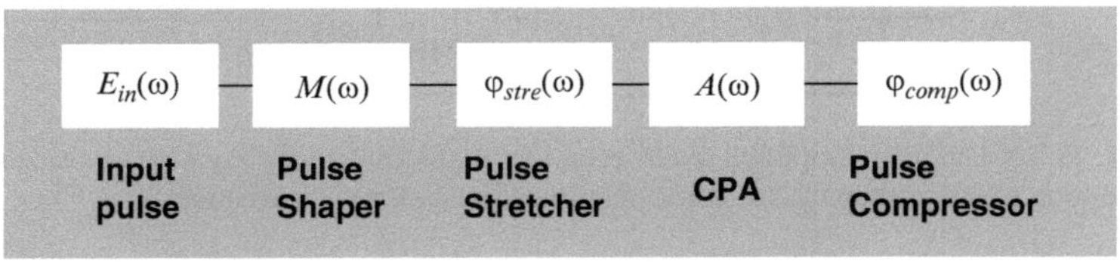

Fig. 5.25 Transfer functions in a pulse shaper and a CPA system

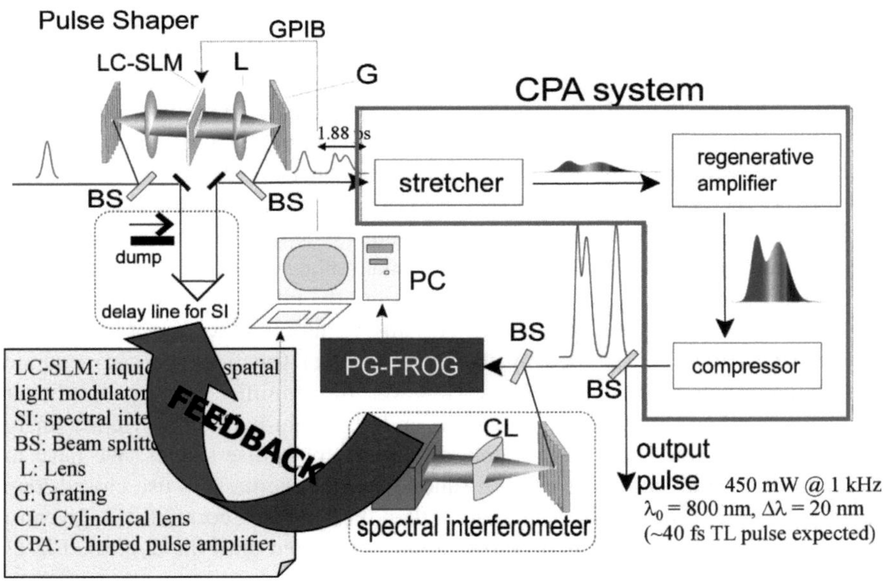

Fig. 5.26 Experimental setup of adaptive pulse shaping with a pulse shaper before an amplifier by referring for FROG traces [14]

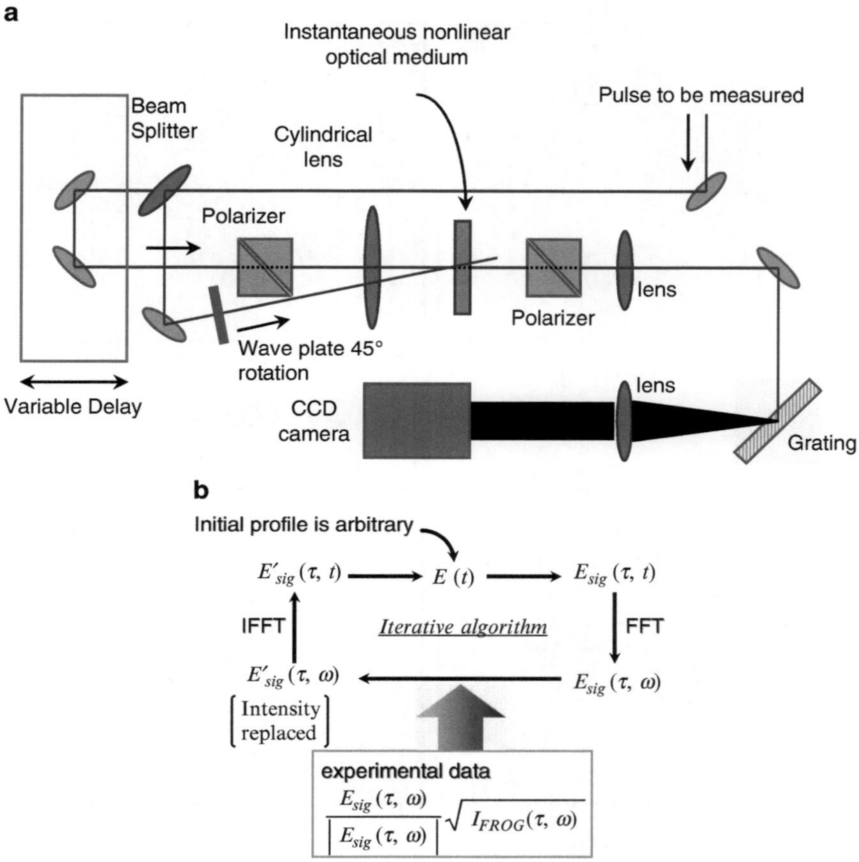

Fig. 5.27 Polarization gate frequency resolved optical gating (PG-FROG) [11]. (**a**) Optical configuration of PG FROG measurement. (**b**) Waveform retrieval algorithm for PG FROG

$$I_{\mathrm{FROG}}(\tau,\omega) = \left| \int_{-\infty}^{\infty} E(t)\,|E(t-\tau)|^2 \exp(i\omega t)\; dt \right|^2. \tag{5.28}$$

To reconstruct temporal intensity and phase from this FROG trace, we need another iterative algorithm, where the algorithm search the complex electric field which makes the following FROG error minimum as an example in Fig. 5.28.

$$G = \left| I_{\mathrm{FROG}}(\tau,\omega) - \left| E_{\mathrm{sig}}(\tau,\omega) \right|^2 \right|. \tag{5.29}$$

When pulse shape is reconstructed from a FROG trace, the two-dimensional trace includes all the amplitude and phase information; thus, the pulse shape itself need not necessarily be calculated for the feedback control. Instead of pulse shape, the

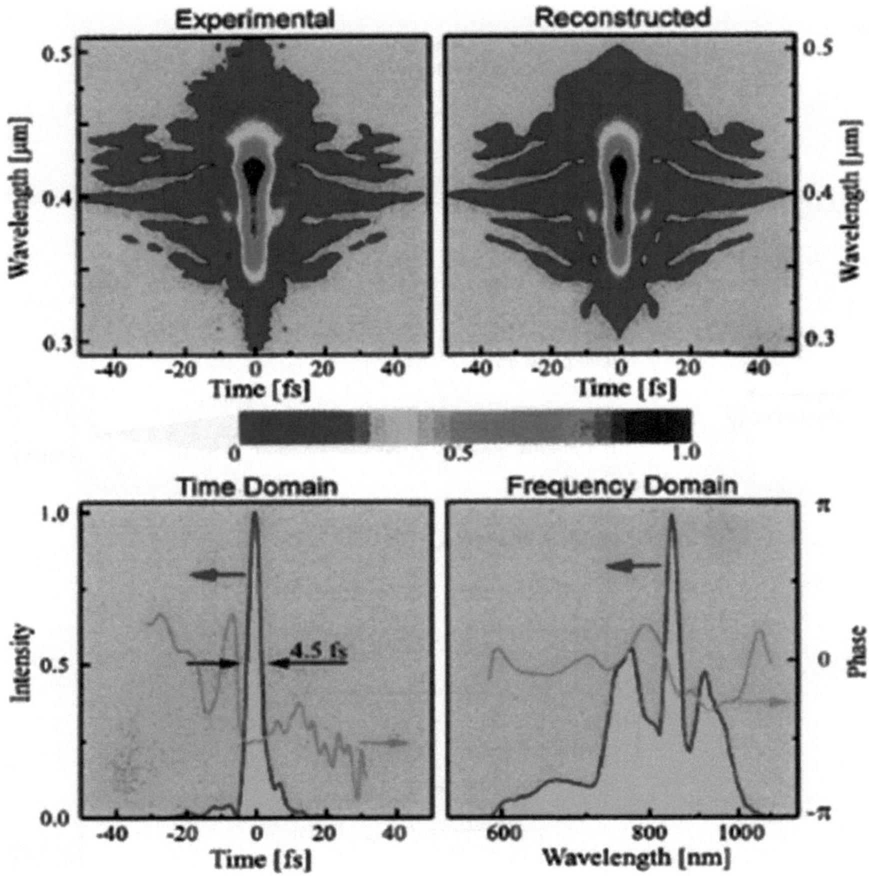

Fig. 5.28 Reconstruction of pulse shape from a FROG trace [11]

two-dimensional image (FROG trace) itself can be used to optimization. This is a great advantage in our scheme.

Figure 5.29 [14] shows some examples of experimental results. The target FROG trace has a very complicated shape. Initially we started with the transform limited laser pulse and then we repeated the experiment so that we finally obtained this complicated FROG trace. We captured the snapshots during the optimization. First, the Fourier transform limited laser pulse was completely destroyed. Then, it gradually approached the desired FROG trace. The final FROG trace is not exactly the same as the target, but we were very satisfied by this shaping performance. Therefore, if we know the amplitude and phase of the desired pulse, we can generate the desired pulse with this feedback scheme.

Another method we have used is TADPOLE pulse measurement, which is a spectral interference measurement (Fig. 5.30) [14]. We can easily reconstruct the both spectral phase and amplitude of the shaped laser pulse without any optimization

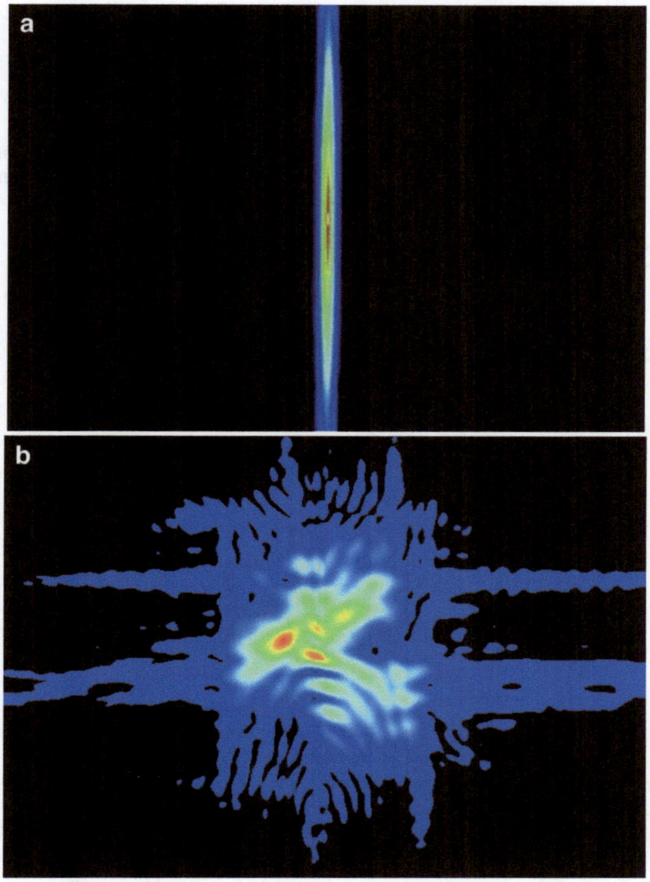

Fig. 5.29 Example of adaptive pulse shaping with a pulse shaper placed before an amplifier by referring for FROG traces (a movie is not available in this text. See ref [14]). (**a**) Initial FROG trace before Optimization. (**b**) Target FROG

algorithm as that in the FROG scheme, although we need an unshaped short reference pulse. Once we obtain the deviation between the target pulse and the measured pulse, we can straightforwardly correct the shaping mask. When repeating this process in a few times, the shaped pulse approaches to the desired pulse.

In Fig. 5.31 [14], the target pulse is the double pulse and each pulse has a different frequency chirp characteristic. After two or three corrections of the pulse shaper, we obtained this pulse that corresponds to the targeted double pulse. The phase is also very consistent with the target.

In Fig. 5.32 [13], these results were obtained using the AOPDF pulse shaper and the iterative shaping using FROG traces. The target pulse was the double pulse with different centre frequency for each pulse; the experimental result is very good.

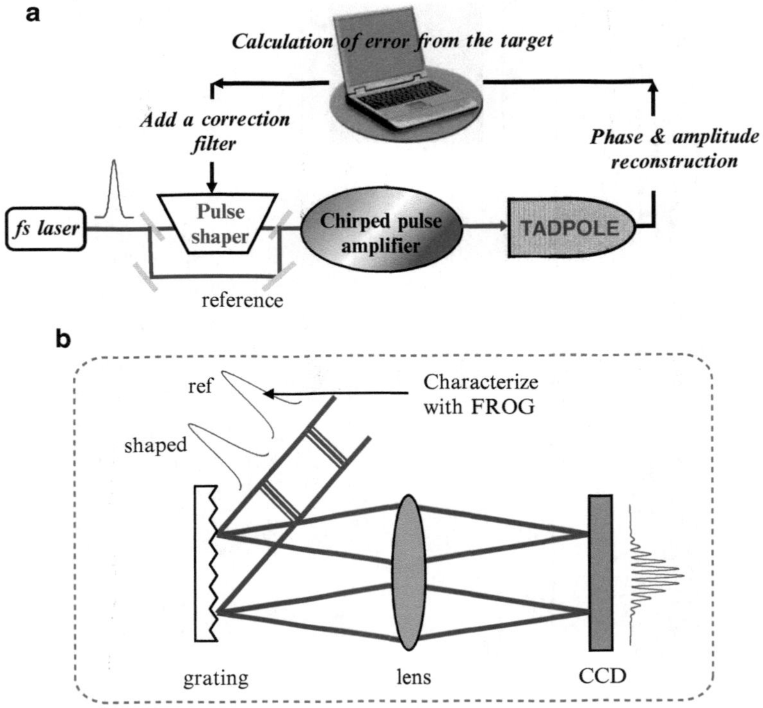

Fig. 5.30 A feedback pulse shaping scheme based on TADPOLE measurements [15]. (**a**) The reference pulse detours the pulse shaper and enters the CPA 600 fs after the shaped signal pulse. (**b**) Schematic setup of spectral interference measurement

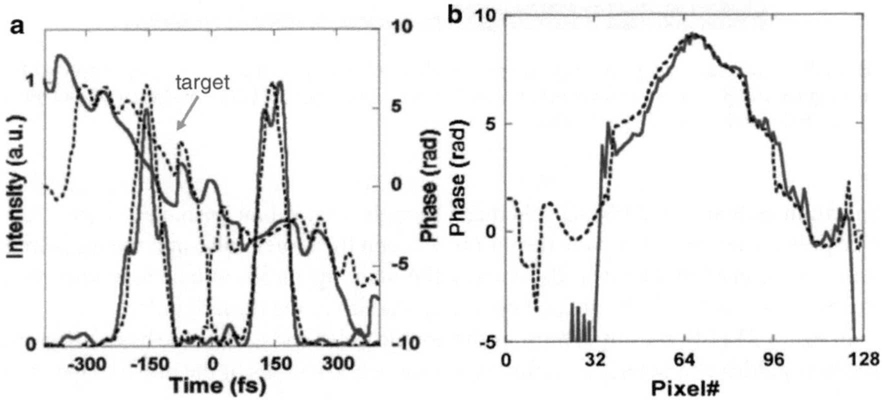

Fig. 5.31 An Example of Accuracy of Shaped Laser Pulses using the Scheme of Fig. 5.30 [15]. (**a**) Target (*dotted curve*) and Shaped pulse (*solid curve*) profile. (**b**) Phase mask profiles of the target pulse and the experimentally obtained pulse

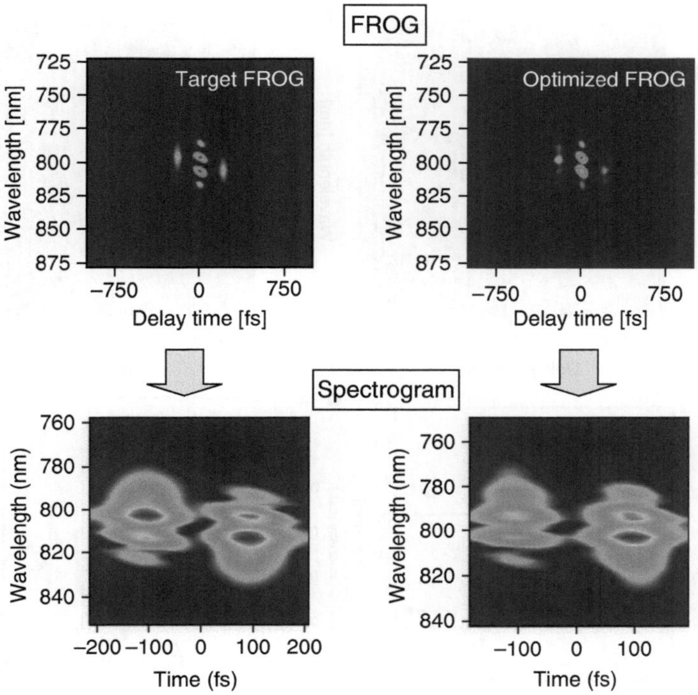

Fig. 5.32 Results of adaptive pulse shaping with an AOPDF by referring for FROG trace [13]

The example in Fig. 5.33 [13] shows our experimental results when we tried to produce the transformed limited pulse. Our target pulse was set shorter than the pulse that was normally obtained from our amplifier. Due the gain-narrowing effect, the output pulse was always slightly longer than the pulse width obtainable from the gain spectrum width of the amplifier. Therefore, we set a slightly shorter target pulse and repeated the iterative shaping using FROG measurements. The results implied that we had successfully obtained the shorter pulse. When we observed the final shaping mask, the central part was well suppressed; therefore, the amplitude mask which can compensate for the gain narrowing effect was automatically constructed during the closed-loop optimization.

By the pulse shape correction scheme using TADPOLE measurements, we can produce more complicated target pulses as seen in Fig. 5.34 [15]. In principle, we can produce any targeted laser pulses given in the pulse shaper time window.

Next we should consider the question: is it enough to apply the pulse shaping technique for various light–matter interactions? The simple answer is no. For example, there is a theoretically calculated optimal pulse for a certain vibrational excitation. When we Fourier transform this temporal shape, the bandwidth is huge and the spectrum ranges from the mid-infrared to the far-infrared (Fig. 5.35). With the present technology, we cannot manage such broad and long wavelengths. Also

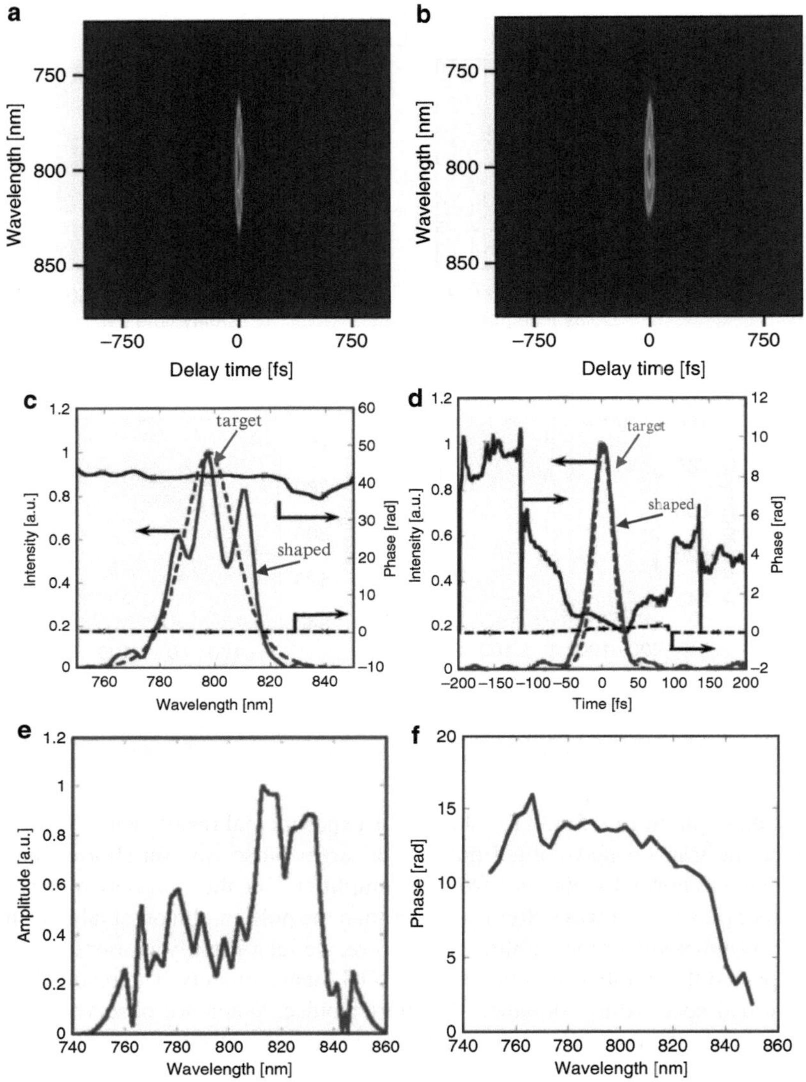

Fig. 5.33 An example of adaptive pulse shaping with an AOPDF by referring for FROG Trace [13]. In this example, the adaptive optimization automatically compensated for gain narrowing. (**a**) Target FROG: 30 fs TL pulse. (**b**) FROG trace of optimized pulse in the experiment. (**c**) Target (*dashed curves*) and optimized (*solid curves*) spectral profiles reconstructed from the FROG traces. (**d**) Target (*dashed curves*) and optimized (*solid curves*) temporal profiles reconstructed from the FROG traces. (**e**) Optimized amplitude function of the AOPDF for shaping a FTL pulse that has been compensated for gain narrowing and residual dispersion. (**f**) Optimized phase function of the AOPDF for shaping a FTL pulse that has been compensated for gain narrowing and residual dispersion

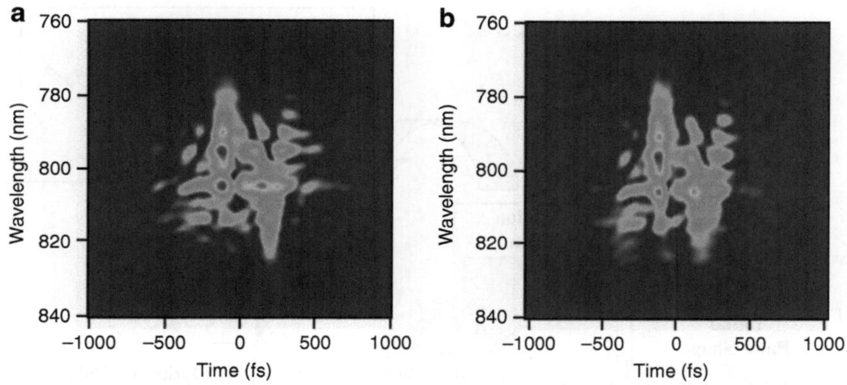

Fig. 5.34 Results of adaptive pulse shaping with an AOPDF by referring for FROG Trace [15]. More complex pulse is shaped as a target. (**a**) Target FROG trace of more complicated pulse. (**b**) FROG trace of the optimized pulse

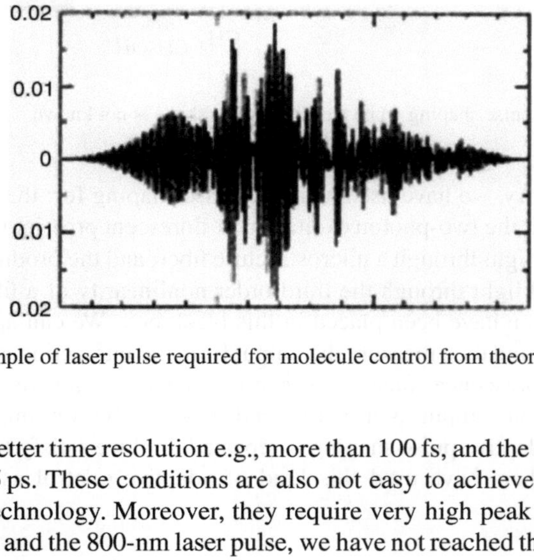

Fig. 5.35 An example of laser pulse required for molecule control from theoretical calculation

we need to get better time resolution e.g., more than 100 fs, and the entire pulse must be longer than 5 ps. These conditions are also not easy to achieve with the present pulse shaping technology. Moreover, they require very high peak intensities. With our pulse shaper and the 800-nm laser pulse, we have not reached these optimization laser pulses thus far.

5.7 How to Generate the Desired Ultrashort Laser Pulse in an Actual Laser System: Case 2: We Do Not Know What the Desired Pulse Shape Is

On the other hand, when we do not know the desired laser pulse shape, we can use a self-learning type adaptive control. Interaction between matter and light is set as a black box; then we can treat any kind of interaction between them (Fig. 5.36).

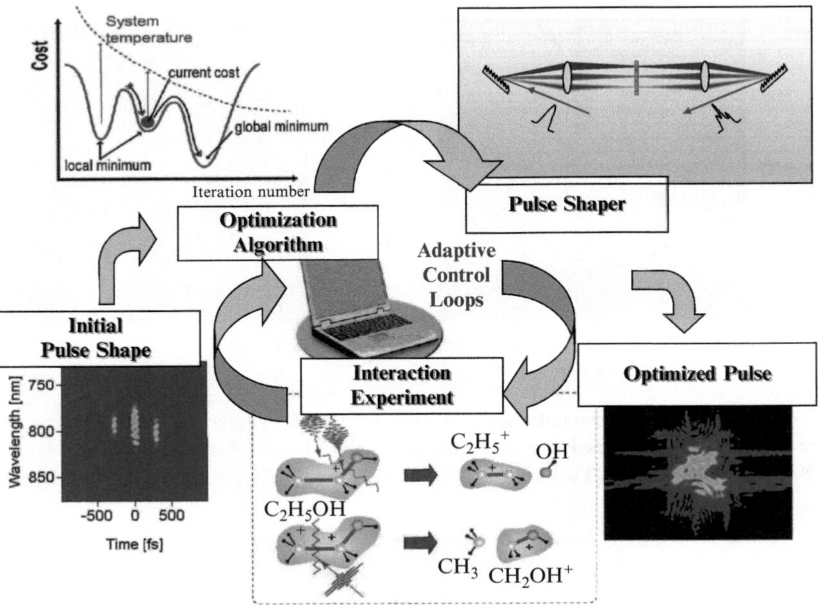

Fig. 5.36 Adaptive pulse shaping when a desired pulse shape is not known

In our laboratory, we have used adaptive pulse shaping for: the dissociated ionization of ethanol; the two-photon excitation of florescent protein; the generation of white continuum light through a microstructure fiber; and the production of photon-number squeezed light through the third order nonlinearity of a fiber. Thus, many kinds of interaction have been placed in this black box. We can apply any type of initial laser pulse and then measure the output from the interaction, such as fragment distributions, fluorescence intensity, spectrum of white light, or squeezing level. Then, by using that output as a control index, a suitable optimization algorithm is used and the adaptive pulse shaping is repeated until we satisfy the output index.

Many researchers have used this kind of adaptive control as summarized in Table 5.3 [16].

5.8 Adaptive Pulse Shaping for Dissociative Ionization of Ethanol Molecules

A few of our experiments are summarized in Figs. 5.37–5.41 [17, 18]. The potential energy surface of a molecule under high laser field is not spectroscopic in nature. There is an AC Stark effect and the light dressed state is temporally varying as a function of the instantaneous laser field. In fact, the vibrational nuclear wave packet cannot follow such a fast change of the potential energy surface, but we may be able to guide the propagation of the nuclear wave packet by shaping this multidi-

Table 5.3 Examples of adaptive pulse control of atom or molecule reactions [16]

fs, shaped pump, GA	$CpFe(CO)_5 \rightarrow Fe(CO)_5^+/Fe^+$, beam
fs, shaped pump, GA	$CpFe(CO)_2Cl \rightarrow FeCOCl^+/FeCl^+$, beam
fs tailored pump, GA	$CpFe(CO)_5 \rightarrow CpFe(CO)_5^+/Fe^+$, beam
fs polarized pump,	$I_2 \rightarrow I + I^*/I + I$, ss jet
Fs, chirped pump–probe	$Na_2 \rightarrow Na_2^+/Na^+$, beam
fs, pump–probe, GA	$CsCl \rightarrow Cs + Cl/Cs^+ + Cl^-$, beam
fs, pump–probe, GA	$CsCl \rightarrow Cs + Cl/Cs^+ + Cl^-$, beam
fs, shaped pump, GA	$CpMn(CO)_3 \rightarrow CpMn(CO)_3^+/CpMn(CO)^+$, beam
fs, shaped pump, GA	$CpFe(CO)_2Cl \rightarrow CpFeCOCl^+/FeCl^+$
fs, shaped pump, GA	$CH_3COCF_3 \rightarrow CH_3^+/CF_3^+$, beam
fs, shaped pump, GA	$C_6H_5COCH_3 \rightarrow C_6H_5CO^+/C_6H_5^+$, beam
fs, shaped pump, GA	$C_6H_5COCH_3 \rightarrow C_6H_5CH_3^+$, beam
ns, IR + UV	HNCO, H + NCO/^{3}NH + CO, beam
fs, shaped pump, GA	$CH_3COCH_3 \rightarrow CH_3CO^+$, beam
fs, shaped pump, GA	$CH_3COCF_3 \rightarrow CF_3^+$, beam
fs, shaped pump, GA	$C_6H_5COCH_3 \rightarrow C_6H_5CO^+/C_6H_5^+$, beam
fs, shaped pump, GA	$C_6H_5COCH_3 \rightarrow C_6H_5CH_3^+$, beam
fs, shaped pump, GA	$CH_2BrCl \rightarrow CH_2Cl^+/CH_2Br^+$, beam
fs, shaped pump, GA	$CpFe(CO)_2Cl \rightarrow CpFeCOCl^+/FeCl^+$, beam
fs, shaped pump, GA	$CpFe(CO)_2Br \rightarrow CpFeCOBr^+/FeBr^+$, beam
fs, shaped pump, GA	$CpMn(CO)_3 \rightarrow CpMn(CO)_3^+/CpMn(CO)_2^+$, beam

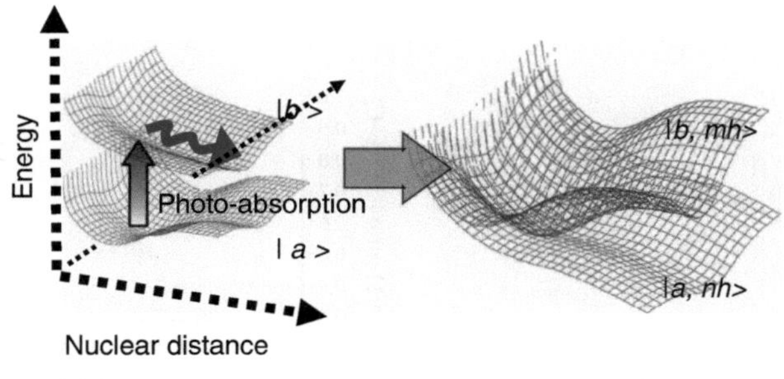

Fig. 5.37 Light dressed potential energy surface in high intensity light field

mensional potential energy surface using the shaped laser pulse. This is the idea of molecule control using a high field laser pulse.

In our experiment, we used ethanol molecules [17, 18]. Its skeleton structures are C–C and C–O units, but the dissociation energies of the C–C and C–O bonds are approximately equal to 3.4 eV. At the leading edge of the ultrashort laser pulse, the parent ions are generated at the first singly charged state via multiphoton ionization. Then, we observe the dissociation of singly ionized molecules and also the doubly charged ion formation followed by the Coulomb explosion.

We used the adaptive pulse shape control in the experiment [18]. In this case, the control index is the ratio corresponding to C–O bond dissociation relative to

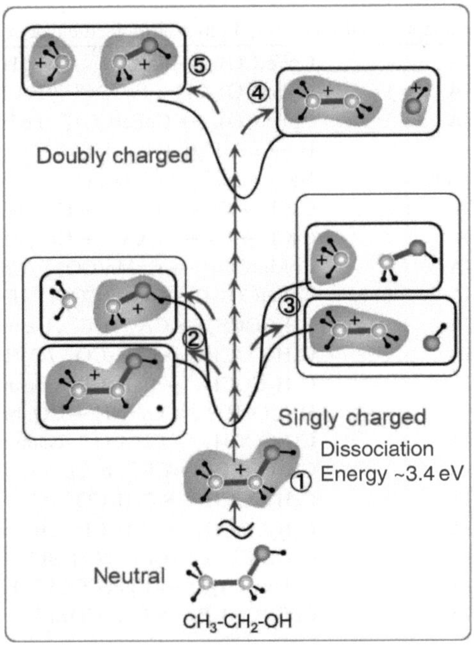

Fig. 5.38 Reaction pathways of dissociative ionization of ethanol

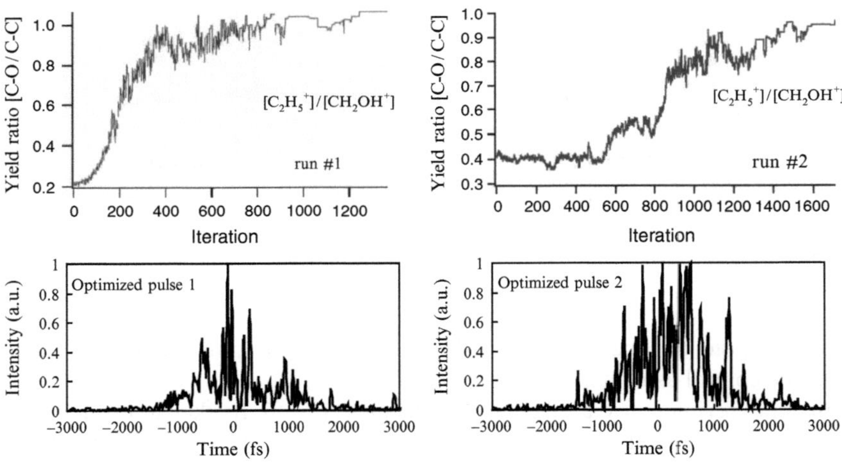

Fig. 5.39 Adaptive control of ethanol referring to $[C_2H_5^+] / [CH_2OH^+]$

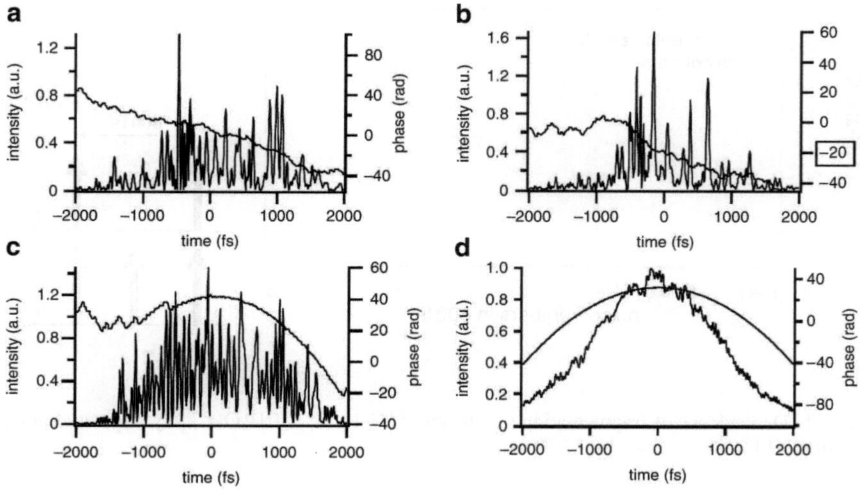

Fig. 5.40 Four pulses that can achieve the same $[C_2H_5^+]/[CH_2OH^+]$ ratio [18]

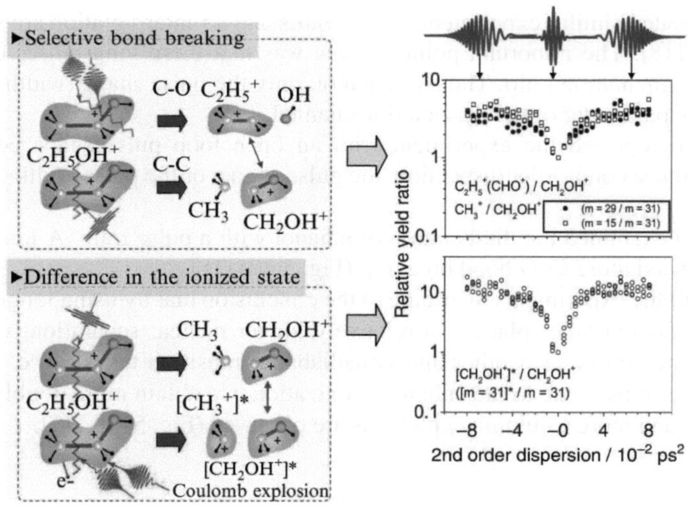

Fig. 5.41 Dependence of fragmentation of ethanol on laser pulse chirp [17]

C–C bond dissociation. Using the transform limited pulse, most of the dissociations are found to be at the C–C bond. Further, by using this ratio for the control index, we performed adaptive control. Starting from the transform limited pulse, we found this index gradually increased and saturated. When we measured the shaped laser pulse at the end of control, we obtained the result of Fig. 5.39. When we changed the parameter of the optimization algorithm, this evolution was slightly different. But

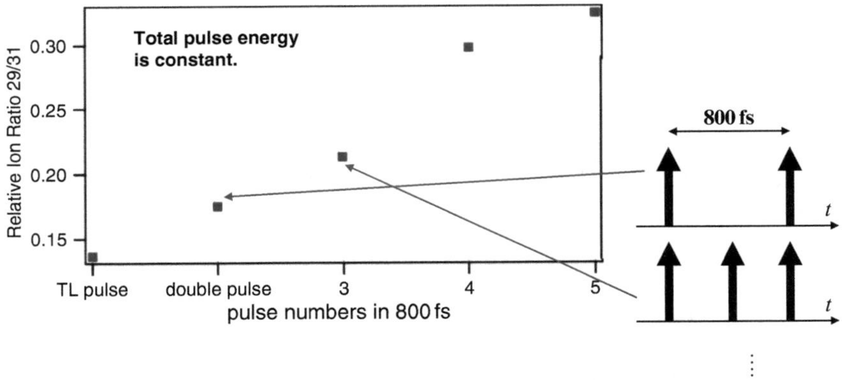

Fig. 5.42 Dependence of fragmentation of ethanol of $[C_2H_5^+] / [CH_2OH^+]$ by sequential pulse excitation [18]

finally we reached the same index value. The pulse shapes are shown in Fig. 5.39. The fine structure was a random modulation without any regularities.

We repeated similar experiments. The results and an interpretation are shown in Fig. 5.40 [18]. The important point we saw was that these long pulses created a similar fragmentation ratio. Thus, in ethanol, only the pulse shape (width) was the controlling parameter of the dissociation channel.

We also repeated the experiment with an open loop pulse shape control by changing the second-order dispersion, the pulse shape, or the pulse width (Fig. 5.41 [17]).

We also examined the dissociation of ethanol with a pulse train. A longer pulse train produced more C–O bond breaking (Fig. 5.42) [18].

From all the experiments, we came to the conclusion that by using longer pulses, C–O dissociation takes place at a relatively longer nuclear separation, where the dressed states cross each other and nonadiabatic transition takes place. Also, by using longer pulses due to the enhanced ionization, we obtain more doubly ionized molecules and more Coulomb explosions are observed (Fig. 5.43) [17].

5.9 Adaptive Pulse Shaping of Two-Photon Excited Fluorescence Efficiency

The final next example is from a biochemistry-type experiment. We observed the two photon excited fluorescence of fluorescent protein and adaptively controlled the two photon fluorescence efficiency (Fig. 5.44). The two photon fluorescent intensity was maximized by the transformed limited pulse because the peak intensity was the maximum. However, regarding the efficiency defined by the fluorescence

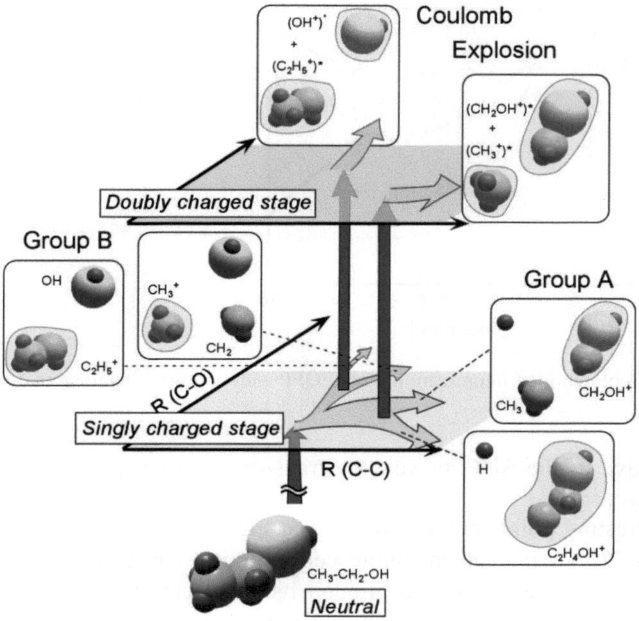

Fig. 5.43 Summary of dissociative ionization of ethanol by chirped laser pulses [17]

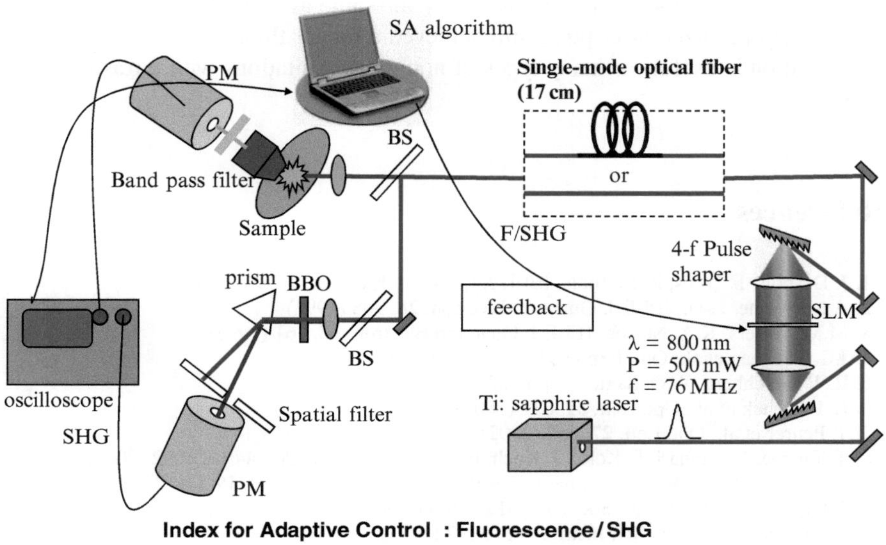

Fig. 5.44 Experimental setup of adaptive control of two-photon pumped fluorescence with femtosecond laser pulses through an optical fiber

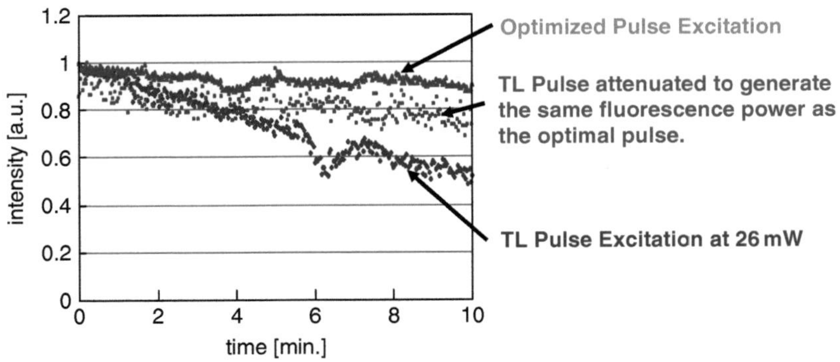

Fig. 5.45 Discoloration of fluorescent protein EGFP with optimized femtosecond laser pulse

power relative to the SHG power, the transform limited pulse was no more an optimal pulse.

The interesting points are summarized in Fig. 5.45. The dots for the TL pulse excitation at 26 mW indicate the fluorescence lifetime curve of the protein. Usually, when excitation of the protein is continued, its fluorescence dies. However, by using the optimized laser pulse, the lifetime was much longer than when excited by a transform limited pulse. The peak intensity was lower for the optimized pulse than the transform limited pulse. Therefore, initially we attenuated the transform limited pulse so that the fluorescence intensity was the same as that of the optimized laser pulse. The optimized laser pulse still achieved a longer fluorescence lifetime. This optimization is useful for many types of nonlinear excitations with ultrashort laser pulses.

References

1. J. Desbois, F. Gires, P. Tournois, IEEE J. Quantum Electron. **9**(2), 213 (1973)
2. A.M. Weiner et al., IEEE J. Quantum Electron. **28**, 908 (1992)
3. M.M. Wefers, K.A. Nelson, IEEE J. Quantum Electron. **32**, 161 (1996)
4. J.C. Vaughan et al., Opt. Express **14**(3), 1314 (2006)
5. R.W. Gerchberg, W.O. Saxton, Optik **35**, 237 (1972)
6. L. Polachek et al., Opt. Lett. **31**, 631 (2006)
7. T. Feurer et al., Opt. Lett. **27**, 652 (2002)
8. T. Tanabe, F. Kannari, F. Korte, J. Koch, B. Chichkov, Appl. Opt. **44**(6), 1092 (2005)
9. T. Tanabe, H. Tanabe, Y. Teramura, F. Kannari, J. Opt. Soc. Am. B. **19**(11), 2795 (2002)
10. J. Paye, A. Migus, J. Opt. Soc. Am. **B12**, 1480 (1995)
11. R. Trebino et al., Rev. Sci. Instrum. **68**, 3277 (1997)
12. F. Verluise et al., Opt. Lett. **8**, 575 (2000)
13. K. Ohno, T. Tanabe, F. Kannari, J. Opt. Soc. Am. B. **19**(11), 2781 (2002)
14. T. Tanabe, K. Ohno, T. Okamoto, M. Yamanaka, F. Kannari, Jpn. J. Appl. Phys. **43**(4A), 1366 (2004)
15. T. Tanabe, M. Yamanaka, T. Okamoto, F. Kannari, IEEE J. Select. Top. Quantum Electron. **10**(1), 221 (2004)

16. M. Dantus, V.V. Lozovoy, Chem. Rev. **104**, 1813 (2004)
17. R. Itakura, K. Yamanouchi, T. Tanabe, T. Okamoto, F. Kannari, J. Chem. Phys. **119**(8), 4179 (2003)
18. H. Yazawa, T. Tanabe, T. Okamoto, M. Yamanaka, F. Kannari, R. Itakura, K. Yamanouchi, J. Chem. Phys. **124**, 204314 (2006)

Chapter 6
Nonlinear Interaction of Strong XUV Fields with Atoms and Molecules

Katsumi Midorikawa

Abstract We introduce advanced techniques in generating intense short-wavelength light in the XUV region. Through the new experimental results obtained when we irradiated atoms and molecules with such intense XUV pulses, we discuss the ultrafast dynamics of atoms and molecules in the attosecond time domain.

6.1 Introduction

There are various high-intensity phenomena occurring in atoms and molecules in a strong optical field (Fig. 6.1). When laser radiation with an intensity of greater than $10^{12}\,\mathrm{W/cm^2}$ is focused, alignment or structural changes in molecules can be seen. With a further increase of the intensity to around 10^{14}–$10^{15}\,\mathrm{W/cm^2}$, tunneling ionization occurs and that results in an explosion in the case of molecules. Further increase of the intensity to around $10^{18}\,\mathrm{W/cm^2}$ produces hard X-rays and high energy particles. These pictures of Fig. 6.1 are drawn based on the interaction with atoms and molecules with intense visible or infrared radiation.

What happens when atoms and molecules are exposed to XUV radiation at the intensity of 10^{12} to $10^{15}\,\mathrm{W/cm^2}$? The answer to this question is the main theme of this presentation.

An interaction of a strong field with atoms and molecules should be qualitatively different when a wavelength is changed from the infrared to the XUV, because the driving force of nonlinear interaction is ponderomotive energy U_{p}, which is proportional to $I\lambda^2$. So when the wavelength is changed, the ponderomotive energy changes greatly. Even if the XUV radiation is at a focused intensity of $10^{15}\,\mathrm{W/cm^2}$,

K. Midorikawa
Laser Technology Laboratory, RIKEN Advanced Science Institute, 2-1 Hirosawa, Wako, Saitama 351-0198, Japan

K. Yamanouchi (ed.), *Lectures on Ultrafast Intense Laser Science 1*, Springer Series in Chemical Physics 94, DOI 10.1007/978-3-540-95944-1_6,

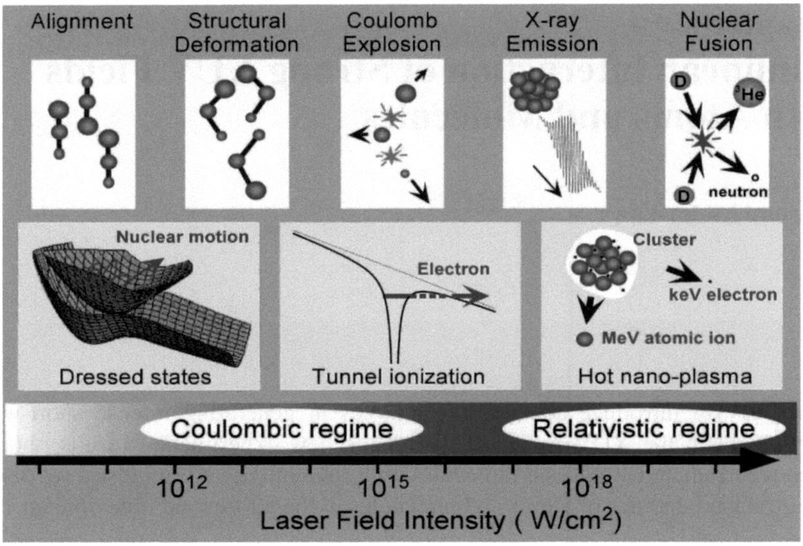

Fig. 6.1 Various phenomena in atoms and molecules in strong optical field [19]. Reproduced by courtesy of Prof. K. Yamanouchi

the ponderomotive energy is much smaller than the ionization potential of atoms or molecules, $U_p \ll I_p$. This means that the Keldysh parameter γ, defined as:

$$\gamma = \frac{\omega_{\text{laser}}}{\omega_{\text{electron}}} = \sqrt{\frac{I_p}{2U_p}} \tag{6.1}$$

becomes $\gamma > 1$. Therefore, when XUV radiation is used, it is hard to cause the tunneling process. Also as the cross section of nonresonant two-photon ionization $\sigma^{(2)}$ is proportional to λ^6 [5], even when the XUV radiation is focused on atoms and molecules at an intensity of around $10^{12}\,\text{W/cm}^2$ the nonlinear process is hard to be observed. To induce nonlinear or high-intensity phenomena in the XUV region, we need much higher intensity compared to that required for visible or infrared radiation.

Another interesting phenomenon observed by using XUV radiation is generation of attosecond pulses. When infrared or visible radiation is used, we can easily obtain picosecond or femtosecond radiation. We have used such ultrashort pulses to investigate the atomic motion in molecules. However, to investigate the motion of an electron in atoms or molecules, we need much shorter pulses reaching down to attoseconds. In order to generate such an ultrashort pulse, we need XUV or soft X-ray radiation.

Why do we need such a short wavelength to generate attosecond pulses? We can take a Ti:sapphire laser field as an example to answer this question (Fig. 6.2). Since a single cycle oscillation of the 800 nm radiation takes 2.7 fs, that is the shortest

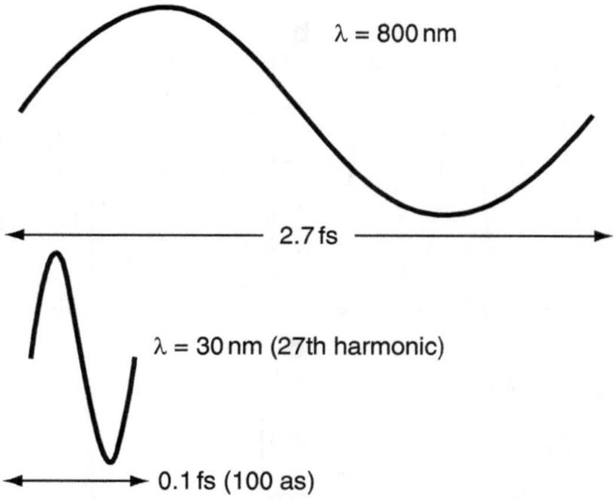

Fig. 6.2 One optical cycle is 2.7 fs for 800 nm and 0.1 fs for 30 nm (27th harmonic)

pulse attainable when we use the Ti:sapphire laser. To obtain a much shorter pulse, we need a shorter wavelength radiation as shown in the figure. If we use a 30-nm radiation that corresponds to the 27th harmonic of the Ti:sapphire laser, the oscillation period is one-27th of the fundamental pulse. This radiation has a potential to generate a 0.1 fs pulse.

A second requirement to be satisfied is bandwidth. The theory of Fourier synthesis of harmonics describes that by adding harmonics in phase a train of ultrashort pulses can be generated as seen in Fig. 6.3. The final pulse width $\Delta\tau$ is inversely proportional to the bandwidth of the radiation $\Delta\nu$ because of the uncertainty principle, $\Delta\nu \cdot \Delta\tau \geq 1$. So we need both large bandwidth and short wavelength to generate attosecond pulses.

To satisfy such requirements, we use high-order harmonics for the generation of attosecond pulses; high-order harmonic generation has already been discussed in this school. Here, I will repeat briefly the generation process. When high-intensity Ti:sapphire laser pulses are focused into a gas cell or jet in a vacuum chamber, the strong-field interaction with atoms or molecules generates high harmonics. In a typical spectrum for high harmonics, intensity rapidly falls off within the first several harmonics and then shows a plateau where all the harmonics have almost the same intensity; that is a result of the interaction of atoms and molecules with high-intensity radiation. Finally the spectrum ends with sharp cutoff.

When we combine several harmonics in phase, we can generate a train of attosecond pulses. Then, the next problem is how to measure the attosecond pulses. Usually, when we measure femtosecond laser pulses in the visible or infrared regions, we use an autocorrelator (Fig. 6.4). In the autocorrelator, the measured pulse is split by a beam splitter and by moving the system mirrors we can introduce a time delay between the two split pulses before focusing them onto a nonlinear medium. But when we implement this method for the XUV radiation to measure attosecond pulses, there are several problems. The first is that we do not have a beam

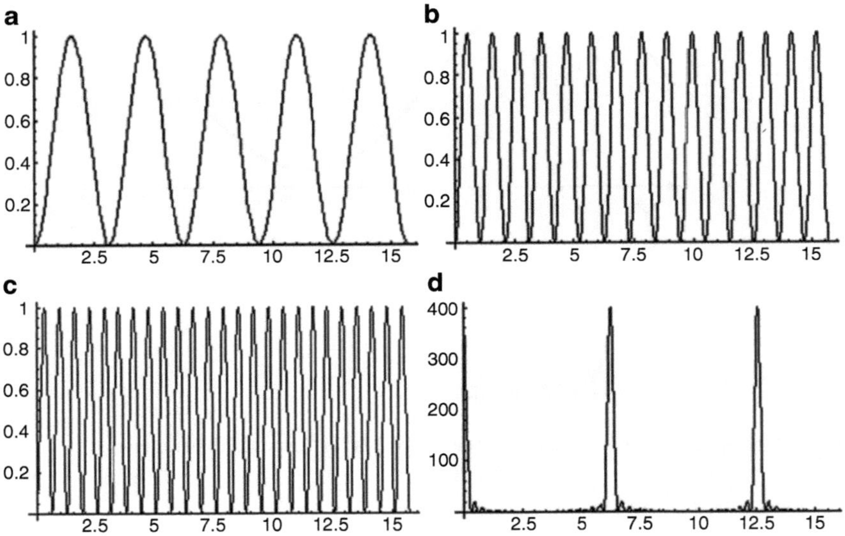

Fig. 6.3 Fourier synthesis of ultrashort pulse train by adding harmonics in phase. (**a**) $N = 1,\ \omega$, (**b**) $N = 1,\ 3\omega$, (**c**) $N = 1,\ 5\omega$, (**d**) $N = 20,\ \phi = \text{constant}$

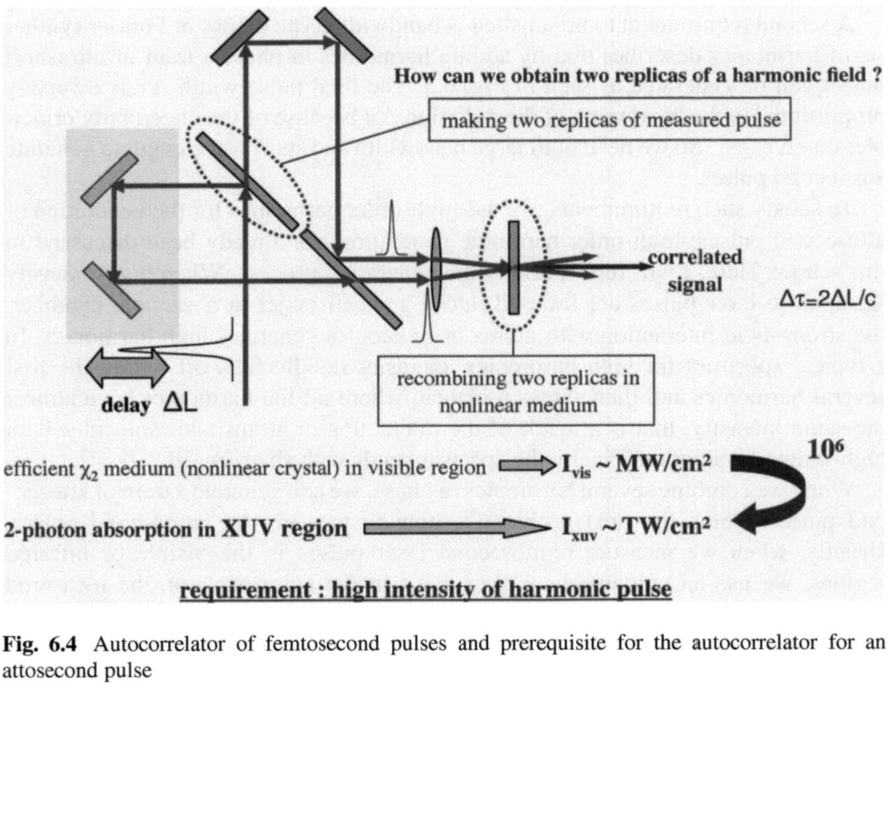

Fig. 6.4 Autocorrelator of femtosecond pulses and prerequisite for the autocorrelator for an attosecond pulse

splitter in the XUV region. The second is that when we use this type of configuration we need a lot of mirrors and they reduce the harmonic intensity considerably. This is because in the XUV region, the reflectivity of mirrors is typically less than 30%.

The third problem is also serious – there is no nonlinear crystal in the XUV region. As mentioned above, to obtain a two- or three-photon process in the XUV region, we need a very high-intensity because the cross section of the nonresonant two-photon ionization is proportional to λ^6. Some researchers are using two-color above-threshold ionization (ATI) scheme developed by a French group, but the temporal shapes of attosecond pulses could not be obtained directly with the two-color ATI process. The two-color ATI process is based on an interference between two harmonics. When we measure autocorrelation, which is an already established technology in the femtosecond region for visible lasers, we can directly obtain temporal shapes of harmonic pulses.

For the rest of this paper, attention is directed to how to generate intense high harmonics in the XUV region for observing the pulse shapes by using the autocorrelation technique. After the generation, we measure the focusing property of the high harmonic, because we must evaluate how much intensity we can achieve. Using such high-intensity, we observe the nonsequential double ionization, which is a typical two-photon process observed in the XUV region. Using this two-photon process, we characterize XUV pulses, and then, an application of such a high-intensity XUV attosecond pulse to a molecular process will be described. That work is a Coulomb explosion of molecules by XUV attosecond pulses, which was done in collaboration with Professor Yamanouchi's group.

6.2 Generation of High-Power High-Order Harmonics

When we consider high-order harmonic generation process, a single atom response can be understood by solving the time-dependent Schrödinger equation:

$$\mathrm{i}\frac{\partial}{\partial\tau}\psi(\vec{r},z,t) = \left[-\frac{1}{2}\nabla^2 + V_{\mathrm{eff}} + \vec{r}\cdot\vec{E}(z,t)\right]\psi(\vec{r},z,t). \tag{6.2}$$

One important result obtained from the theoretical calculation is the maximum harmonic order, which is given as the sum of the ionization potential and three times the ponderomotive energy:

$$E_{\mathrm{max}} = I_{\mathrm{p}} + 3.2U_{\mathrm{p}}. \tag{6.3}$$

The underlying physics is well known, so only a brief explanation is necessary (Fig. 6.5). When we focus the high-intensity laser onto atoms and molecules with an intensity comparable to the Coulomb field inside the atoms or molecules, which corresponds to 10^{14}–$10^{15}\,\mathrm{W/cm^2}$, tunneling ionization occurs and a freed electron is captured and accelerated by the optical field. After the acceleration when the optical

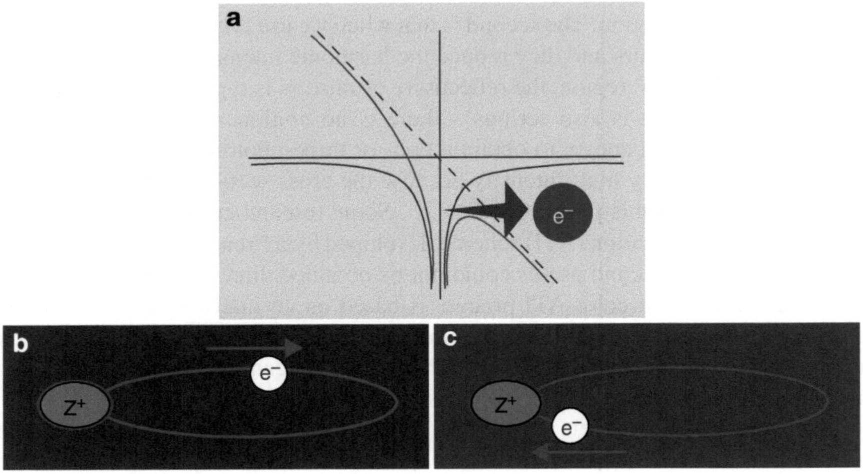

Fig. 6.5 Schematic drawing of the three-step model of the high-order harmonic generation. (**a**) Tunneling ionization by strong optical field. (**b**) acceleration by optical field. (**c**) recombination.

field changes its sign in the second half-cycle, the freed electron returns to the parent ion and recombines with it. At this instance this kinetic energy of electrons is converted into a harmonic photon and the maximum harmonic photon energy is expressed by a simple equation (6.3), and as predicted with the time-dependent Schrödinger equation.

When we focus the Ti:sapphire laser at an intensity of 10^{14}–$10^{15}\,\mathrm{W/cm^2}$, the electron kinetic energy reaches 100 eV which corresponds to 10-nm radiation.

The time-dependent Schrödinger equation should be useful to predict the generation of high-order harmonics, but the calculation is very time consuming. In order to include propagation effects, we need a much simpler calculation method based on the strong field approximation model. In this model, we consider only two states – the ground and continuum. In the continuum state, electrons behave like classical particles and their motion is expressed by the quasi-classical action term $S(p_s, t, \tau)$ as:

$$\begin{aligned}\vec{r}(t) = \mathrm{i}\int_0^\infty \mathrm{d}\tau \left(\frac{\pi}{\nu + \mathrm{i}\tau/2}\right)^{3/2} \mathrm{d}^*\left(\vec{p}_s - \vec{A}(t)\right) \exp\left(-\mathrm{i}S(\vec{p}_s, t, \tau)\right) \\ \times \vec{\varepsilon}(t-\tau)\cdot\vec{d}\left(\vec{p}_s - \vec{A}(t-\tau)\right)\exp(-\Gamma t) + c.c.\end{aligned}, \qquad (6.4)$$

from which we obtain the harmonic phase factor equal to $\exp[-\mathrm{i}S(p_s, t, \tau)]$. Consequently, the atomic dipole phase is proportional to the focused intensity as:

$$\Phi_\mathrm{i} = -S \approx -\alpha_\mathrm{i} I. \qquad (6.5)$$

This is different from the conventional nonlinear optical process. In the conventional nonlinear process like the generation of the second and third harmonics, the dipole phase does not depend on the intensity. But the interaction of the high-intensity pulse makes the high harmonic dipole phase strongly dependent on the intensity. When we tightly focus the fundamental laser pulses, the intensity changes rapidly around the focused region. It means there is a change of the dipole phase along the optical axis. Therefore, to achieve phase matching for efficient generation of high-order harmonics, we need to carefully consider such effects.

In order to avoid such a difficulty, we proposed the use of two laser-guiding techniques several years ago, one of which is using a hollow fiber and the other is using self-guiding. In the self-guiding scheme [17, 18], the self-focusing originating from the change in the nonlinear refractive index,

$$n(r) = n_0 + n_2 I(r), \tag{6.6}$$

balances the plasma defocusing represented as:

$$n(r) = (1 - N_e(r)/N_c)^{1/2}. \tag{6.7}$$

This is shown in Fig. 6.6b. Since both the pump phase and intensity are almost constant in the guided region, we do not need to consider the dipole phase change associated with the intensity change along the propagation axis.

Using these techniques, we have obtained a large enhancement of harmonics (Fig. 6.7). In the usual high harmonic spectral distribution, the harmonics rapidly decrease within the first several orders and then reach a plateau, but in our techniques, harmonics in the phase-matched region are enhanced strongly. These techniques work well with a pump energy less than a few mJ. When we try to increase the harmonic energy, we need a higher pump intensity or higher pump energy. But, in turn, if we focus the Ti:sapphire laser of several mJ or more onto the fiber, it easily destroys the fiber and it also changes the propagation conditions. Therefore, these techniques work well within an energy of a few mJ, and we need further developments to increase the harmonic energy.

Under the phase-matched condition, the evolution of the harmonic field is simply described by an equation that includes terms for the dipole, phase-matching factor, and medium absorption:

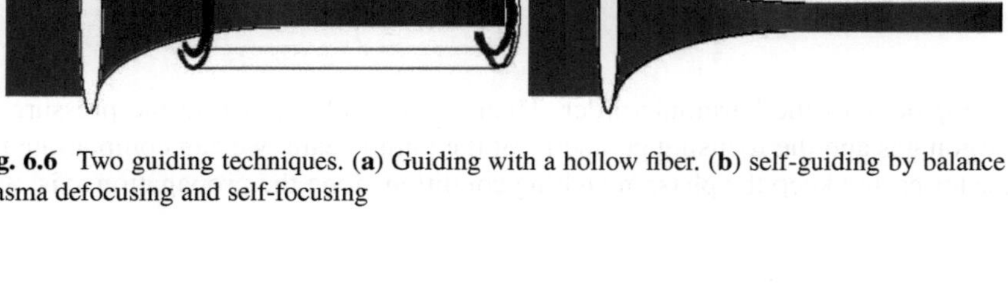

Fig. 6.6 Two guiding techniques. (**a**) Guiding with a hollow fiber. (**b**) self-guiding by balance of plasma defocusing and self-focusing

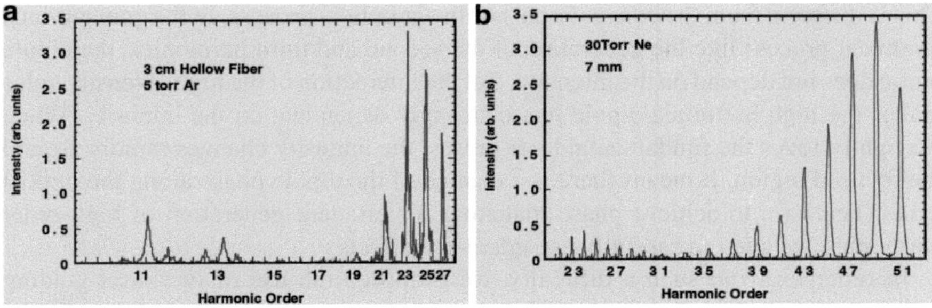

Fig. 6.7 High-order harmonics obtained by the hollow fiber scheme (**a**) and the self-guiding scheme (**b**) [16–18]

$$\frac{\mathrm{d}}{\mathrm{d}z}E_{\mathrm{q}} = \mathrm{i}N\left|D_{\mathrm{q}}\right|\exp(\mathrm{i}\Delta kz) - \frac{\alpha}{2}E_{\mathrm{q}}. \tag{6.8}$$

By integrating this equation along the propagation direction, we can obtain the total photon number N_{q} as [1]:

$$N_{\mathrm{q}} = AN^2|D_{\mathrm{q}}|^2\frac{1+\exp(-\alpha L) - 2\cos(\Delta kL)\exp(-\alpha L)}{(\alpha/2)^2 + (\Delta k)^2}. \tag{6.9}$$

From this equation, we can easily obtain the optimized conditions leading to the most efficient medium condition as $= L_{\mathrm{med}} > 3L_{\mathrm{abs}},\ L_{\mathrm{coh}} > 5L_{\mathrm{abs}}$.

This has led us to develop a new pumping geometry, in which laser pulses are loosely focused. For this, the geometrical phase is determined by the focusing property of the pump pulse. This geometrical phase, the so-called Gouy phase, is compensated by medium's dispersion to keep the phase matching condition,

$$\Delta k_{\mathrm{gouy}} + \Delta k_{\mathrm{neu}} + \Delta k_{\mathrm{nonl}} = 0. \tag{6.10}$$

When using Ar as a medium, the dispersion of this medium is negative, i.e.,

$$\Delta k_{\mathrm{neu}} = -1.08p\ \mathrm{cm}^{-1}(\mathrm{Argon}), \tag{6.11}$$

and the geometrical phase has a positive dispersion as:

$$\Delta k_{\mathrm{gouy}} = q\frac{1}{z_0\left(1+\frac{L^2}{z_0^2}\right)}, \tag{6.12}$$

where q denotes the harmonic order. Then by carefully adjusting the pressure of an argon gas and the focusing geometry of the pump beam, we can compensate for these terms and keep the phase-matching condition along the propagation axis.

After the phase matching is satisfied, the total output energy becomes proportional to the emission cross section, or the emission area of the harmonics and the square of the product of the interaction length and gas pressure as:

$$E_{\mathrm{HH}} \propto AL^2P^2, \tag{6.13}$$

where A, L, and P denote the emission area of high-order harmonics, the interaction length, and the gas pressure, respectively.

But the product of the pressure and the interaction length, *PL*, is limited by the absorption of the medium, which is automatically determined by the medium as long as the harmonic wavelengths are fixed.

Consequently, in order to increase the total harmonic energy we must increase the emission area of the harmonics. To do that, we increased the focusing length of the pump beam. For the developed geometry, as in Fig. 6.8, the Ti:sapphire laser pulses are focused with a 5-m-long focusing-length lens to the gas cell. In this gas cell, the interaction length is 10 cm, which is much longer than the typical harmonic generation geometry ever reported. Because, as just mentioned, since the total PL product is limited, the gas pressure is decreased when the interaction length is increased, so that the PL product is kept optimum.

This geometry gives us some advantages over conventional one. First, we can ideally focus the laser beam because of the low pressure and long interaction length; this allows automatic selection of a single mode of the pump propagation. Second, we can linearly increase the pump energy if we increase the focal length. We checked whether this long focusing geometry works, and some results are shown in Fig. 6.9. When we changed the focusing length of the lens, the emission area changed and the output energy was found to be linearly proportional to the focus area as predicted. With the increase in the focusing length, the third advantage can be seen; the beam divergence of high-order harmonics decreases with the increase of the interaction length. Consequently, we have obtained the output energies of 5 μJ at 62 nm in Xe, 0.3 μJ at 30 nm in Ar, and 40 nJ at 13 nm in Ne.

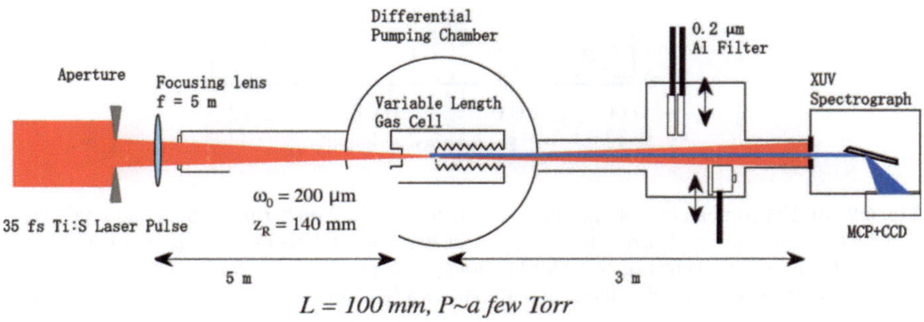

Fig. 6.8 Experimental setup for high-harmonic generation using a loosely focusing geometry

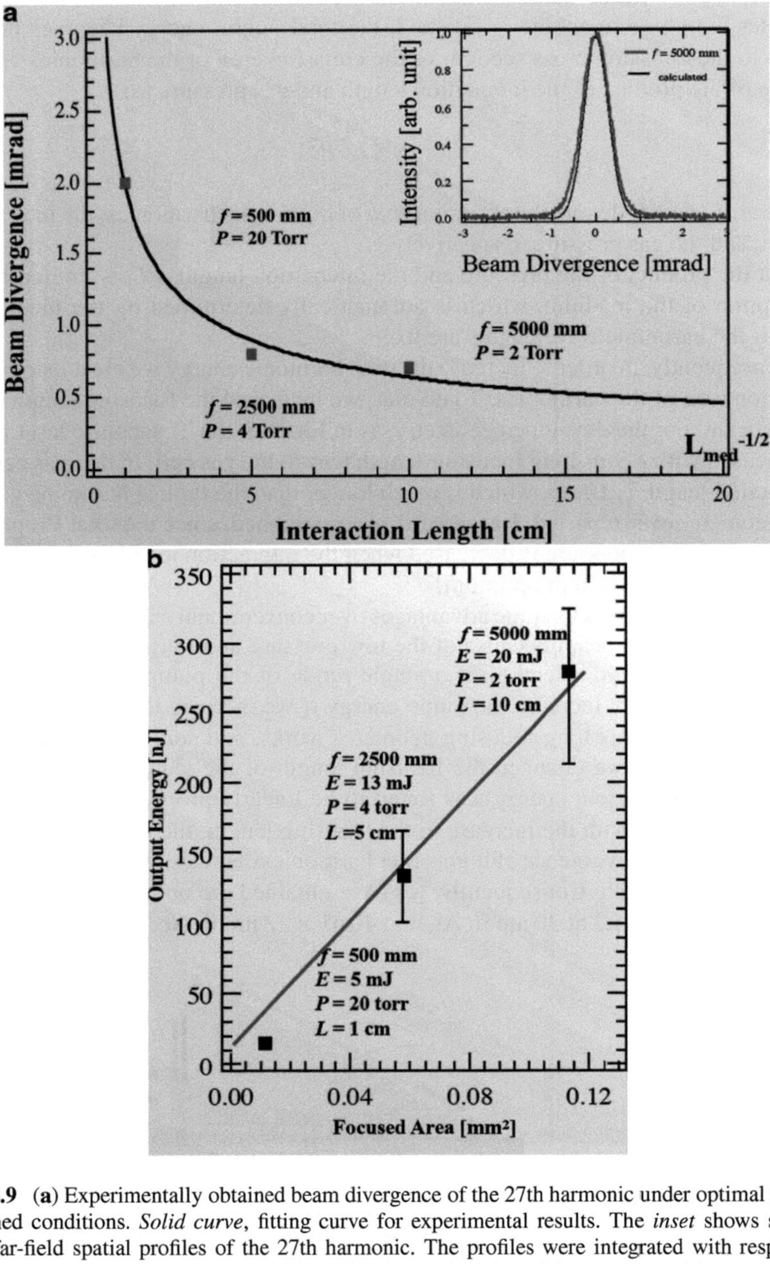

Fig. 6.9 (**a**) Experimentally obtained beam divergence of the 27th harmonic under optimal phase-matched conditions. *Solid curve*, fitting curve for experimental results. The *inset* shows single-shot far-field spatial profiles of the 27th harmonic. The profiles were integrated with respect to wavelength. (**b**) Energy yield of the 27th harmonic as a function of the spot area of the pump pulse at the focus

6.3 Spatial Properties of High-Order Harmonics

Figure 6.10a, b shows the harmonic spectral distribution and the 2D spatial profile of the 27th harmonic beam, respectively. In this experiment, we set a phase matching point around the 27th harmonic. The conversion efficiency was typically 10^{-4} to 10^{-5} and the output energy was more than 300 nJ when Ar was as a medium [12]. As a result of the phase matching, only five harmonics were observed around the 27th signal. By changing the position of the spectrometer slit step by step, we were able to obtain the 2D spatial structure of the 27th harmonic output (Fig. 6.10b).

When Xe is adopted, the orders of the generated harmonmics are lower than in the case of Ar, because the ionization potential of Xe is much lower than that of Ar. However, the large nonlinearity of Xe helps us to improve the conversion efficiency by one order of magnitude. Indeed, by using Xe, we obtained an output of about 5 μJ at around 60 nm.

Using Ne as medium, we obtained the harmonics energy of the order of 40 nJ at wavelength of 13 nm with perfect spatial coherence (Fig. 6.11). This 13-nm radiation is very useful for the next generation EUV lithography process, and we are planning to use it for the metrology of EUV optics [14].

We investigated the focusing property of the 27th harmonic, in order to know the focused intensity (Fig. 6.12). Since the generation efficiency of our system is typically 10^{-5}, a very strong pump radiation propagates co-linearly with the harmonic outputs. Therefore, in order to observe and use high harmonics, we must remove this strong pump radiation coaxially propagating with the harmonic beam. Usually, a thin metal filter is used for this purpose, but that works only for a few mJ pumping. In our case the pump radiation is much stronger, and consequently, the harmonics signals are also very intense, so that thin filters are easily destroyed. Therefore, we

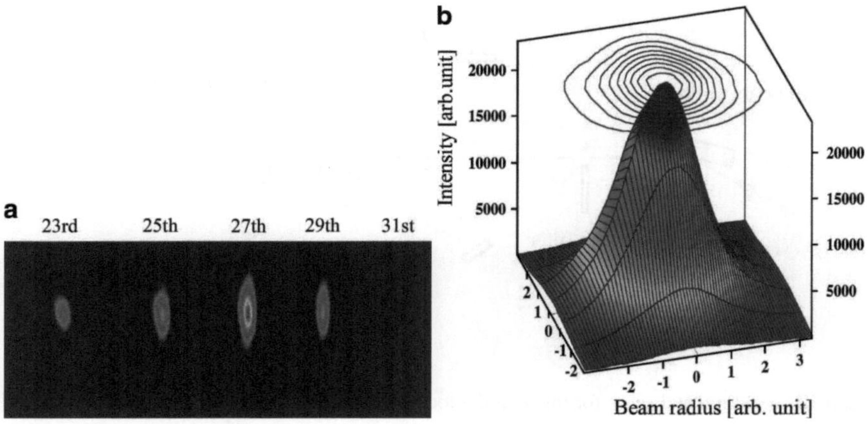

Fig. 6.10 (**a**) High-order harmonic spectrum measured by the spectrometer. (**b**) Spatial profile of 27th harmonics

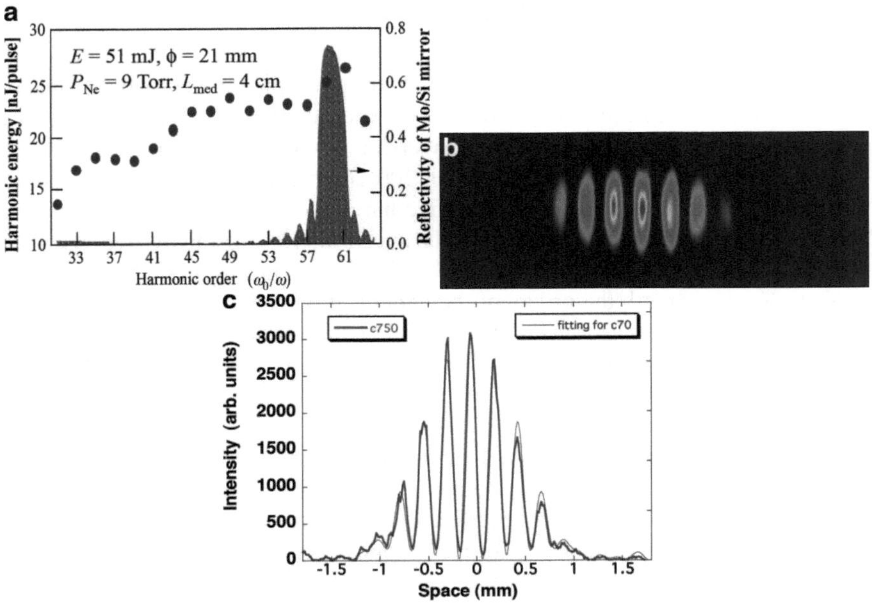

Fig. 6.11 (**a**) 59th harmonic energy and the reflectivity of Mo/Si mirror. (**b**) CCD image of interference pattern of 13-nm harmonic at 60% of beam diameter. (**c**) Intensity profile of interference pattern of 13-nm harmonic at 60% of beam diameter

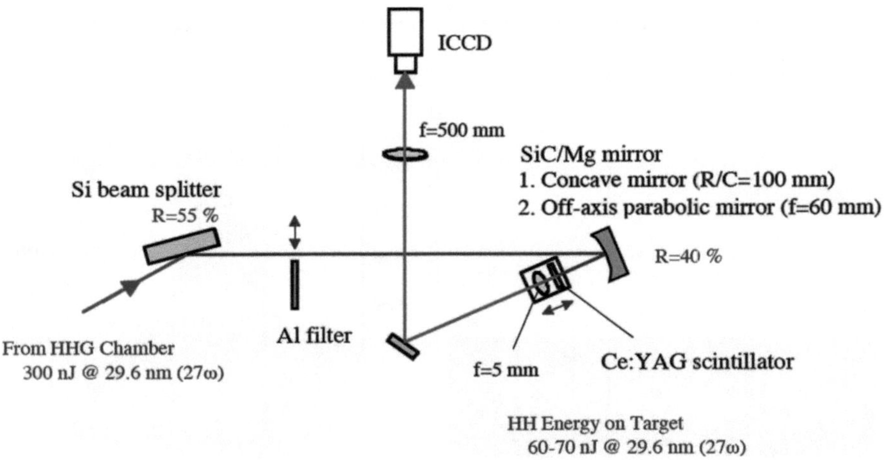

Fig. 6.12 Experimental setup for imaging the focused profiles of high-order harmonics

developed a new beam splitter that is a bulk silicon plate and durable for high-power pulses [15].

Silicon and silicon carbide have very large refractive indices in the visible region, but most of the materials have a refractive index around unity in the XUV region. When silicon (Si) or silicon carbide is used, the Brewster angle at 800 nm becomes very large and is about 75°, while a Brewster angle of the 30-nm radiation is 45°. By setting the silicon or silicon carbide at the proper incident angle, the pump radiation is almost transmitted or absorbed by the material, while the XUV light is reflected efficiently.

After the pump radiation is removed with the silicon plate, high-order harmonics in the range of several orders are reflected, and one single harmonic order is selected and focused by a multilayer mirror. For the measurement of the focusing property of the 27th harmonic, SiC/Mg multilayer mirror was used to select and focus it to the scintillator which converts the XUV radiation to visible radiation. After that the visible image is transferred by the image-relay optics and it is detected by a CCD camera.

Figure 6.13 shows spot size as a function of the distance from the focusing mirror with a focal length of 60 mm and off-axis angle of 24°. For our system, the M^2 value, which is a measure of the quality of laser beams, was measured to be 1.4, indicating that the beam has almost perfect Gaussian-like property.

It is known that the scintillator detector sometimes shows a saturation tendency, we cross-checked the results by observing the ablation pattern produced by the harmonic beam irradiation. We focused the 27th harmonic radiation onto a gold film target, placed just at the focus point (Fig. 6.12), by removing the scintillator and using a thin metal filter. We clearly observed a single shot ablation pattern (Fig. 6.14). By scanning with an atomic force microscope, we obtained a focus spot size of about 1 μm which is almost the same as that obtained in the scintillator experiment, and we determined a focused intensity of the order of 10^{14} W/cm^2 at 42 eV [6].

6.4 Characterization of Attosecond Pulses by PANTHER

Figure 6.15 shows double ionization by optical field. Horizontal is single photon energy and vertical is the optical field intensity. Many researchers have already observed the double ionization process using optical radiation at the X-ray or visible region. One extreme achieved by low-intensity X-ray radiation is the single-photon double ionization. When a very high-energy photon is absorbed by atoms or molecules, a large and sudden potential change associated with ionization occurring in the molecules and atoms produces doubly ionized atoms and molecules. This process was investigated by synchrotron radiation and is well-understood as the shake-off process. The other extreme is a very high-intensity but one photon energy is less than a few electron volts. This situation has already been achieved by the Ti:sapphire laser technology. In this case, multiple photons induce the tunneling process and produce a free electron. This electron can be captured again by the

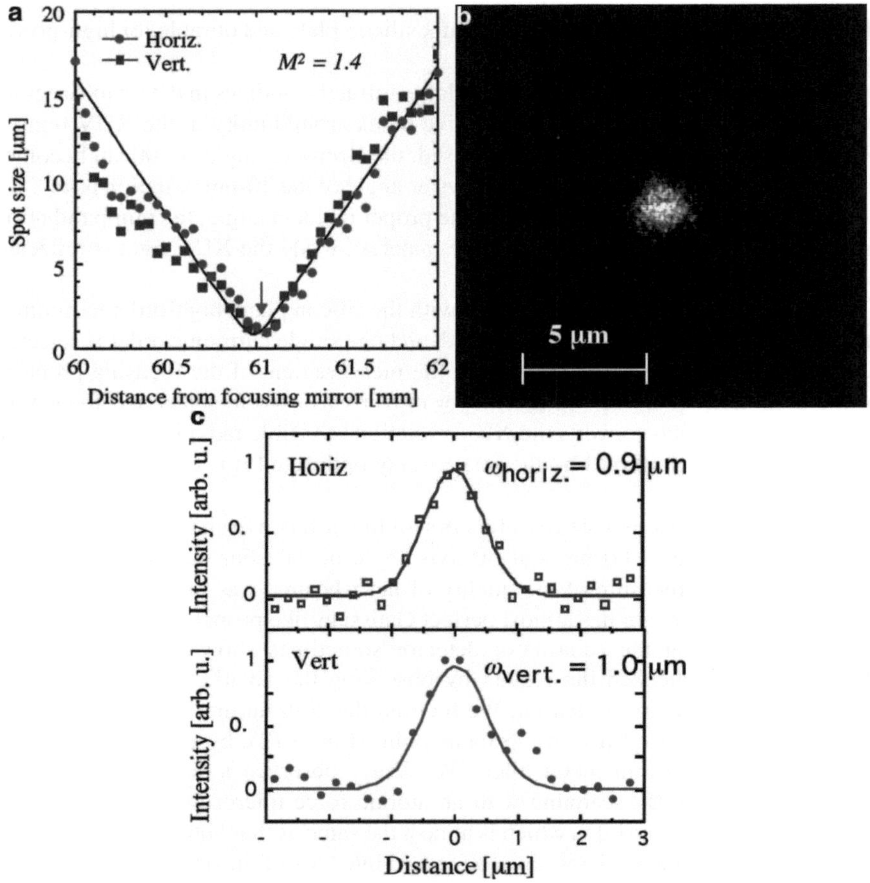

Fig. 6.13 (**a**) Focused beam spot size of the 27th-harmonic wave as a function of distance from the focusing mirror. The *circles* and *squares* correspond to the horizontal and vertical directions, respectively. The best-fit curve is also shown by the *solid curve*. (**b**) Typical focal spot image for the 27th-harmonic beam. (**c**) The horizontal and vertical profiles of the 27th harmonic beam

optical field and returns to the ion core. This process, called re-scattering, is similar to the harmonic generation process and produces a secondary electron. This process is a multiphoton double ionization. Our major interest lies in what happens between these two extreme regions, or in other words, what kind of processes would occur in the intermediate case between a single-photon double ionization by X-ray radiation and a multiphoton double ionization by visible radiation. In Fig. 6.15, a two-photon double ionization by XUV photons is depicted as a typical nonlinear process in the XUV region. The occurrence of this process was predicted more than 20 years ago, but no report had been made about this process before. We observed it for the first time in 2005.

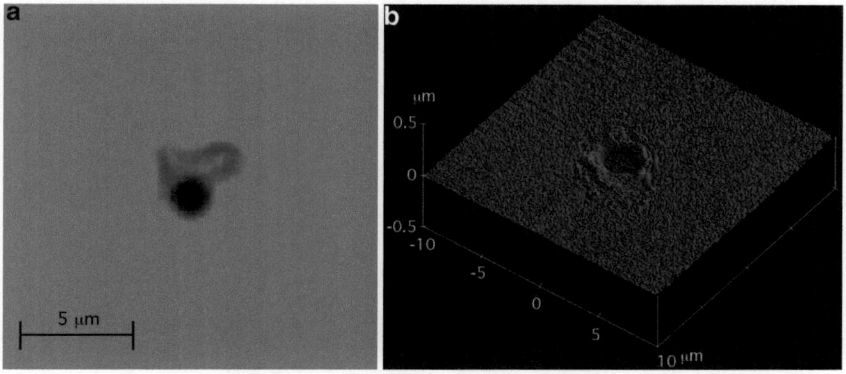

Fig. 6.14 (**a**) Optical microscope image and (**b**) atomic-force microscopy image of an ablation pattern on an Au target

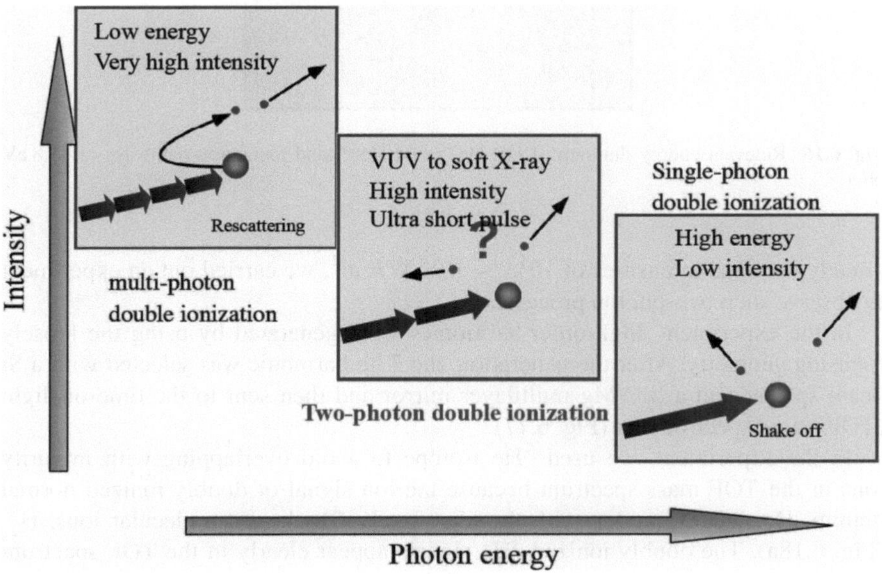

Fig. 6.15 Double ionization in atoms and molecules

Figure 6.16 shows the relevant energy levels of the two-photon process in He when the photon energy is 42 eV that correspond to the 27th harmonic [4, 10]. There are a direct two-photon process and a sequential process, in which after the single-photon absorption, singly ionized ions absorb two photons. These processes compete with each other when the harmonics interact with He atoms. The yields of these processes were predicted before, and it was argued that an intensity of $10^{13}\,\mathrm{W/cm^2}$ would be sufficient to observe the two-photon process. As we have

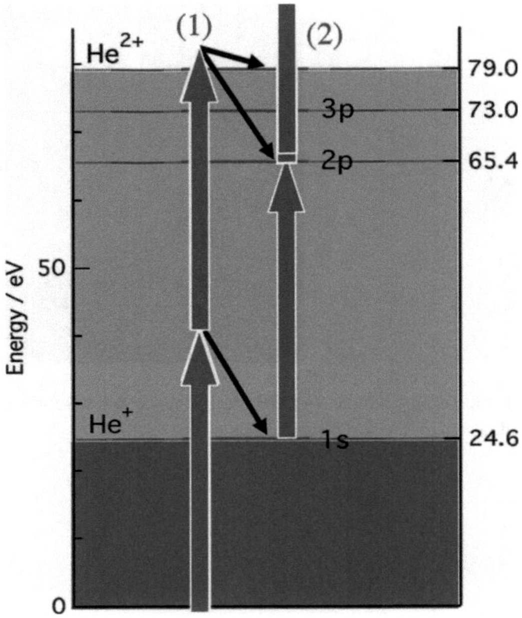

Fig. 6.16 Relevant energy diagram of He, He^{+}, and He^{2+} and ionization pathways of 41.8 eV photon

already obtained intensities of $10^{13} \sim 10^{14}\,W/cm^2$, we carried out an experiment to observe such two-photon processes.

In the experiment, high-order harmonics were generated by using the loosely focusing geometry. After the generation, the 27th harmonic was selected with a Si beam splitter and a SiC/Mg multilayer mirror and then sent to the time-of-flight (TOF) mass spectrometer (Fig. 6.17).

In the experiment, we used ^3He isotope to avoid overlapping with impurity ions in the TOF mass spectrum because the ion signal of doubly ionized normal helium, He, would overlap with the ion signal of hydrogen molecular ions, H_2^+ (Fig. 6.18a). The doubly ionized ^3He signals appear clearly in the TOF spectrum and are indicative of the occurrence of the two-photon process [2, 7]. As there are two or three competing processes, we had to check whether this is truly the direct two-photon double ionization. Then, we measured the intensity dependence. When the sequential process occurs, a total of three photons are needed, that is, a one-photon ionization followed by a two-photon absorption. From the intensity dependence, such processes can be clearly distinguished. Since the He^{+} ion yield is proportional to the intensity of the 42 eV harmonic, its signal is regarded as intensity of the harmonic. Plotting the yield of doubly ionized He^{2+} ion as a function of the He^{+} ion yield (Fig. 6.18b), we obtain a slope of 2.0, indicating that this is the true nonsequential two-photon process. The result is also supported by some theoretical calculations.

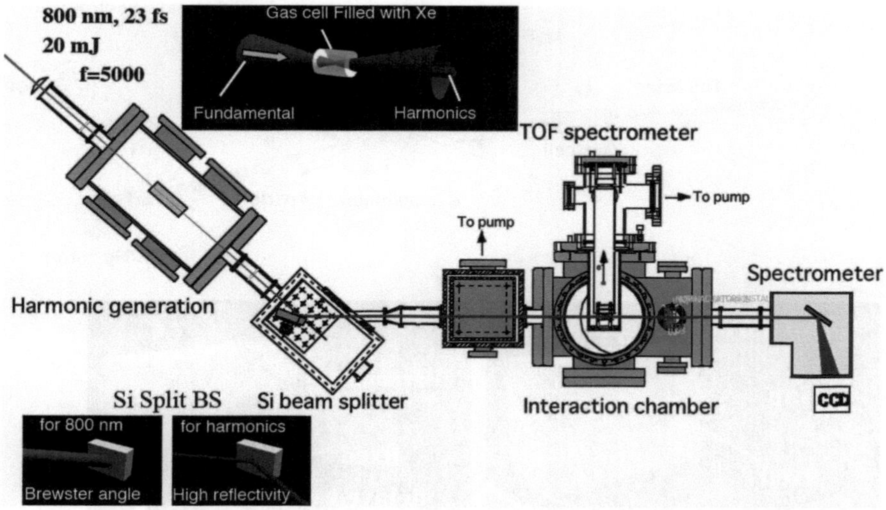

Fig. 6.17 Experimental setup for observing two-photon double ionization in He by the 27th harmonic pulse

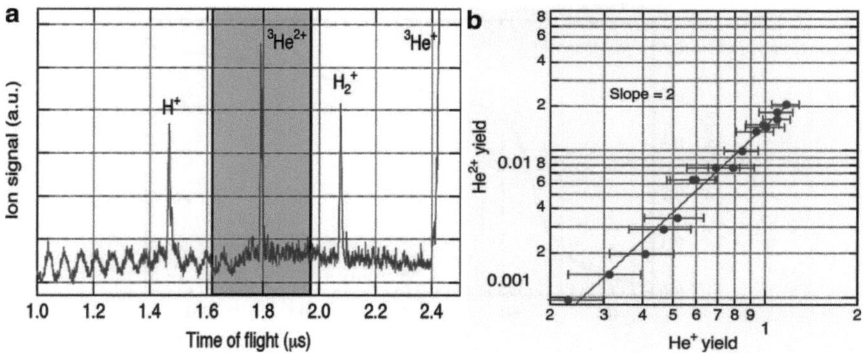

Fig. 6.18 (**a**) The time-of-flight ion spectrum of the two-photon double ionization experiment. (**b**) Doubly charged He ion yield as a function of the 27th harmonic intensity

Next, we turned our attention to how to measure the temporal width of the attosecond pulses or the XUV pulses. For this, we need to construct an autocorrelator (Fig. 6.19). In our autocorrelator, two pieces of the beam splitter were used side by side, and then, the harmonics generated are directed to the boundary between the two splitters where the harmonic beam is spatially divided into two. One of the beam splitter plates is placed on a movable stage. By changing the position of this plate, we can introduce a time delay to one of the two split harmonic beams. Then, the two beams are focused at one point with a spherical multilayer mirror.

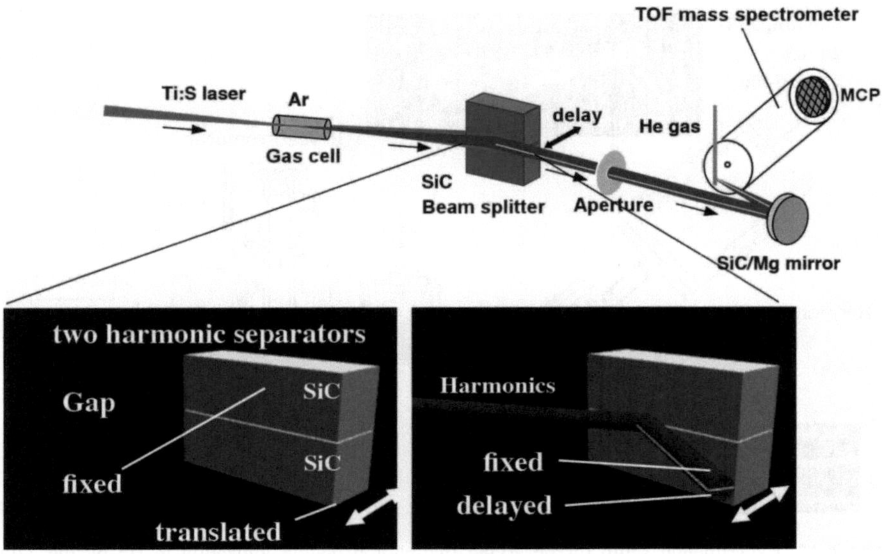

Fig. 6.19 Autocorrelator for XUV attosecond pulse characterization

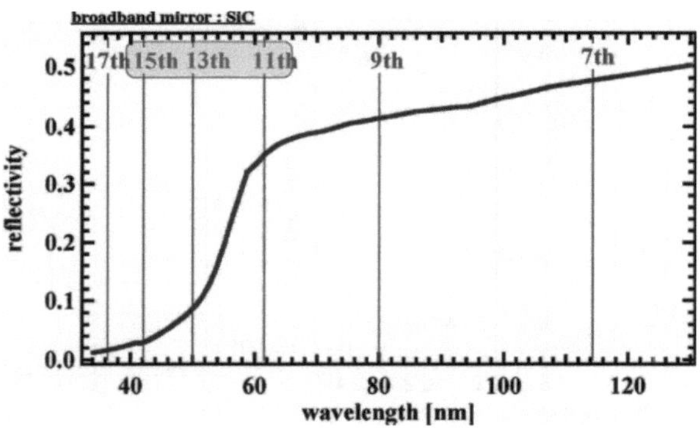

Fig. 6.20 Reflectivity of SiC mirror

The superiority of this autocorrelator lies in the fact that we use only one optical component to split the harmonic beam and to introduce the temporal delay; we can minimize loss of the higher harmonic energy. This is especially of importance in the XUV region. For the beam splitter, we use a flat mirror, which is much easier to be fabricated than a spherical mirror, although its surface quality is crucial.

For the attosecond pulse generation, we need to combine several harmonics. However, in the 27th harmonic area, that is, in the 30-nm region, high reflective multilayer mirrors with sufficient bandwidth were not available, so we changed the wavelength to 60 nm, where we can use silicon carbide mirrors (Fig. 6.20). In order to simplify the analysis of the two-photon ionization process, we limited the number of harmonics with a Sn filter. The total number is actually three from 11th to 15th.

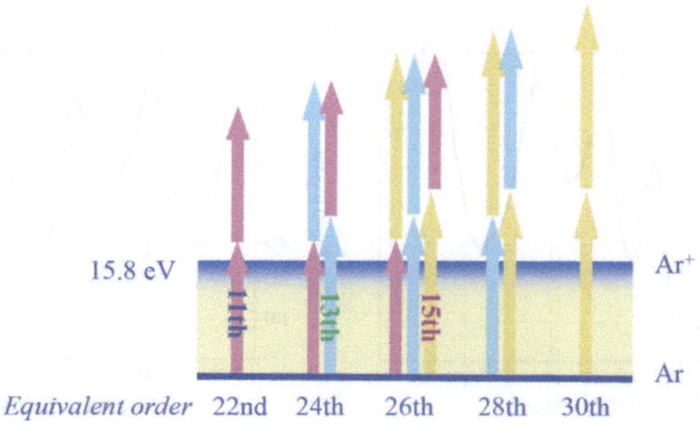

Fig. 6.21 Schematic of two-photon interactions between the harmonic photons and Ar atom

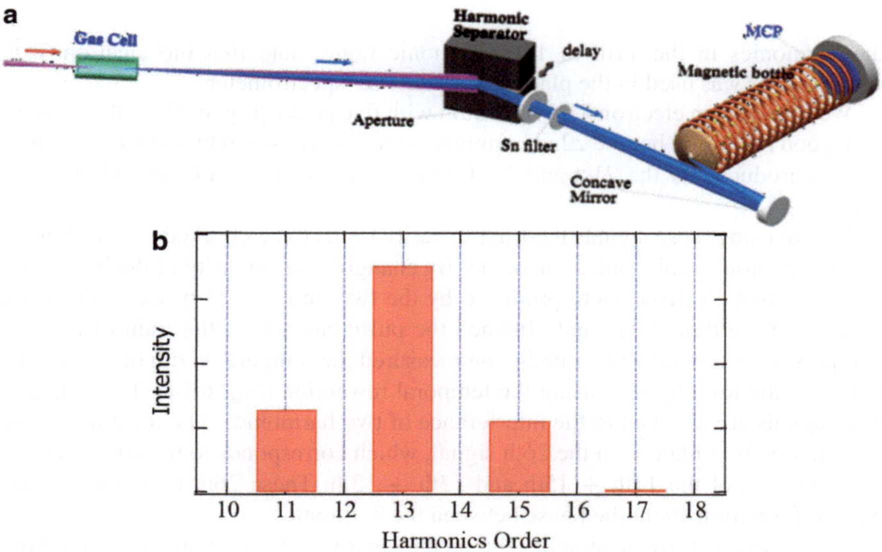

Fig. 6.22 (**a**) Autocorrelation measurement for an attoseocnd pulse train, and (**b**) high-order harmonic distribution after Sn filter

In the double ionization experiment, we observed ion signals which gave information on the intensity of the harmonics, but in this case we used electron signals. In this experiment we used the two-photon ATI process, which is another kind of nonlinear optical process (Fig. 6.21). The advantage of detecting the electron signal is that it provides information on the harmonic phase. To completely reconstruct the pulse shape we need information on the phase between the harmonics. After some calculations we can reconstruct more precisely the temporal shape of the harmonics by measuring the intensity and phase of the harmonics.

The experimental setup (Fig. 6.22a) is similar to the measurement of the pulse width of the 27th harmonic [7] except that we used Xe to efficiently generate

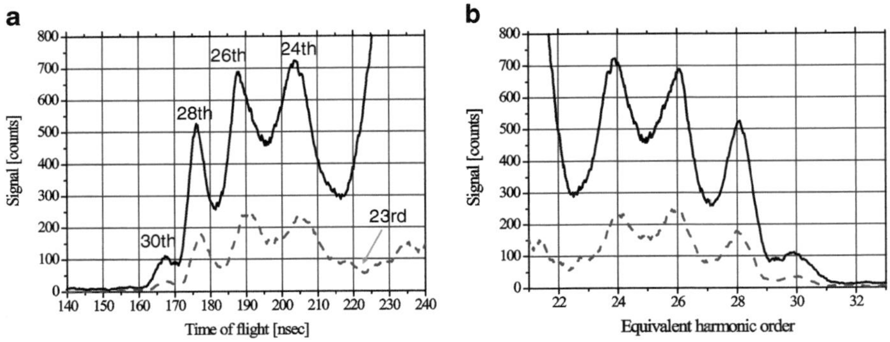

Fig. 6.23 Typical ATI spectrum of Ar at fixed delay

the harmonics in the 11th to 15th harmonic region and that the electron TOF spectrometer was used in the place of the ion TOF spectrometer.

We obtained an electron TOF spectrum with five peaks (Fig. 6.23) with a reasonably good S/N ratio, but the 22nd harmonic signal was covered by the large electron signals produced by the 21st and 23rd harmonics, which slightly leaked from the filter.

Before using these signals their nonlinearity needs to be checked as was done in the two-photon double ionization of He. By changing the intensity of the harmonics we found that electrons were produced by the two-photon ATI process. Then, we introduced the time delay and obtained the pulse envelop of the train of attosecond pulses for around 15 fs. Finally we measured the temporal width of attosecond pulses in the train by increasing the temporal resolution (Fig. 6.24). The 24th and 28th signals are ascribed to the interference of two harmonics. The most important information is contained in the 26th signal, which corresponds to the combinations of harmonics of the 11th + 15th and 13th + 13th. These 26th harmonic signals carry information about the phase between the harmonics.

As described above, to simplify the spectral analysis we used only three harmonics, the 11th, 13th and 15th, for the demonstration of our method (Fig. 6.24) [8]. The relative intensity of the two-photon transitions are written as:

$$I_{22}(\tau) \propto I_{11}^2, \tag{6.14a}$$

$$I_{24}(\tau) \propto 2I_{11}I_{13}\{1 + \cos(2\omega_f\tau)\}, \tag{6.14b}$$

$$I_{26}(\tau) \propto \left\{I_{13} + 2\sqrt{I_{11}I_{15}}\cos(2\omega_f\tau)\right\}^2 - 8I_{13}\sqrt{I_{11}I_{15}}\sin^2(\Delta\Phi)\cos(2\omega_f\tau), \tag{6.14c}$$

$$I_{28}(\tau) \propto 2I_{13}I_{15}\{1 + \cos(2\omega_f\tau)\}, \tag{6.14d}$$

$$I_{30}(\tau) \propto I_{15}^2, \tag{6.14e}$$

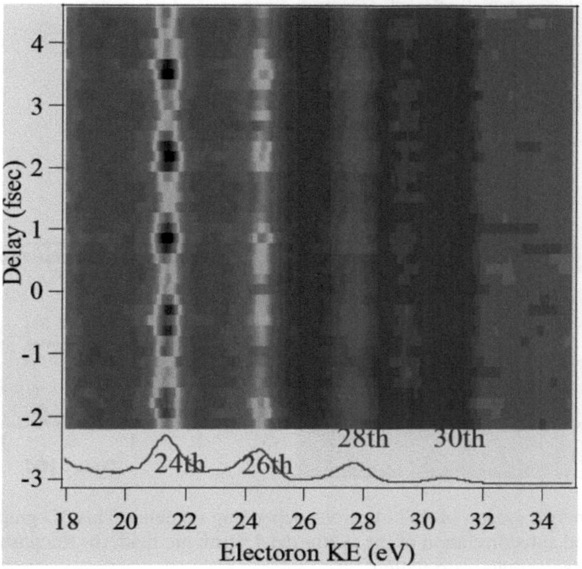

Fig. 6.24 Temporal delay dependence of two-photon ATI electron spectra

where $\Delta\Phi$ is expressed as:

$$\Delta\Phi \equiv (\phi_{11} + \phi_{15})/2 - \phi_{13}, \tag{6.15}$$

and the group delay dispersion, ϕ double dots, at the 13th harmonic is expressed using $\Delta\Phi$ as:

$$\ddot{\phi} = \Delta\Phi / \left(2\omega_{\mathrm{f}}^2\right). \tag{6.16}$$

From this spectral analysis we were able to determine the pulse width as 450 as. In the research field of ultrafast optics, there is a custom to name a new measurement method after an animal, so we named our method PANTHER which stands for Photoelectron Analysis with Nonresonant Two-photon ionization for Harmonic Electric field Reconstruction (Fig. 6.25).

6.5 Autocorrelation Measurement of Attosecond Pulses by Molecular Coulomb Explosion

Next we tried to use these attosecond pulses for observing ultrafast dynamics of molecules. When we focus intense femtosecond Ti:sapphire laser pulses to molecules like N_2, we can easily produce doubly ionized molecules which Coulomb explode due to the static Coulomb force. By measuring the momentum and energy

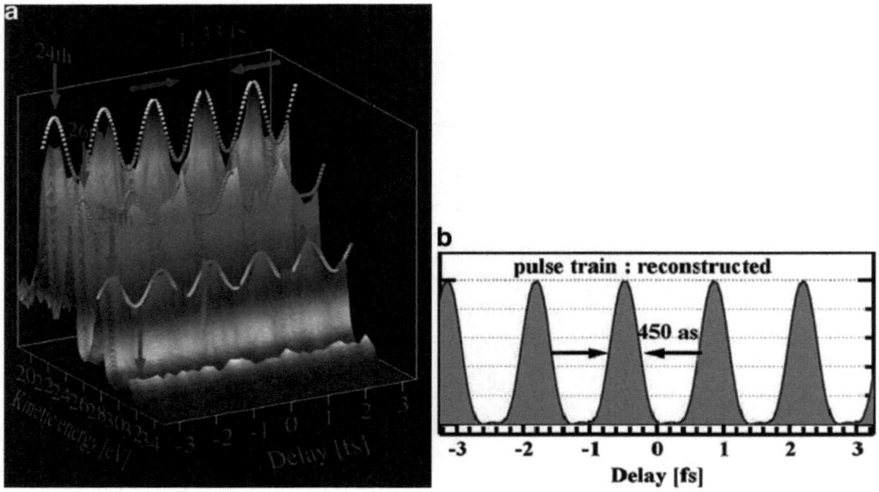

Fig. 6.25 (**a**) Energy spectra of ATI electrons depending on delay. This 3D graph corresponds to the mode-resolved autocorrelation of the synthesized harmonic field. (**b**) Reconstructed pulse train from the mode-resolved autocorrelation

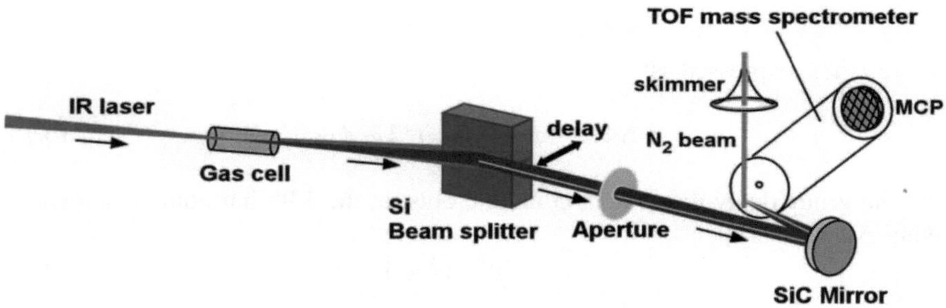

Fig. 6.26 Schematic drawing of the experimental setup for the autocorrelation measurement of attosecond pulses by molecular Coulomb explosion

of the ions generated by the Coulomb explosion using a 2D detector, we can estimate the internuclear distance of a molecule just before the Coulomb explosion by simple calculation because the ionization proceeds much faster than the motion of nuclei.

In order to observe the Coulomb explosion using the attosecond pulses, we used a setup similar to that for the ATI experiment, but we changed the target medium from atoms to molecules and we detected ion signals using the ion TOF mass spectrometer as shown in Fig. 6.26 [11]. In the experiment, the loose focusing condition ($f = 5$ m) was adopted for achieving geometrical phase matching, and the Si beam splitter was used to set a delay between two harmonic pulse beams and to remove fundamental light (less than 0.1% is reflected).

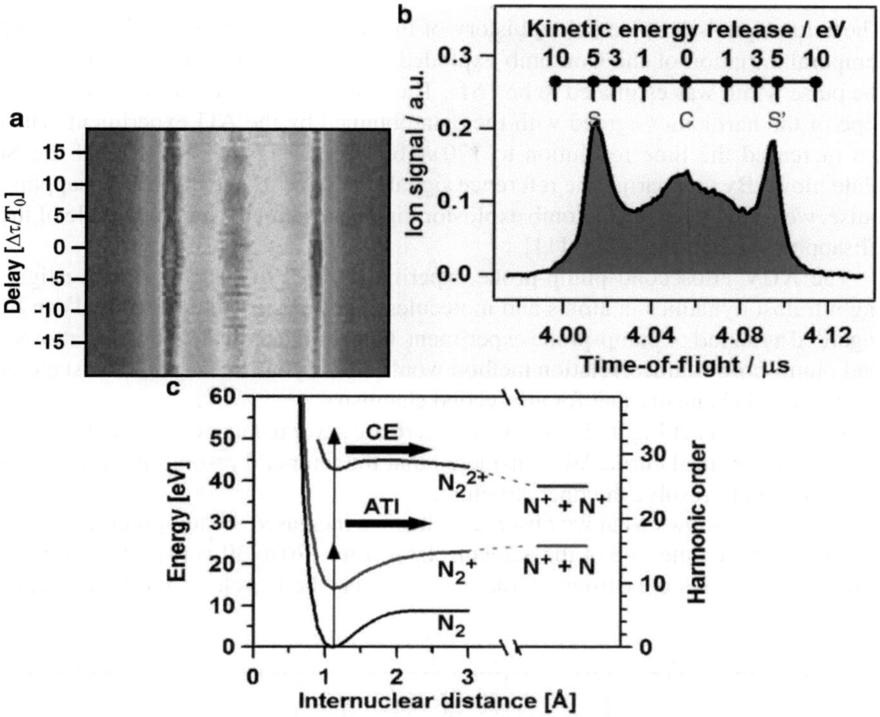

Fig. 6.27 (**a**) Autocorrelation trace of the time-of-flight mass spectra of N_2 recorded as a function of the delay Δt. (**b**) Enlarged view of the time-of-flight mass spectrum of N_2 at $\Delta t = 0$. Side peaks are backward and forward components generated from the Coulomb explosion of $N_2^{2+} \rightarrow N^+ + N^+$. (**c**) Potential energy curves for N_2, N_2^+ and N_2^{2+}

To observe ion signals, we used N_2, because it has a large cross section around the 13th harmonic (Fig. 6.27c). After the two-photon absorption, doubly ionized nitrogen molecules are produced, and then, they undergo a Coulomb explosion,

$$N_2^{2+} \rightarrow N^+ + N^+. \tag{6.17}$$

When the Coulomb explosion occurs, ions toward the forward direction reach the detector faster than those toward the backward direction do. Consequently, a double-peak structure appears in the TOF mass spectrometer signals as shown in Fig. 6.27b. Conversely, the appearance of the two peaks can be regarded as an evidence of the Coulomb explosion. Between these two side peaks, there was a central signal produced by a dissociative ionization process,

$$N_2^{+} \rightarrow N + N^+. \tag{6.18}$$

These ion signals provide a time history of the harmonics. Figure 6.27a shows the temporal variation of the Coulomb exploded ion signals. From the measurement, the pulse width was estimated to be 16 fs. The temporal structure of the pulse envelope of the harmonics agreed with the data obtained by the ATI experiment. Then we increased the time resolution to 170 as by decreasing the step size of the Si plate move. By comparing the reference signal produced by the 800 nm Ti:sapphire pulse, we found that the Coulomb explosion signals appeared every half cycle of the Ti:sapphire signal (Fig. 6.28) [11].

The XUV attosecond pump-probe experiments are very useful for investigating ultrafast dynamics in atoms and molecules. The autocorrelation method can be regarded as a kind of pump-probe experiment. Our results indicated that this attosecond pump-probe autocorrelation method would be very useful for the next stage of femtosecond chemistry and for attosecond chemistry.

When we look at Fig. 6.28 into detail, there are several data points which deviate from the theoretical curve. We considered that the temporal resolution of 170 as is not sufficient to resolve the finer structure.

Figure 6.29 shows what we observed when we increased the temporal resolution by changing the time step of the autocorrelator from 170 to 30 as [8]. There are two different structures appearing alternatingly at every half-cycle of the fundamental

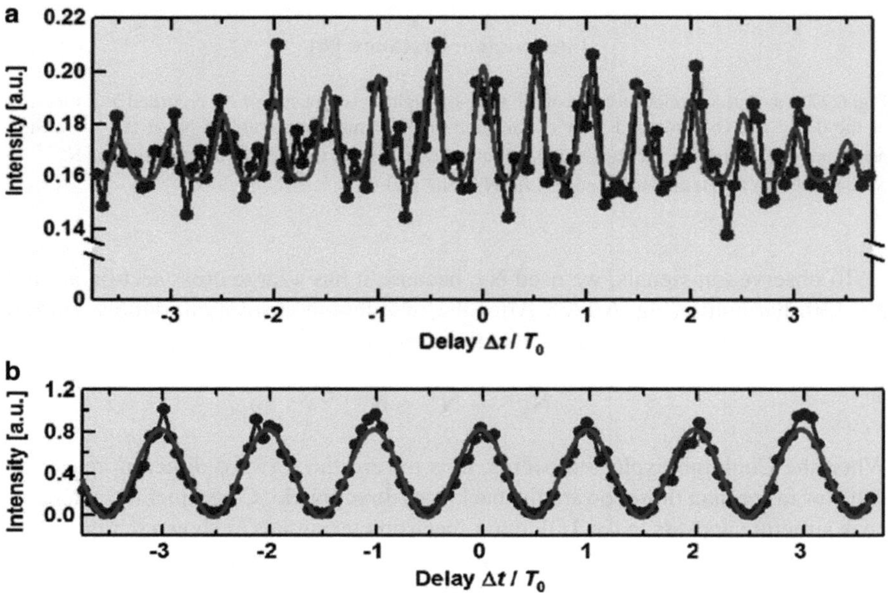

Fig. 6.28 (**a**) Ion signal intensities of the two side-peak areas as a function of Δt and (**b**) autocorrelation trace by detecting the fragment ions N^+ generated by the intense fundamental light field at 800 nm.

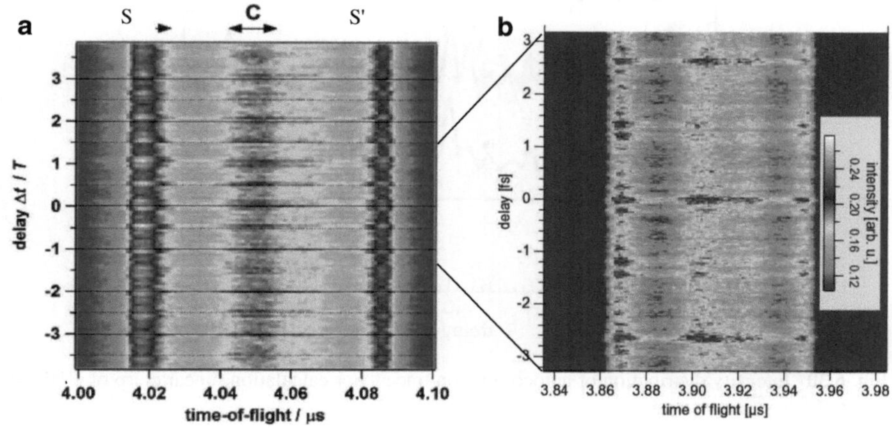

Fig. 6.29 Temporal variation of the ion spectrum by translating the delay between the two replicas of the harmonic fields. Sampling at every 30 as, and the delay is scanned from −3 to 3 fs. The gray scale of the intensity is shown in the *inset*

field. One structure has a peak at the center and two side lobes, but the other structure has only two peaks without a central peak. The origin of this structure is understood if we reconsider the generation process of high-order harmonics. In the three-step model of high-order harmonic generation, the electron escapes near the peak of the fundamental optical field and returns back to the parent ion in the second half-cycle, then it recombines to produce the harmonics. In the next half-cycle the electric field changes its sign and again, an electron escapes at a peak and returns, but the direction is opposite. Therefore, between the first and second processes, the harmonic fields change their phase by π radians. Accordingly, what we observed is a temporal variation of the electric field itself, resulting in the autocorrelation trace of the electric field of the attosecond pulse. Usually, in the visible and infrared regions, researchers use interferometric autocorrelation to observe the electric field of femtosecond laser pulses. We achieved a similar direct electric field measurement in the XUV region using attosecond pulses.

Figure 6.30 shows the 1D-interferometric autocorrelation trace. The calculated curve was obtained by taking into account the focusing geometry in which the total number of harmonics contributing was five. This calculation is not so simple, because when we use the harmonics from the 11th to 19th, their beam spot sizes are considerably different depending on their order. This autocorrelation result could not be obtained by a conventional autocorrelator. Since, in this all reflective autocorrelator, all the energy is reflected into the focus, the variation of spatial intensity distribution should be considered to reconstruct the original electric field in the calculation. Therefore, we must carefully include the change in spot size depending on the order of the harmonics. From the measured curve we were able to obtain a very important feature of the harmonics, that is, the phase relation among attosecond pulses. This result will be useful for frequency metrology in the near future. It

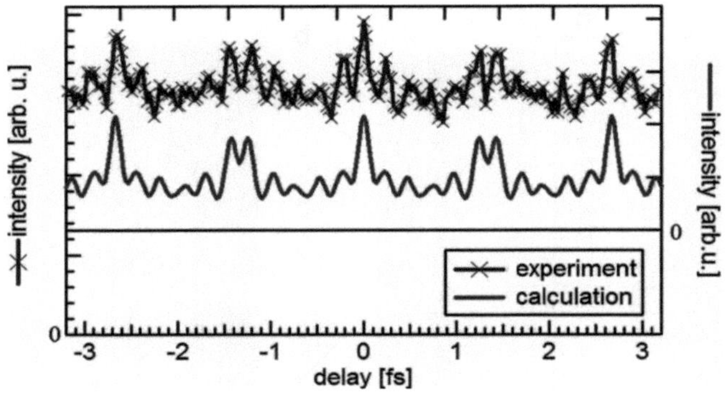

Fig. 6.30 Intensity distribution of autocorrelation traces. For calculation, linear chirp of GDD $= 1.3 \times 10^{-32}\ \mathrm{s}^2$ is assumed

should be also noted that an almost single optical cycle pulse has been achieved in the attosecond pulses. The number of cycles measured was only 1.3 cycles. Without such extremely small cycles of the electric field, we could not observe this kind of autocorrelation signals.

6.6 Summary

In summary, we have developed an intense high harmonic source by using a loosely focusing geometry that produced high enough power to induce a nonlinear optical process in the XUV region. Using such intense harmonics, we observed nonlinear two-photon processes, such as two-photon double ionization in He and ATI in Ar in the XUV region. Then we used them for direct characterization of the temporal properties of the XUV attosecond pulses. Furthermore, we have achieved a Coulomb explosion in a molecule in the attosecond time scale, and demonstrated for the first time the measurement of interferometric autocorrelation using the Coulomb explosion signals.

References

1. E. Constant, D. Garzella, P. Breger, E. Mével, Ch. Dorrer, C. Le Blanc, F. Salin, P. Agostini, Phys. Rev. Lett. **82**, 1668 (1999)
2. H. Hasegawa, E. Takahashi, Y. Nabekawa, K. Ishikawa, K. Midorikawa, Phys. Rev. A **71**, 023407 (2005)
3. K. Hoshina, A. Hishikawa, K. Kato, T. Sako, K. Yamanouchi, E. Takahashi, Y. Nabekawa, K. Midorikawa, J. Phys. B: At. Mol. Opt. Phys. **39**, 813–829 (2006)
4. K. Ishikawa, K. Midorikawa, Phys. Rev. A **65**, 043405 (2002)
5. S. Klarsfeld, A. Maquet, J. Phys. B **12**, L553 (1979)

6. H. Mashiko, A. Suda, K. Midorikawa, Opt. Lett. **29**, 1927 (2004)
7. Y. Nabekawa, H. Hasegawa, E. Takahashi, K. Midorikawa, Phys. Rev. Lett. **94**, 043001 (2005)
8. Y. Nabekawa, T. Shimizu, T. Okino, K. Furusawa, H. Hasegawa, K. Yamanouchi, K. Midorikawa, Phys. Rev. Lett. **96**, 083901 (2006)
9. Y. Nabekawa, T. Shimizu, T. Okino, K. Furusawa, H. Hasegawa, K. Yamanouchi, K. Midorikawa, Phys. Rev. Lett. **97**, 153904 (2006)
10. T. Nakajima, L. Nikolopoulos, Phys. Rev. A **66**, 041402 (2002)
11. T. Okino, K. Yamanouchi, T. Shimizu, K. Furusawa, H. Hasegawa, Y. Nabekawa, K. Midorikawa, Chem. Phys. Lett. **432**, 68 (2006)
12. E. Takahashi, Y. Nabekawa, T. Otsuka, M. Obara, K. Midorikawa, Phys. Rev. A **66**, 021802 (2002)
13. E. Takahashi, Y. Nabekawa, M. Nurhuda, K. Midorikawa, JOSA B **20**, 158–165 (2003)
14. E. Takahashi, Y. Nabekawa, K. Midorikawa, Appl. Phys. Lett. **84**, 4 (2004)
15. E. Takahashi, H. Hasegawa, Y. Nabekawa, K. Midorikawa, Opt. Lett. **29**, 507 (2004)
16. Y. Tamaki, O. Maya, K. Midorikawa, M. Obara, IQEC '98 (1998)
17. Y. Tamaki, J. Itatani, Y. Nagata, M. Obara, K. Midorikawa, Phys. Rev. Lett. **82**, 1422 (1999)
18. Y. Tamaki, Y. Nagata, M. Obara, K. Midorikawa, Phys. Rev. A **59**, 4041 (1999)
19. K. Yamanouchi, Science **295**, 1659 (2002)

Chapter 7
Ultrafast X-Ray Absorption Spectroscopy Using Femtosecond Laser-Driven X-Rays

Hidetoshi Nakano

Abstract We introduce recent advances in the techniques of generating laser-driven ultrashort x-ray pulses. We demonstrate that the intensities of the generated x-ray pulses can be enhanced by arranging a prepulse prior to the main laser pulse and by the nanostructure prepared on the surface of the target materials. Through the pump-and-probe EXAFS measurements, we probe in real time the melting process of the solid Al induced by the intense laser pulses having the fluence below the ablation threshold.

7.1 Introduction

This presentation begins with a description of x-ray generation by femtosecond lasers, and then moves on to describe the properties of these plasma x-rays and techniques to enhance their conversion efficiency, which includes the prepulse technique mentioned by Professor Nakamura. Some experimental results of the time-resolved absorption measurement of the optically excited materials are detailed and the spatiotemporally resolved absorption measurement system is also briefly mentioned.

With recent developments of high-power femtosecond laser technologies, ultrafast x-ray sources that are normal laboratory size can be achieved which produce high-order harmonics or laser-produced plasma x-rays. High-order harmonics are coherent ultrafast radiation; however, it was difficult to obtain high energy photons. Although these harmonics were initially very weak and were not very efficient for practical use, several methods have overcome such difficulties. Now, they are useful for diagnosing ultrafast responses of materials since they can provide access even in the attosecond regime. We have used laser plasma x-rays as a probe since they can cover a wide range of energy regions from several tens of electron volts

H. Nakano
NTT Basic Research Laboratories, NTT Corporation, 3-1 Morinosato Wakamiya, Atsugi, Kanagawa 243–0198, Japan

K. Yamanouchi (ed.), *Lectures on Ultrafast Intense Laser Science 1*, Springer Series in Chemical Physics 94, DOI 10.1007/978-3-540-95944-1_7,

to kilo-electron volts. However, this emission is incoherent due to its emission mechanism. The pulse duration should be limited by laser pulse duration at least at the femtosecond level in the kilo-electron volt regime. The pulse duration in the soft x-ray or UV region is typically around several picoseconds to 10 ps.

Such laser-based x-ray sources naturally emit short pulses; another important feature is that such sources emit a photon flux that is precisely synchronized to the incident laser pulse. Therefore, such sources can be used as a dynamic probe to observe the responses of the materials induced by the femtosecond laser pulse by using a pump-probe-type experiment. These ultrashort bursts of electrons and ions (quantum emissions) allow direct observation of changes in materials caused by not only the laser but also the electrons or ions.

Figure 7.1 compares peak brightness of typical x-ray sources. The figure includes data on the brightness obtained from Spring-8 – one of the largest synchrotron facilities – in terms of the peak brightness with that achieved by laser-generated x-rays. The peak brightness is typically considered as the time average along with the repetition rate; here, it is divided by the pulse duration and thus this graph of sources indicates a very high peak brightness can be achieved.

Heat or photoelectrons are emitted when a material absorbs x-rays, i.e., x-rays are incident on the material. Of the emission, some part is diffracted by the periodic structures, and some part passes through as transmitted x-rays. The spectrum of transmitted x-rays has some areas exhibiting a sharp change in the absorption, known as absorption edges, which indicate the ionization potential of the atoms of a particular species present in the material. Therefore, by carefully measuring such edges, the constituents of the material can be determined.

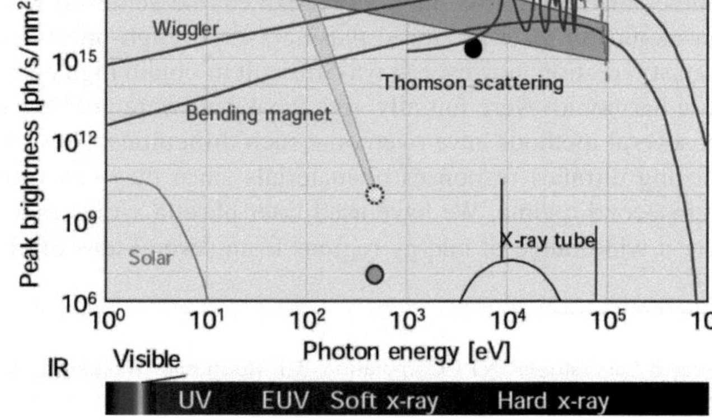

Fig. 7.1 Peak brightness of x-ray source

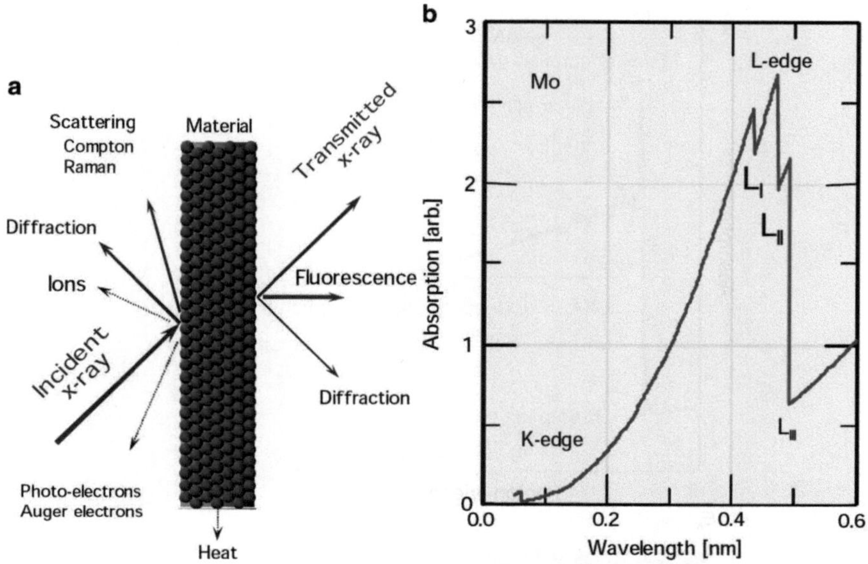

Fig. 7.2 (**a**) Phenomena induced by x-ray absorption, and (**b**) an example of x-ray absorption spectrum

A close observation of the plasmic spectra reveals fine structures near the absorption edges known as the x-ray absorption fine structure (XAFS), indicated here in Fig. 7.2. The XAFS is comprised of two parts: the x-ray absorption near-edge structure (XANES), which appears very close to the absorption edge and is sensitive to the electronic state of the atoms; and the extended XAFS (EXAFS), which is distant from the absorption edge and reveals the local arrangement among the nearest neighboring atoms. The EXAFS is an oscillating structure and by observing it, we can determine how far the nearest neighboring atoms are placed from the absorbing atom.

As is well known, and is shown in Fig. 7.3, x-rays are absorbed by materials through the photoelectric effect and are also absorbed by atoms; the latter causes the promotion of co-electrons from the atom into the continuum state. This leaves the atom with an empty electronic level at the core. The electron ejected from the atom is called a photoelectron, and it plays an important role in the absorption of fine structures.

There is interference between the wave packet of the photoelectrons directly emitted from the absorbed atom and the wave packet scattered from the nearest neighboring atoms; therefore, if such an interference pattern is observed in the spectral domain, we can determine the coordination of the atoms.

The interference pattern can be written as:

$$x(k) = \sum_{\mathrm{j}} \frac{N_{\mathrm{j}} f_{\mathrm{j}}(k) \mathrm{e}^{-2k^2\sigma_{\mathrm{j}}^2}}{kR_{\mathrm{j}}^2} \sin[2kr_{\mathrm{j}} + \delta_{\mathrm{j}}(k)]. \tag{7.1}$$

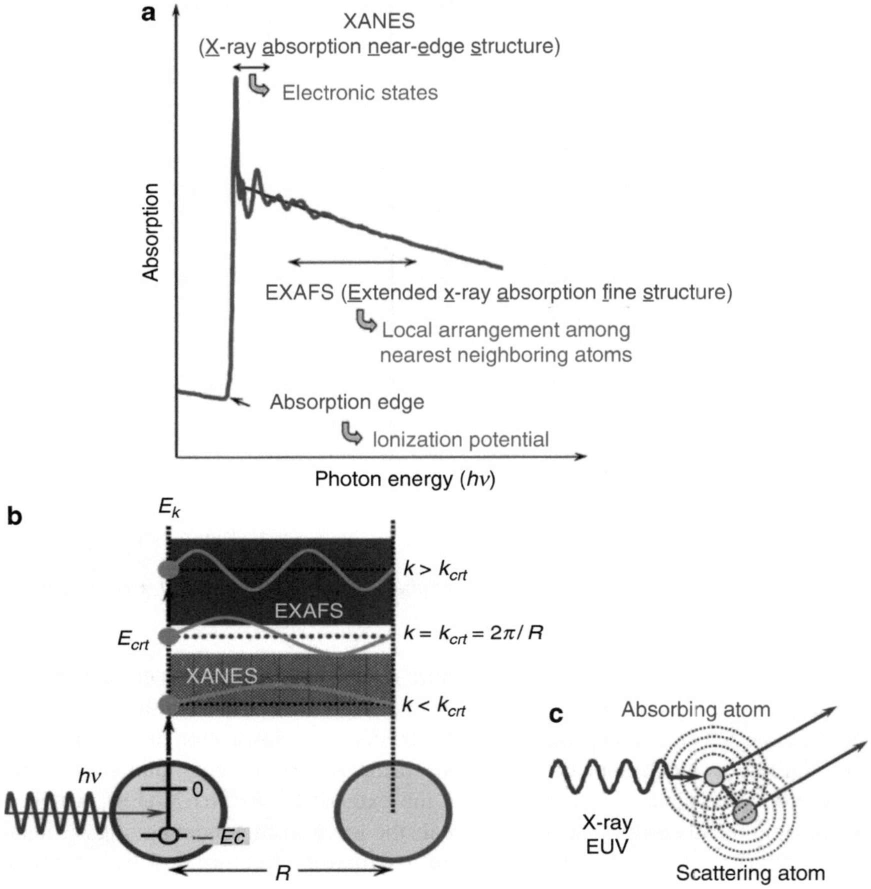

Fig. 7.3 Rich information included in x-ray absorption spectrum (**a**), and cartoon of fine structures in x-ray absorption spectrum (**b, c**)

To emphasize the oscillating structure, this function is sometimes weighted by the square or cube of k, where k stands for the wave number (cf. equation of Fig. 7.4).

A careful observation of the absorption spectrum reveals the absorption edges and some moderate absorption structure in the pattern, which is recorded by the interference between the photoelectron waves of some of the absorbed atoms and scatter from the neighboring atoms. The EXAFS information provides the local atomic coordination distance and the number of atoms. The XANES information is sensitive to the formal valence state and the coordination environment Such information is a complement to x-ray diffraction. Targets of study can be any random system including crystals, amorphous structures, and liquid states, although the extracted information is limited to one dimension. XAFS is now widely used to determine material structures in a random K system. Prepulses can potentially be used as probes to obtain dynamic response information.

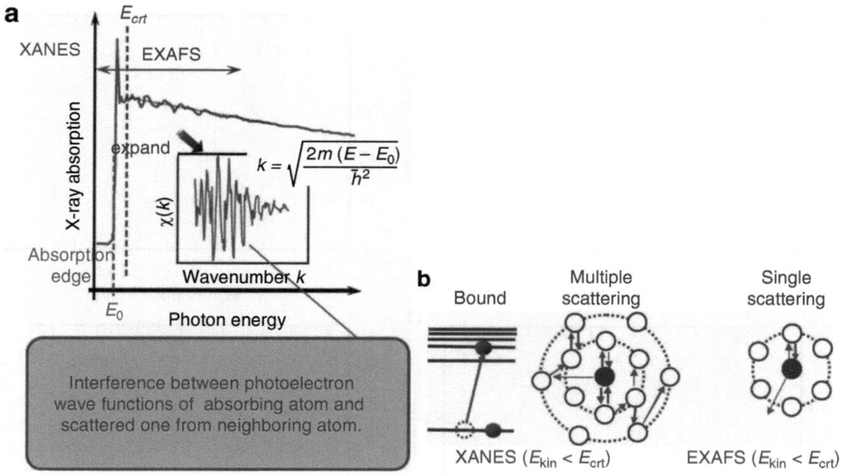

Fig. 7.4 XAFS: x-ray absorption fine structures. (**a**) Absorption spectrum. (**b**) Scattering of photoelectrons: origin of fine structures in x-ray absorption spectrum

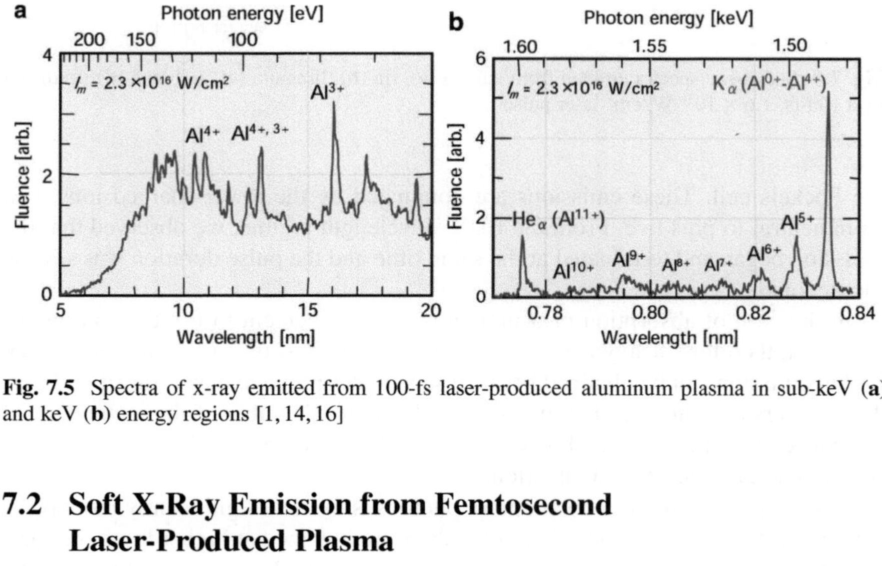

Fig. 7.5 Spectra of x-ray emitted from 100-fs laser-produced aluminum plasma in sub-keV (**a**) and keV (**b**) energy regions [1, 14, 16]

7.2 Soft X-Ray Emission from Femtosecond Laser-Produced Plasma

Figure 7.5 shows example spectra emitted from a 100-fs laser produced aluminum plasma. The peak intensity of the laser pulse incident on the target was 10^{16} W/cm^2. The left-hand spectrum shows the EUV regime emission near 100 eV which we used in later experiments and the right-hand spectrum shows the keV regime emission from the K-shell. The contrast ratio of the intense laser to the prepulse was 10^{-6} while that to the ASE level was significantly lower. Here, the prepulse is the chirp pulse caused by the regenerative amplifier, which cannot be suppressed by

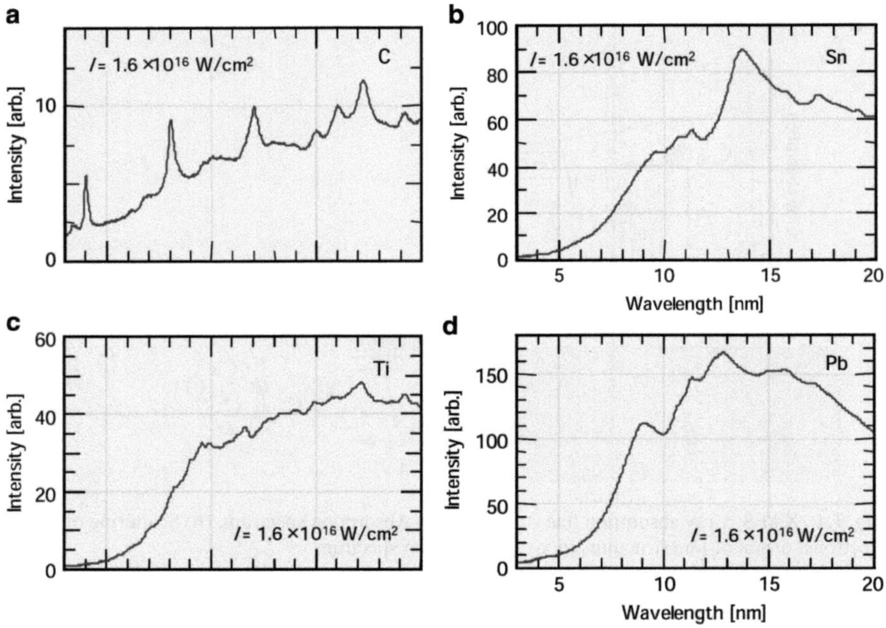

Fig. 7.6 Soft x-ray spectra emitted from silicon (**a**), tin (**b**), titanium (**c**), and lead (**d**) irradiated by a 100-fs, 1.6×10^{16} W/cm^2 laser pulse

the Pockels cell. These emissions are dominated by the lower-charged ions, i.e., from neutral to plus five. From the EUV wavelength regime, we observed that the emission began and terminated at the same time and the pulse duration was several picoseconds.

In the case of absorption measurements, it is convenient to use the continuous spectrum; therefore, heavy metals are used as the target (Fig. 7.6). Sometimes, even higher emission can be obtained because such materials have many transition lines. The conversion efficiency is a function of the atomic number of the materials and the conversion efficiency is observed to gradually increase with the atomic number of the material with small undulation.

The simplest method to obtain more photons is to use a significantly more powerful laser. Of course, higher photon numbers are obtained at higher laser energies. An alternate inexpensive method is to insert a weak prepulse immediately before the intense laser pulse in order to modify the interaction between the targets with the intense laser pulse. A third method is to prepare an array of nanostructures on the target surface; this implies that nanotechnology enhances x-ray emissions.

We can briefly describe how the prepulse function works by using Fig. 7.7. In the case of a clean main pulse, the strong main pulse directly interacts with the solid material. The interaction volume is very limited and most of the pulse is reflected back by the solid density plasma. Therefore, the conversion efficiency is very low.

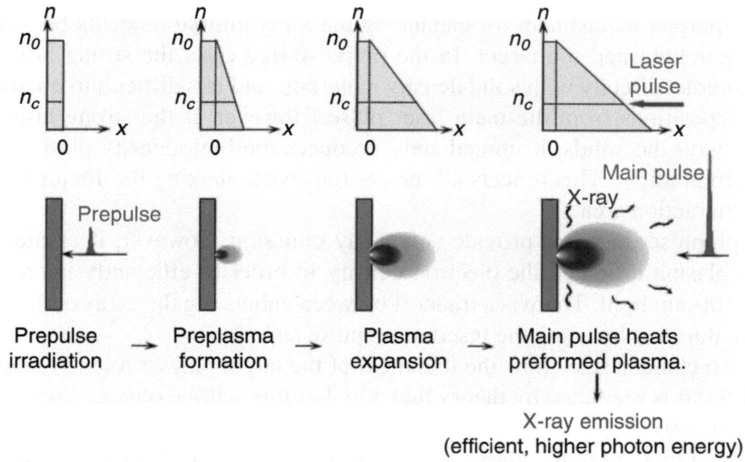

Fig. 7.7 The role of the prepulse in the x-ray generation

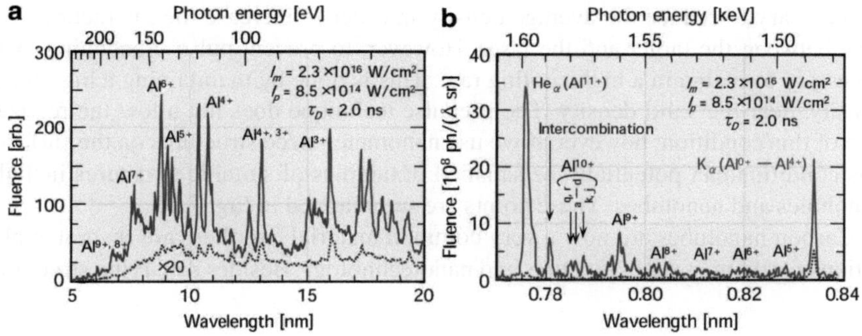

Fig. 7.8 Enhanced x-ray emission from Al plasma created by a pair of 100-fs laser pulse in sub-keV (**a**) and keV (**b**) energy regions [1, 16] (H. Nakano 1996)

However, if a weak prepulse is inserted before the intense laser pulse, a preplasma forms. This preplasma soon expands and matches the physical density or the laser emission of the main pulse. Then, the preplasma is hit with an intense laser pulse to obtain strong x-ray emissions. By controlling certain prepulse parameters, we can control the plasma parameter that interacts with the main pulse, and thus control the x-ray emission.

Using the prepulse, we were able to enhance the x-ray emission. In Fig. 7.8, the left side shows the emission spectrum near 100 eV and the right hand side shows the occasional emission from the aluminum. In both regions, the x-ray emission was significantly enhanced and very strong line emissions are observed here, which indicates the interaction between the low-density plasma. We obtained the highest enhancement of greater than 100 times from an Al plasma formed by a pair of 100-fs laser pulses.

An important mechanism for enhancing the x-ray emission occurs between the strong laser light and the target. In the prepulse-free case, the strong laser pulse should interact directly with solid density materials, and it is difficult to obtain large energy depositions from the main laser pulse. However, if the strong laser pulse interacts with the solids, it immediately produces the high-density plasma with a very sharp gradient. This reflects all the photons back, making it difficult to obtain a large interaction area.

The prepulse does not provide soft x-ray emission; however, it creates some gaseous plasma to match the electron density in order to efficiently interact with emitted 800-nm light. There is a tradeoff between enhancing the x-ray emission and the pulse duration by using the inserted prepulse technique.

We also characterized how the thickness of the target plays a role in obtaining a higher flux; it is predicted by theory that a higher flux will be obtained with a thick target (Fig. 7.9).

The third method noted above for increasing conversion efficiency is modification of the target structures by introducing nanostructures. To overcome the marginal interaction volume between the femtosecond laser burst and the target material, it is necessary to reduce the average density in order to increase the interaction volume between the target and the laser. However, to prevent pulse broadening, it is necessary to maintain a high cooling rate. This is done by maintaining a high local density, near the solid density. The prepulse technique does not allow the realization of this condition; however, if we use nanometer-sized structures on the surface, this condition can potentially be satisfied. Examples of suitable structures include nanoholes and nanotubes. These points are summarized in Fig. 7.10.

Carbon nanotubes are now a very common material and there are several applications for them in nanostructures and nanotechnology. Besides satisfying structural

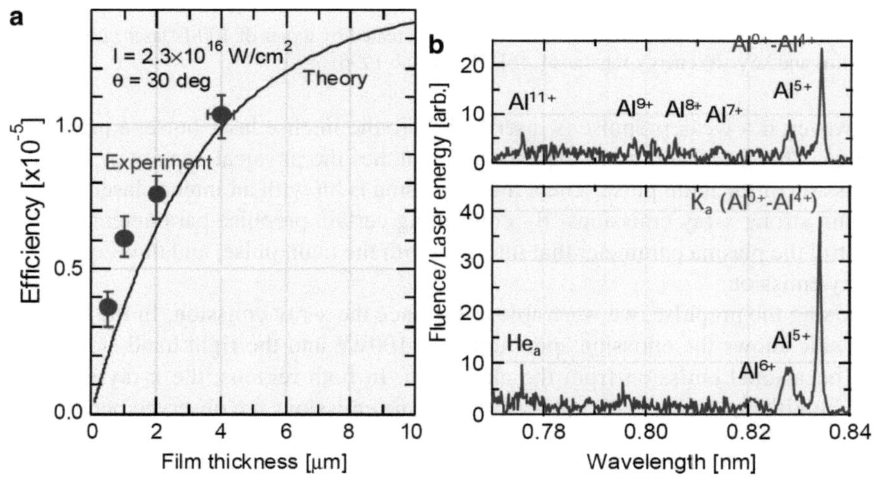

Fig. 7.9 Al-Kα emission as a function of target thickness [3]

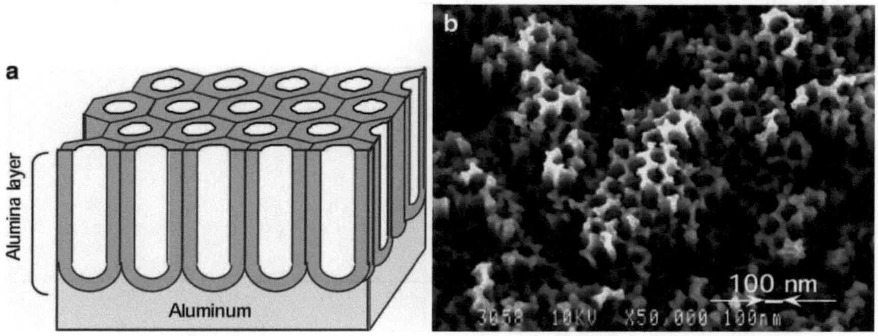

Fig. 7.10 Target with nanostructured surface. (**a**) Schematic of porous aluminum target structure. (**b**) SEM image of porous aluminum

Fig. 7.11 Enhanced water-window soft x-ray emission by carbon nanotube [2]

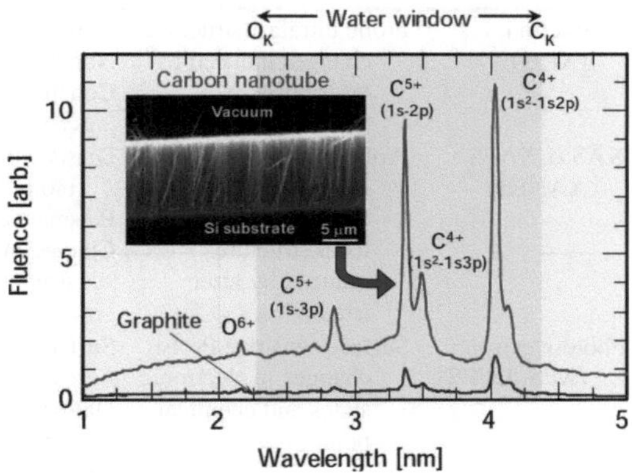

features of thickness, volume and depth by the nanotubes, carbon ions have strong transition lines in the water-window region so carbon can be used to observe biological species. Therefore, we focused on carbon as a target to enhance the water-window region soft x-rays.

The gray spectra of Fig. 7.11 can be obtained by using a properly designed carbon nanotubes structure, a ten times greater photon flux can be obtained at the water-window region as compared to the solid case using graphite. However, there is again a tradeoff between the photon flux and the x-ray duration. There is broadening up to several tens of picoseconds, although the pulse duration is shorter than that in the prepulse case.

We thought we might be able to enhance spectrum emission using carbon nanotubes which are well known as efficient electron emitters. We placed the carbon nanotube structure on a smoothed surface to obtain energetic electrons for line emission. If the laser is incident on the carbon nanotube, hot electrons are emitted and they enter the smoothed surface and hit the core electrons to create Kα emission. For example, in Fig. 7.12, comparing just a planar silicon wafer with a silicon wafer having the carbon nanotubes structure, we achieved a significantly higher intensity from the latter.

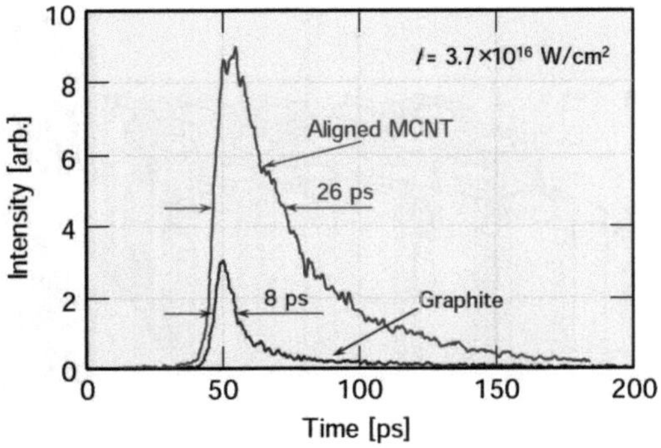

Fig. 7.12 SiKα emission from Si covered by carbon nanotube [2]

Table 7.1 Applications of ultrafast laser-driven x-ray sources

	Feature	Examples	Sources
Diffraction (XRD)	Probe ultrafast structural changes induced by laser pulse irradiation	Damage, shock Nonthermal melting Coherent phonon	Monochromatic hard x-ray (K_α-line) pulse
XAS (EXAFS, XANES)	Probe changes in electronic states, chemical bond, and local structures induced by laser irradiation	Dissociation of molecules Plasma diagnosis Change in energy structure	Broadband x-ray pulse (Dense emission lines, Bremsstrahrung)
Photo-electron: (XPS, EUPS)	Surface sensitive Probe changes in electronic states and chemical bond	Surface (5–75 Å) dynamics Dissociation of molecules Dynamics of core levels	(quasi-) monochromatic x-ray pulse

7.3 Time-Resolved XAFS Measurement of Optically Excited Silicon

Table 7.1 summarizes what can be probed using various ultrafast laser-driven x-ray sources.

As an application example, we show the results for a low-intensity case which means that we are below the damage threshold where only the electronic state is modified. Figure 7.13 shows a schematic diagram of our setup. We used a titanium sapphire laser that operates at 10 Hz. Its pulse duration was 100 fs and the energy was approximately 50 mJ; this output was divided into two parts by the beam and 80% of it was diffracted to create the plasma at the x-ray source. The emitted broadband soft x-ray pulse was focused on to the sample by using concave mirrors. After passing through the sample, the soft x-ray pulse was collected by the toroidal mirror onto the entrance slit of a flat-field grazing-incidence spectrograph. We set the unequally grooved grating with nominal groove number to 1,200 per millimeter.

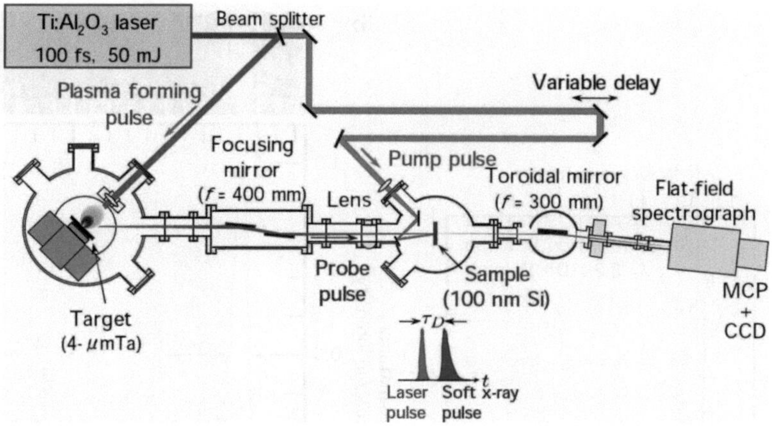

Fig. 7.13 Setup for time-resolved soft x-ray absorption measurements

The remaining part of the laser energy was sent into the variable delay line and was then focused onto the sample. It is necessary to overlap the laser pulse with the soft x-ray pulse probe, which has a focal spot size around 300 μm, in space and time. Therefore, in this experiment, we used a very tiny sample size, 100 μm in diameter and 100 nm in thickness, which limits the effective aperture.

Further, by changing the variable delay line, relative delays can be adjusted between the laser pulse and the soft x-ray pulse. For our purpose, the soft x-ray source must provide a continuous spectrum over a wide enough wavelength range with enough photons, but without complicated fine structures. Figure 7.14 shows a large number of photons were obtained around 100 eV, and if we divided the transmission spectrum by the emission spectrum, we obtained the transmission spectrum of the sample which agreed with the theoretical value. Therefore, this source is suitable for low intensity experiments. The pulse duration was around 10 ps at 100 eV; therefore, our system can resolve 10 ps phenomena to observe the responses.

Figure 7.15 shows a typical transmission spectrum. In this graph, we accumulated 45 shots, but such a spectrum can be observed in a single shot. The time interval between the laser and soft x-ray pulses was almost 0 where the maximum change was observed. The black curve was the transmission spectrum obtained when the pumping-laser light was blocked. The gray curve was obtained when the sample was irradiated by a laser pulse with an intensity of $3 \times 10^{10}\,\mathrm{W/cm^2}$, which was well below the damage threshold. If we calculate the differential transmission, a significant change is observed near the $L_{\mathrm{II,III}}$ absorption edge. This change only appeared near this particular absorption edge.

A careful observation reveals an undulation in the absorption spectrum corresponding to the EXAFS oscillation. By analyzing this undulation, we obtained the distance of neighboring atoms as 2.33 Å, which agrees with the distance measured by x-ray diffraction. Since we did not observe a clear difference in EXAFS signals induced by the laser pulse irradiation, the interatomic distance did not have much

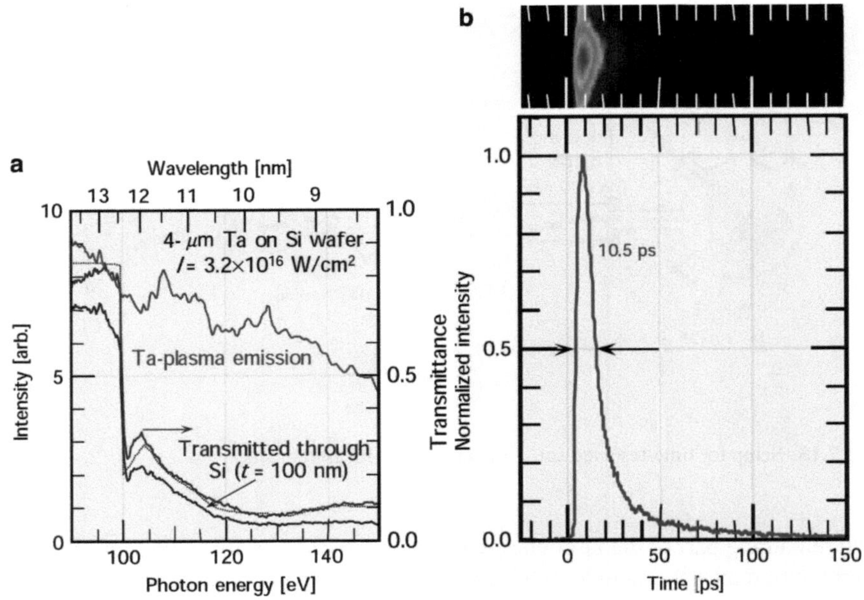

Fig. 7.14 Soft x-ray emitted from a Ta-plasma created by a 100-fs laser pulse [3]. (**a**) Time-integrated soft x-ray spectra. (**b**) Pulse shape of soft x-ray near 100 eV

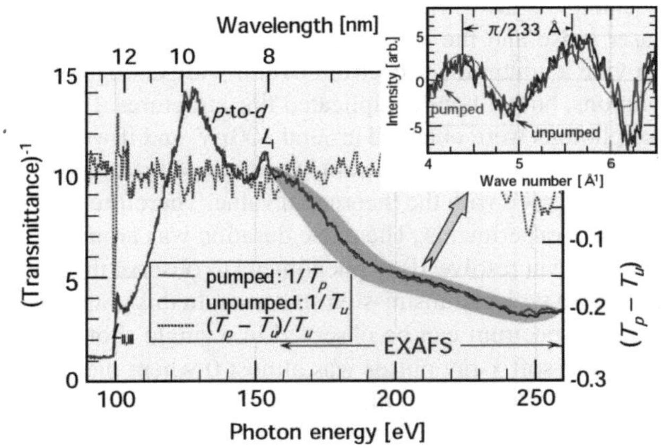

Fig. 7.15 Soft x-ray absorption in a sample [3]

effect. However, for a detailed discussion, we need to undertake measurements in the broader energy region to improve accuracy.

We observed that the laser irradiation caused a downward shift of about 0.2 eV in the absorption edge (Fig. 7.16). Because the $L_{II,III}$ absorption edge is sharp, a sharp dip near the edge appeared in the differential transmittance as the edge shifts.

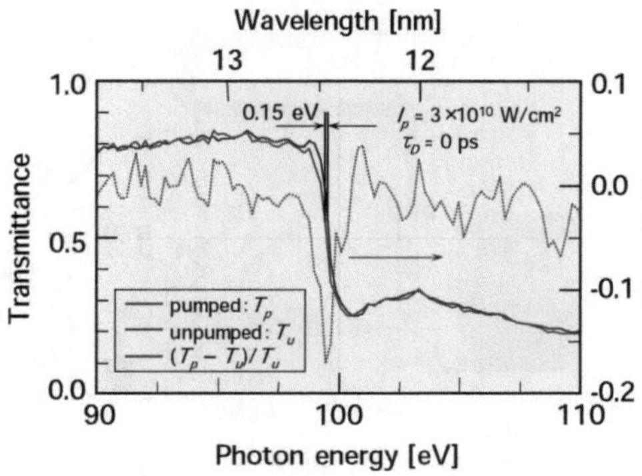

Fig. 7.16 Change in soft x-ray transmittance caused by a laser pulse irradiation [3]

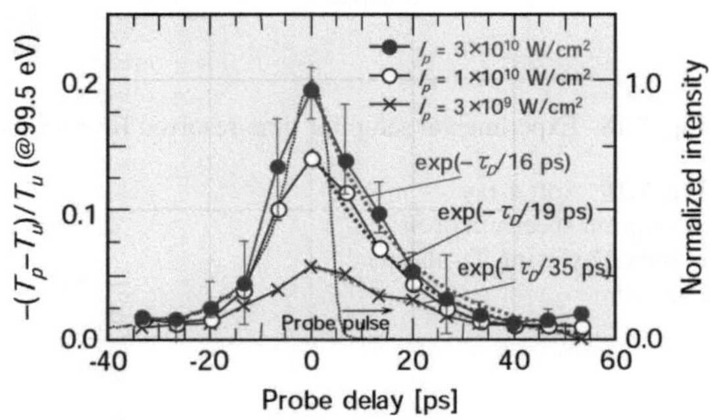

Fig. 7.17 Dependence of differential transmittance at 99.5 eV on a probe pulse delay [3]

Figure 7.17 shows the temporal evolution of the differential transmission.

On the negative delay side, the dependence of the relative increase in absorption on the delay was well fitted to the pulse shape of the soft x-rays. Recovery was observed on the positive side. The depth of the differential transmittance and the recovery time constant strongly depend on the incident laser pulse intensity; therefore, this change in the transmittance is a result of the nonlinear effect.

It is well known that in highly excited semiconductors, the band edge shifts downwards due to the renormalization of the band energy.

If the laser field is high, the optically induced electrons and holes become dense enough to act as a plasma. Because of the strong Coulomb interaction among laser-excited carriers, the energy of photo-induced electrons and holes in their respective bands is reduced, resulting in band-gap reduction.

We assume that our observations are related to this effect. The electron–hole plasma density of $20/\text{cm}^3$ is expected to reduce the band-gap by approximately 0.2 eV, which agrees with our observations. The Auger recombination time of such high-density plasma is expected to be several picoseconds which also agrees with our observations. Therefore such an effect can potentially be observed by using soft x-ray absorption spectroscopy.

In the next example of the absorption measurements, we were above the damage threshold, so we would destroy the sample at each laser shot. We needed to measure the irreversible process.

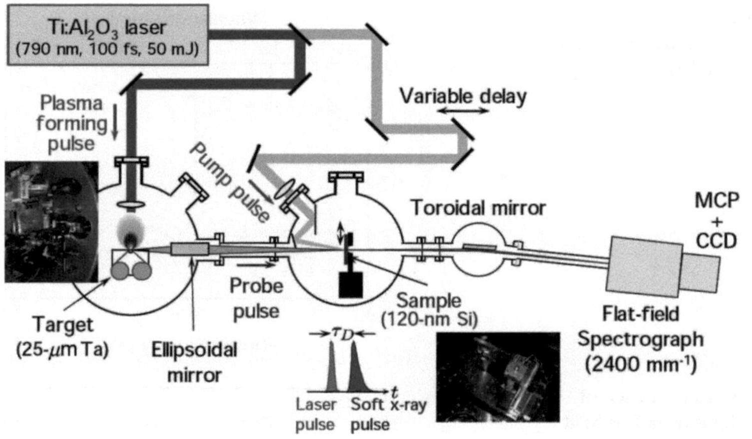

Fig. 7.18 Experimental setup for time-resolved EXAFS measurements

Fig. 7.19 Soft x-ray absorption spectrum of Si obtained by using Ta-plasma emission

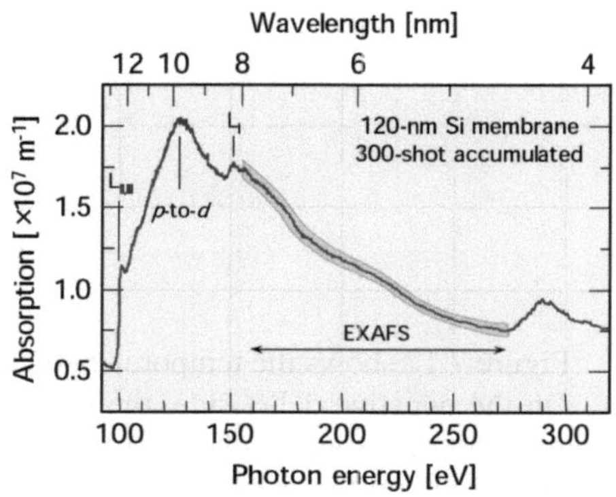

For this purpose, we thus modify our experimental setup from that shown previously (Fig. 7.13) when we were below the damage threshold. The modified set-up is shown in Fig. 7.18. We placed the sample on a translation stage to expose fresh surface at each laser shot. In this case, we must tightly focus the laser beam on the sample. The fluctuations of the source point might affect the overlapping between the laser pulse and the soft x-ray probe pulse on the sample.

In the soft x-ray absorption spectrum of Fig. 7.19, we focus on the fine structure and therefore we had to accumulate more shots to obtain a good signal-to-noise ratio. In the case of silicon, we were able to observe several typical absorption structures near the edges of the sample as well as the EXAFS oscillation. By analyzing the oscillation structures, a very good sinusoidal oscillation was observed which could be treated by Fourier transform. From this we obtained the mean value of the interatomic distance for silicon of 2.32 Å which agreed very well with the literature values.

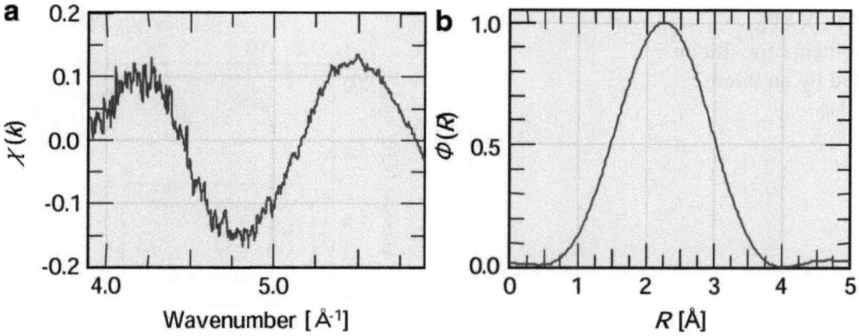

Fig. 7.20 Analyzing EXAFS signal. (**a**) EXAFS function extracted from Fig. 7.19. (**b**) Radial distribution function

Figure 7.20 shows that the structure near the absorption edge collapsed and the EAXFS oscillations become unclear when the sample was exposed to an intense laser pulse.

Next, we compared three cases using silicon (Figs. 7.21, 7.22, and 7.23). In the first case, the probe soft x-ray pulse is transmitted through the sample well before the laser pulse is incident. In this case, no big difference was observed by the laser pulse irradiation. In the second case, if the laser pulse and the soft x-ray pulse overlap on the sample, the laser pulse irradiation caused a significant decrease in the absorption near the $L_{II,III}$ edge and the sharp structure collapsed while keeping the overall absorption almost unchanged. Well after the laser pulse irradiation, the overall absorption dropped clearly over the entire energy region. In the third case, the structure near the $L_{II,III}$-edge was completely modified by the strong laser pulse irradiation. This structural change in absorption spectrum indicated that the electronic structure of the sample was changed from the semiconductor state to the metallic state.

At the delay time of −330 ps, the EXAFS spectra did not show a big difference (Fig. 7.24). The laser pulse irradiation caused a decrease in the oscillation amplitude and a slight peak shift to lower wave number side when the soft x-ray pulse and the laser pulse overlapped on the sample. By analyzing the EXAFS spectra at this time delay, we obtained the atomic distance of about 2.43 Å, while it was 2.32 Å for the unperturbed case. This distance was in good agreement with the literature values of 2.45–2.5 Å for liquid silicon at 1,700–1,800 K. The expanded interatomic distance could be explained by laser-induced rapid melting [11–13], which happened within the soft x-ray pulse duration of 7 ps.

The oscillation structure in the EXAFS spectrum almost disappeared after the laser pulse irradiation as shown for the time delay of 1,670 ps. Taking into account the decrease in the overall absorption, the disappearance of the oscillation structure suggested further disordering silicon due to a phase transition into the gaseous phase. Actually, long after the laser irradiation, the laser-heated area was completely blown off.

Fig. 7.21 XAFS measurements for silicon irradiated by an intense laser pulse

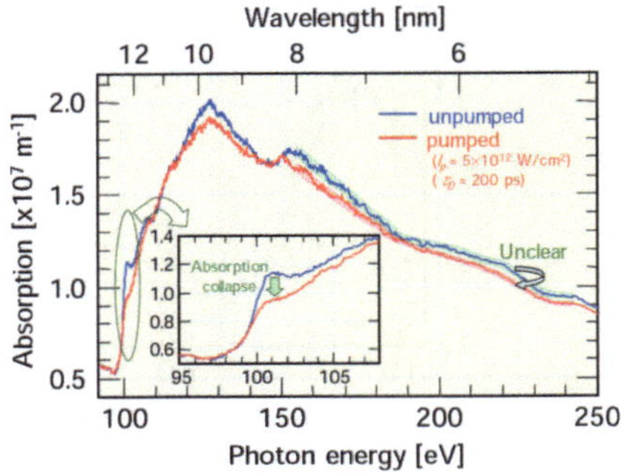

Fig. 7.22 Absorption spectra of the silicon sample at various time delays [4]

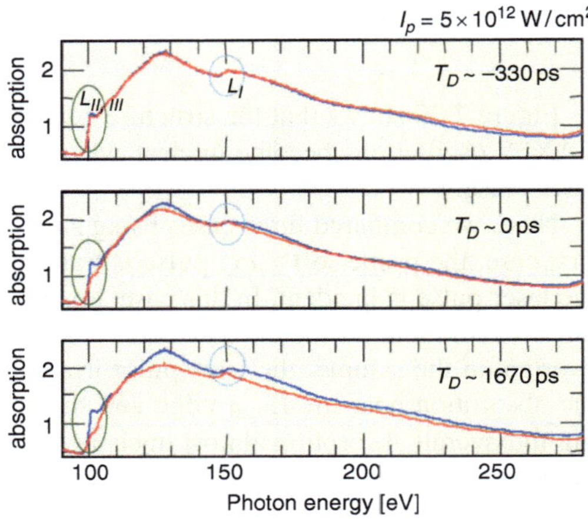

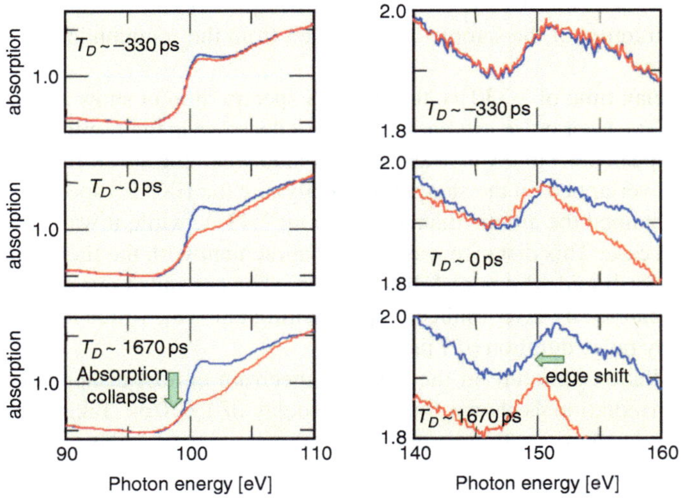

Fig. 7.23 Near-edge (L_I, $L_{II,III}$) absorption spectra of Si at various time delays

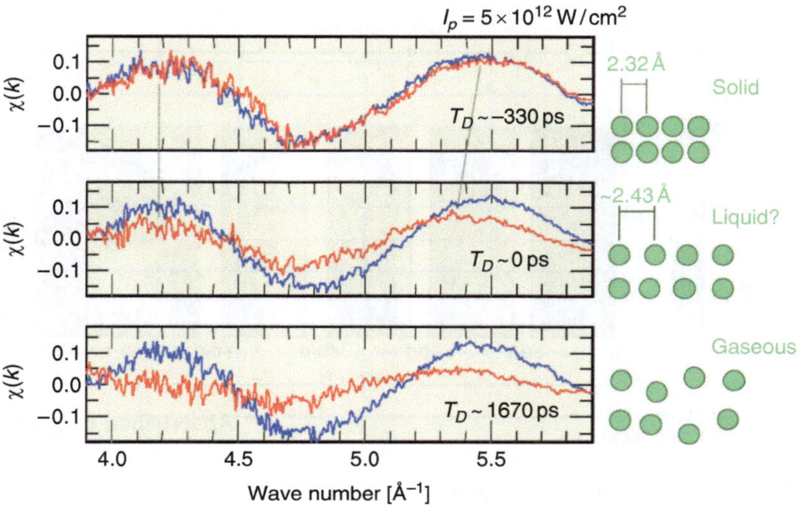

Fig. 7.24 EXAFS spectra at various time delays [4]

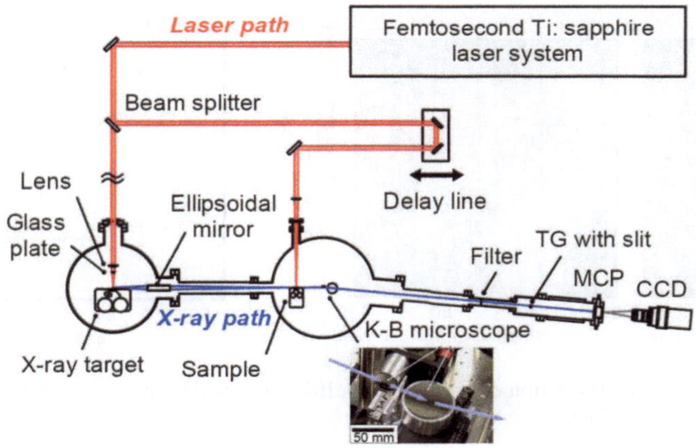

Fig. 7.25 Experimental setup for probing ablation plasma [9]

7.4 Spatiotemporally Resolved XAS

Figure 7.25 shows our experimental setup for probing ablation plasma. The ablation plasma temporarily changes its nature with space and time. Therefore this was a good example to test our system. The ellipsoidal mirror was used at the grazing incidence angle, and in such a case, deposition of small particles does not affect it so much. Debris might be a serious problem when we use a high-quality multilayer mirror at near-normal incidence. Further, the laser ablation particles distribute more

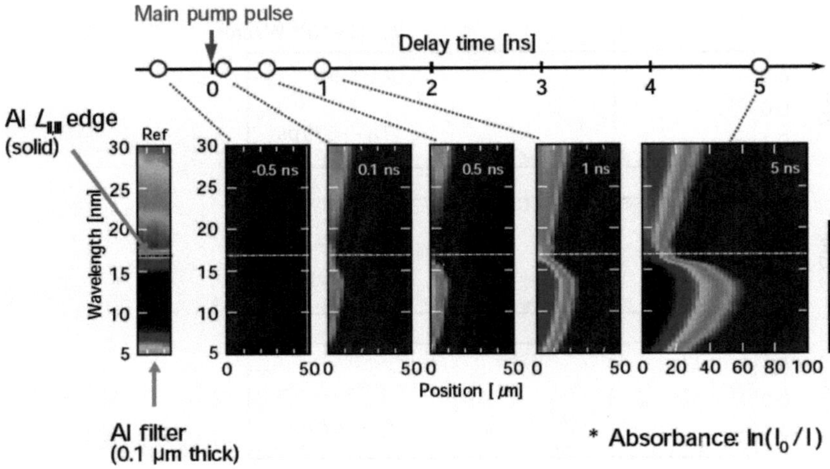

Fig. 7.26 Spatiotemporally resolved spectra

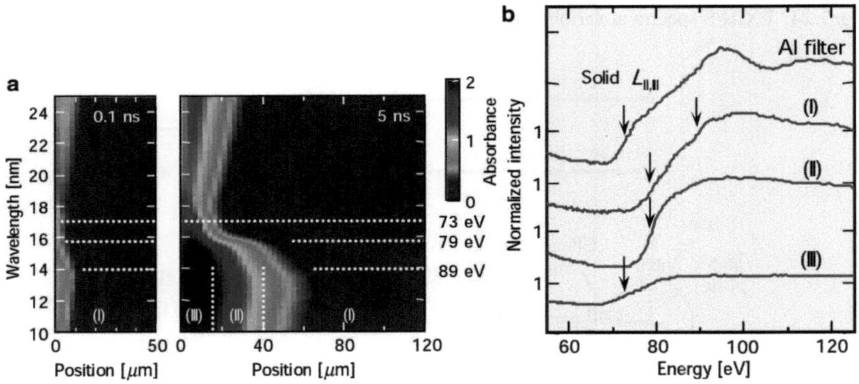

Fig. 7.27 Change in absorption edge position [9, 10]. (**a**) Absorbance images near Al-$L_{II,III}$ edge. (**b**) Shifted Al-$L_{II,III}$ edges

in the normal direction of the sample. Our system has a spatial resolution around 10 μm.

Figure 7.26 reproduces spatially and temporally resolved absorbance spectra taken at different times after the pump pulse struck the sample. A careful observation of these absorption edges allows the type of particle that was there to be determined (Fig. 7.27). The outermost region should be dominated by neutral particles and the mirror might be in the liquid state, while the nearest part might be in the dense condensed state.

By modifying the system to obtain higher spectral resolutions, such as by changing the transmission grating, we can more precisely determine what kind of species are distributed there (Fig. 7.28).

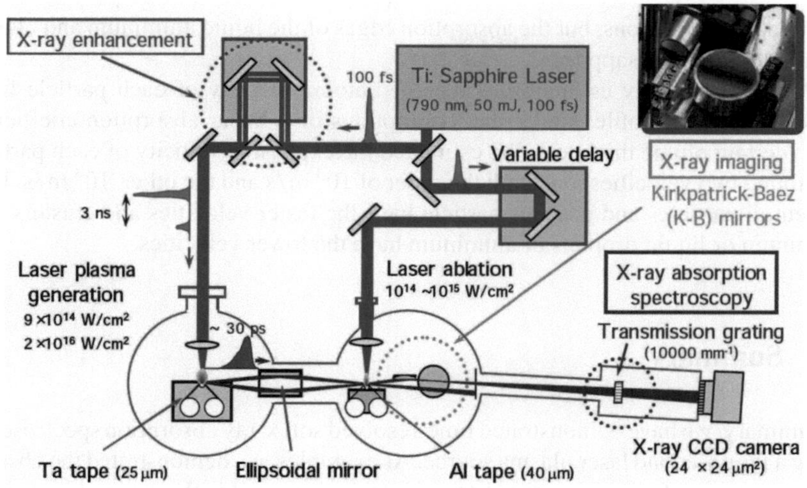

Fig. 7.28 Improved system for Spatiotemporally resolved XAFS measurements

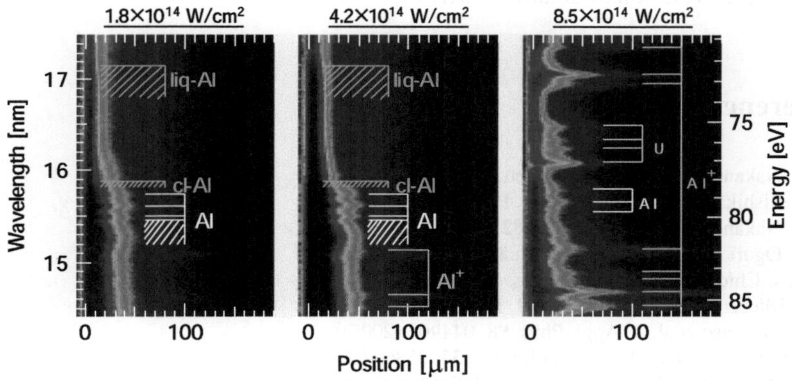

Fig. 7.29 Dependence of absorbance images at 3-ns time delay on laser peak intensity [15]

Figure 7.29 shows the peak intensity dependence of the absorbance image at the time delay of 3 ns. By comparing these absorbance spectra to previous reports [5–10], we concluded that the plume consisted of four components. At the lowest intensity, there was a clear absorption edge near 72 eV, so this edge corresponded to liquid aluminum, and a small edge-like structure was observed near 78 eV, which might correspond to an aluminum cluster because the absorption edge shifted significantly towards high energy. At the moderate intensity, three absorption lines appeared near 79 eV, and these absorption lines corresponded to neutral aluminum atoms. At the highest intensity, the absorbance image drastically changed. This spectrum consists of absorption lines, which could be explained by the singly charged aluminum ions, but the absorption edges of the liquid aluminum and cluster aluminum almost disappeared.

We could roughly estimate the averaged atomic density of each particle from these absorption profiles under the assumptions of a linear absorption coefficient and a certain plume thickness. We estimated the expansion velocity of each particle and found two velocities: one with the order of 10^3 m/s and the other, 10^4 m/s. Ions or neutral particles and gaseous particle have the faster velocities and clusters like aluminum or liquid droplets of aluminum have the lower velocities.

7.5 Summary

In summary, we have demonstrated time-resolved soft x-ray absorption spectroscopy using a femtosecond laser plasma source. As examples, we demonstrated the absorption structures of silicon exposed to an intense laser field. We have also demonstrated a spatiotemporally resolved soft x-ray absorption spectroscopic system to observe the evolution of ablating particles.

References

1. H. Nakano et al., Appl. Phys. Lett. **79**, 24 (2001)
2. T. Nishikawa et al., Jpn. J. Appl. Phys. **42**, L990 (2003)
3. H. Nakano et al., Proc. SPIE **5482**, 11 (2004)
4. K. Oguri et al., Appl. Phys. Lett. **87**, 011503 (2005)
5. B.N. Chichkov et al., Appl. Phys. A **63**, 109 (1996)
6. S. Eliezer et al., Phys. Rev. B **69**, 144119 (2004)
7. S. Amoruso et al., J. Appl. Phys. **98**, 044907 (2005)
8. T. Ohyanagi et al., Jpn. J. Appl. Phys. **33**, 2586 (1994)
9. Y. Okano et al., Rev. Sci. Instrum. **77**, 046105 (2006)
10. Y. Okano et al., Appl. Phys. Lett. **89**, 221502 (2006)
11. S.L. Johnson et al., Phys. Rev. Lett. **91**, 157403 (2003)
12. K. Sokolowski-Tinten et al., Proc. SPIE **3343**, 46 (1995)
13. C.W. Siders et al., Science **286**, 1340 (1999)
14. H. Nakano et al., Appl. Phys. Lett. **69**, 2992 (1996)
15. K. Oguri et al., Phys. Rev. Lett. **99**, 165003 (2007)
16. H. Nakano et al., Appl. Phys. B **79**, 469 (2004)

Chapter 8
Quantum Emission and Its Application to Materials Dynamics

Kazutaka G. Nakamura

Abstract This chapter is divided into two parts. In the first part, we introduce quantum emission processes and their mechanisms by referring to our recent studies on the generation of electrons, ions, and X-ray obtained by irradiating solid materials with intense laser pulses. In the second part, we describe the results of time-resolved measurements of the ultrafast dynamics obtained by taking advantage of the quantum emission processes. We introduce two examples: one is the time-resolved measurements of the laser ablation processes through the electron radiography or the imaging, and the other one is the X-ray diffraction of laser-irradiated semiconductors.

8.1 Introduction

When we apply laser irradiation or a high pressure to materials, a phase transition is induced, for example, a change of crystal structures. However, the duration of the phase transition is unknown, and whether any metastable states exist or not is also unknown. So we want to investigate materials dynamics as summarized in Fig. 8.1.

We have used laser pump and X-ray probe techniques to perform ultrafast time-resolved measurements. When a laser pulse is irradiated onto a material, it is heated up and its crystal structures change. When pulse X-rays are directed to the sample after some delay time, a diffraction pattern can be obtained which includes information on transient structures of the material.

In order to apply this technique we need to generate very short pulse X-rays. We have used X-ray pulses generated by quantum emission (Fig. 8.2), which is initiated by femtosecond laser irradiation with power of 10^{16}–$10^{18}\,\mathrm{W/cm^2}$. Its ponderomotive energy is in the keV range.

K.G. Nakamura
Materials and Structures Laboratory, Tokyo Institute of Technology, Yokohama, Japan

K. Yamanouchi (ed.), *Lectures on Ultrafast Intense Laser Science 1*, Springer Series in Chemical Physics 94, DOI 10.1007/978-3-540-95944-1_8,

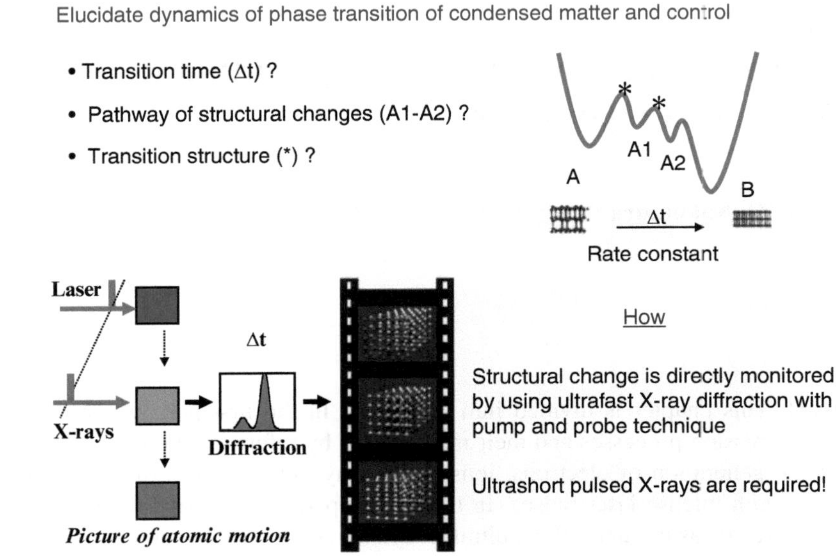

Fig. 8.1 Motivation: materials dynamics

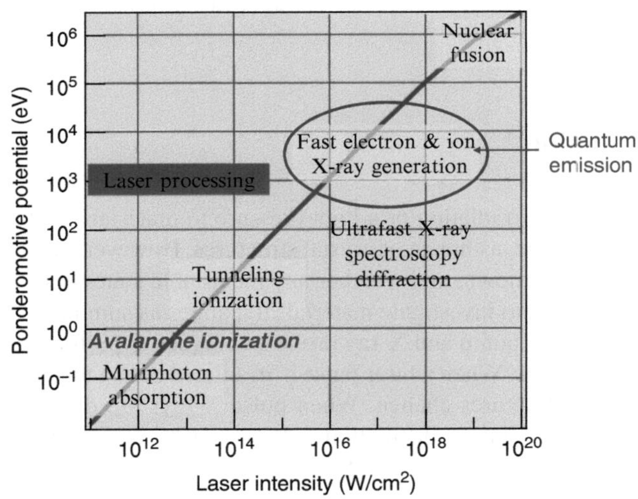

Fig. 8.2 Interaction of laser with solid materials: laser ablation

8.2 Quantum Emission

Quantum emission involves a generation of ultrashort pulsed and high-energy beams of electrons, ions, and photons (X-rays or gamma rays) as indicated in Fig. 8.3. A very thin film metal target is placed in a vacuum chamber. When the intense

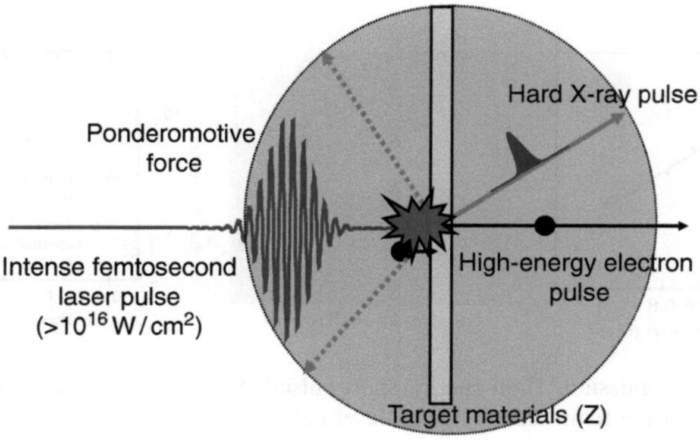

Fig. 8.3 Quantum emissions. High-energy short-pulsed beams of electrons, ions, and photons (X-rays) generated from femtosecond intense laser field ($>10^{16}\,\mathrm{W/cm^2}$)

femtosecond laser beam is irradiated onto the metal target, a preplasma is formed by the prepulse and the electrons in the prepulse are accelerated by the laser field of the main pulse. The acceleration mechanism is due to the ponderomotive potential:

$$\begin{aligned} U_p &= \frac{1}{2} m_{\mathrm{e}} \left\langle v_{\mathrm{osc}}^2 \right\rangle = \frac{eE_0^2}{4m_{\mathrm{e}}\omega^2} = \frac{2\pi e^2 I}{m_{\mathrm{e}} c \omega^2} \\ &= 9.3 \times 10^{-14} I(W/cm^2) \lambda(im)^2 c \end{aligned} \tag{8.1}$$

or wake field depending on the density of the plasma. The accelerated electrons are at very high energy (MeV levels), and they collide with the metal atoms of the target and excite inner shell electrons. The characteristic X-rays are generated. High-energy X-rays up to MeV levels are also generated by Bremsstrahlung of the high-energy electrons. Further, these high-energy electrons ionize absorbed materials on the metal target such as water and hydrocarbons. These generated ions are accelerated to MeV levels by the electric double layer.

Figure 8.4 gives some typical results for quantum emission obtained using our experimental setup. We used a thin copper foil, and irradiated it by the laser beam from the left side. In the photo it is possible to identify the plasma emissions, the high-energy electron and proton beams, generated in the right-side direction and the X-rays isotropically emitted from the source point. The energy spectrum shows that the electrons have energy up to 1.3 MeV and the effective electron temperature is estimated to be about 300 keV. This is comparable to the ponderomotive potential in these intensities. Therefore, in this case, the electrons are accelerated by the ponderomotive force. We also measured the proton energy spectrum. Its shape is slightly different from that of the electron spectrum; however the maximum energy is 1.3 MeV. High-energy X-rays are also generated by Bremsstrahlung and their energy spectrum corresponds to that of the electrons.

X-ray diffraction is usually carried out using hard X-rays at energies around 10 keV. The energy spectrum of these X-rays is in a range of 2–10 keV. Suitable

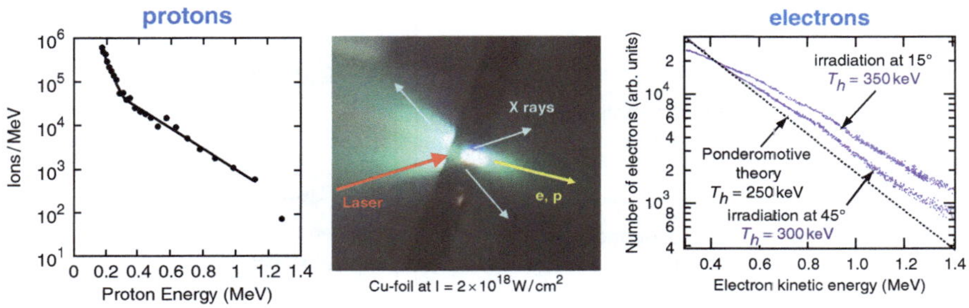

Fig. 8.4 Quantum emission. High-energy short-pulsed beams of electrons, ions and photons (X-rays) generated from femtosecond intense laser field ($>10^{16}$ W/cm^2)

metal targets are copper, nickel, iron, and titanium. The contrast ratio between characteristic X-rays and the white component is very high.

For the copper target, we obtained the copper K_α line at 8 keV and the copper K_β line at 9 keV.

The measured photon number for the copper K_α line was 6.5×10^{10} photons/ 4π sr/pulse. This photon number per pulse is bigger than that from synchrotron radiation. However, the X-rays are emitted isotropically from the point source, so usable X-rays for diffraction measurement were limited. If we set the sample at 10 cm from the source point, then we can obtain about 10^5 photons for one pulse.

We typically use a laser wavelength around 800 nm with prepulse contrast ratio of 10^{-6}. The pulse widths of X-rays depend on the prepulse contrast ratio, so we can generate a shorter pulse by using a smaller contrast ratio. The pulse width of X-rays also depends on the thickness of the target.

It is important to measure the pulse width of the generated X-rays. We measure it by using an X-ray streak camera. In the case of copper K_α X-rays, we have measured a pulse width of 6 ps (Fig. 8.5, right-side spectrum). However, the actual pulse width may be much shorter because the measured pulse width is limited by the time-resolution of the X-ray steak camera. At least, our experiments have shown that X-ray diffraction can be achieved with a time-resolution of 10 ps.

X-rays can also be generated from a gas target. We have used micro-meter-size argon clusters and obtained their X-ray diffraction patterns. Figure 8.6 shows X-ray lines from He-like, B-like, and C-like argon clusters. The photon number of the X-rays from the cluster is about ten times less than that from solid target, but they are also useful for X-ray diffraction experiments. There is no debris when we use clusters.

8.3 Time-Resolved Imaging with Quantum Emission

Figure 8.7 is the setup for the electron energy measurement, which is installed in the vacuum chamber. When the metal target is irradiated with a laser beam, electrons are ejected. Their energy spectrum is measured by using a magnetic type spectrometer.

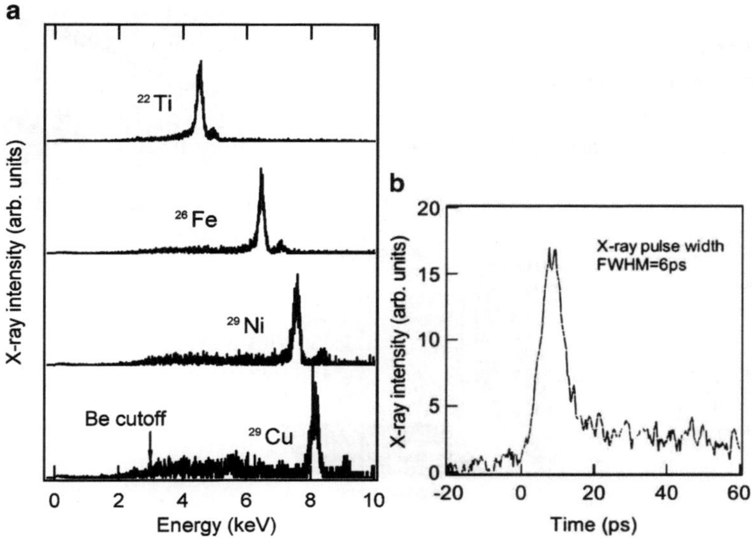

Fig. 8.5 Femtosecond laser induced hard X-ray pulse ($I = 10^{17}\,\mathrm{cm}^2$). The temporal profiles obtained by the X-ray streak camera. The energy extends from 1 to 10 keV are accumulated. $I \sim 10^{17}\,\mathrm{W/cm}^2$ [1,2]. (**a**) Energy spectra; Characteristic X-rays obtained by X-ray CCD camera at power density of $3 \times 10^{16}\,\mathrm{W/cm}^2$. (**b**) Pulse width of CuKa; 6.5×10^{10} photons/4π sr/pulse

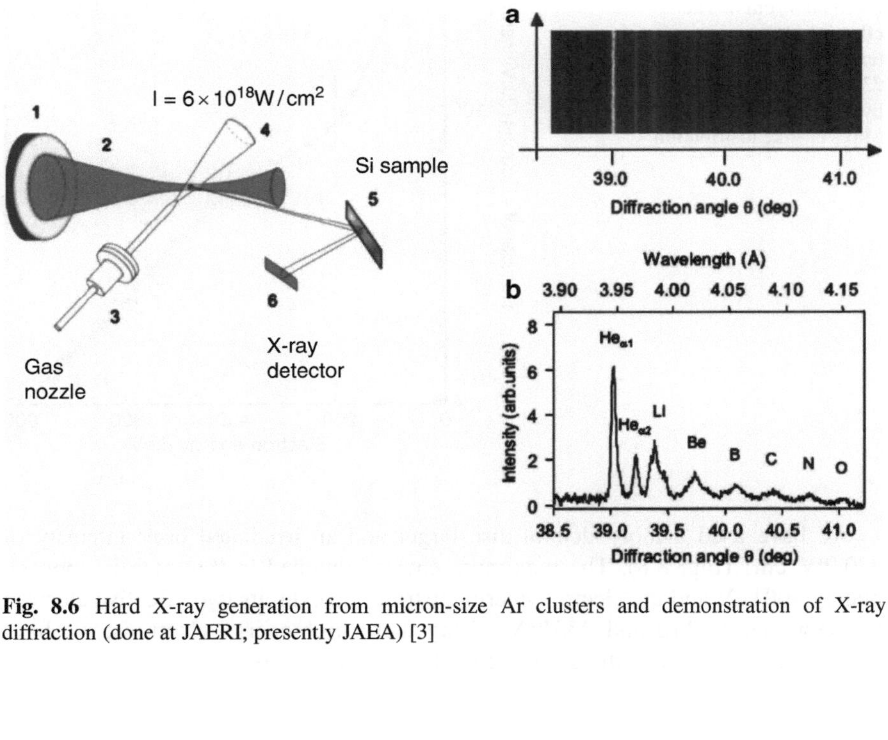

Fig. 8.6 Hard X-ray generation from micron-size Ar clusters and demonstration of X-ray diffraction (done at JAERI; presently JAEA) [3]

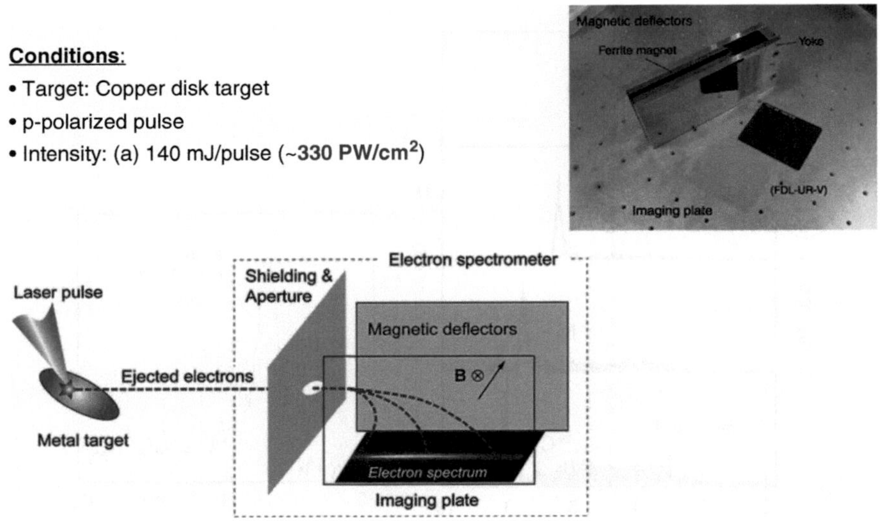

Fig. 8.7 Electron energy measurement

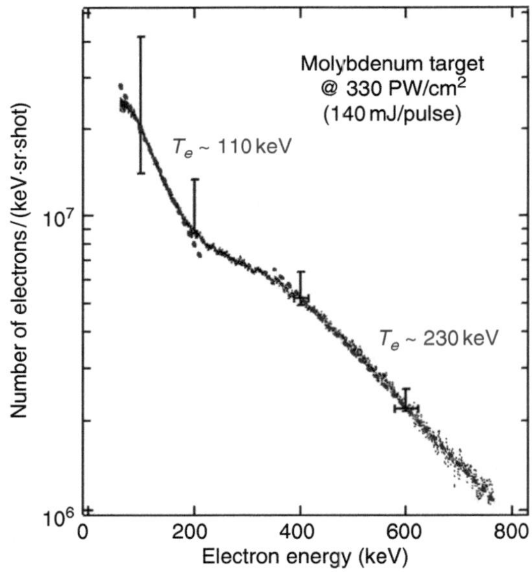

Fig. 8.8 Electron energy distribution from Mo disk target at laser intensity of 140 mJ/pulse ($\sim$330 PW/cm^2). The effectice electron temperatures are 110 and 230 keV, which are obtained by assuming two Maxwell-like distribution

We have used a molybdenum disc target and an irradiated laser intensity of 330 PW/cm^2 (Fig. 8.8). The maximum energy obtained in this case is approximately 800 keV and the electron energy distribution exhibits two effective electron temperatures, at 110 and 230 keV. Using these high-energy electrons, we have demonstrated electron radiography of the laser ablation process.

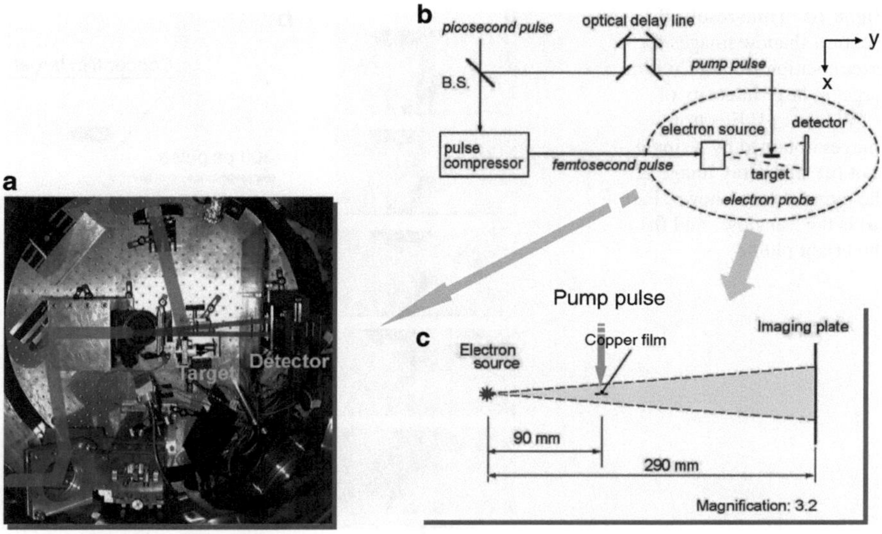

Fig. 8.9 Experimental setup for electron imaging for laser ablation plasma. A picture (**a**), a schematic drawing of the system (**b**) and a configuration between the electron source and the target (**c**) are shown.

Figure 8.9 is a schematic of our experimental setup in which we used a pump-probe method. A picosecond laser beam is divided in two beams. One beam is directed to the pulse compressor and focused onto the metal target to generate electron pulses and the other beam is directed to the optical delay line and the metal target.

Figure 8.10 shows some time-resolved images obtained for a copper film target. When the laser beam was irradiated on the sample, ablation occurred. After 370 ps, the plume could be seen. The plume expanded as the delay time increased and two structures became apparent. These are seen as bright and dark parts in the images. The bright part is the place where electrons concentrate and the dark part is the place where electrons are blocked or absorbed.

The plume expansion was evaluated by using the distance between the edge of the plume and the surface of the sample. Results are shown in Fig. 8.11. The expanding speeds of the bright and dark parts were obtained as approximately 970 and 110 km/s, respectively. Simultaneously, we measured the velocity distribution of the emitted ions by using the time-of-flight measurement. The mean velocity of the ions was about 78 km/s. The expansion of the dark plume image is considered as the expansion of the ions during the laser ablation process.

We suppose that the bright part corresponds to the plume of the expanding electrons (Fig. 8.12) because electrons have a lighter mass than ions, so their expanding speed is much higher than that of ions. When the laser beam is irradiated onto the metal target, it causes a laser ablation and electrons are rapidly ejected from the

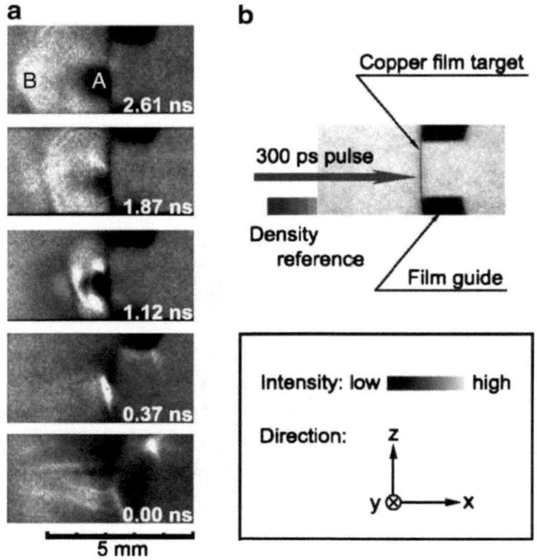

Fig. 8.10 Time-resolved electron shadow images for laser ablation from a Cu film target at laser intensity of 1 TW/cm^2. [4] Electron images obtained by a single shot (**a**) and X-ray image of the target (**b**) are shown. Part (**a**) is the "shadow" and (**b**) is the bright plume

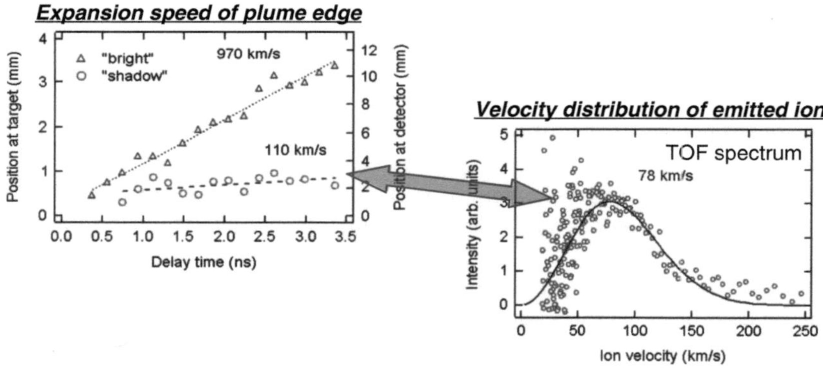

Fig. 8.11 Expanding speeds of plumes and ions. The ion velocity agrees well with the expansion speed of the shadow plume. The shadow plume reflects ions

target, and after that, the ions and electrons undergo expansion. Neutral particles are also generated and they undergo expansion. Usually, they have much lower speed.

This experiment is one demonstration of electron imaging.

As mentioned before, if the intense laser beam is irradiated onto the metal target, high-energy protons are obtained. The protons are considered to be generated from impurities such as water and hydrocarbons on the metal target. If protons are generated from these absorbed species, their amount and energy cannot be controlled. So we examined a double-layered target, composed of PVME (polyvinyl methyl ether) and metal, in order to control the proton amount. We were able to increase the amount of the high-energy protons by 10 or 100 times by using the polymer-coated target. The maximum energy of the protons was also increased.

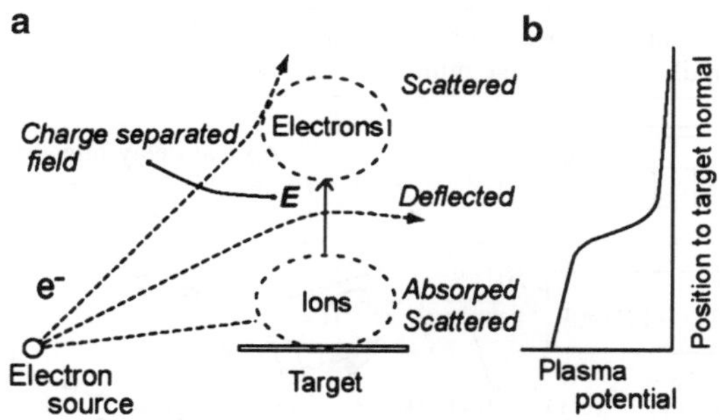

Fig. 8.12 Possible mechanism of electron imaging. "Bright" plume is due to focusing of deflected electrons, and "Shadow" plume reflects ion plume, reflecting the charge-separated field. This is a useful diagnostic tool for laser ablation

Fig. 8.13 A picture of the experimental system for laser plasma X-ray generation

High-energy protons can also be used for dynamic imaging such as imaging of shock compression of laser-irradiated materials.

8.4 Time-Resolved X-Ray Diffraction

In the rest of this chapter, we discuss time-resolved X-ray diffraction (Fig. 8.13) including some examples from laser-irradiated Si, Ge, and GaAs.

Our experiment setup has a table-top laser system with a maximum power of 8 TW. A schematic of the setup is given in Fig. 8.14. The 300-ps laser beam delivered from the amplifier is divided into two beams. One beam is pulse-compressed to 50 fs in the vacuum chamber and focused onto the metal target to generate X-ray pulses. The generated X-rays are extracted in air through a Be window and directed to the sample. Using a symmetrical Bragg diffraction geometry, we detect the diffracted X-rays with a liquid-nitrogen cooled X-ray CCD camera. The other beam

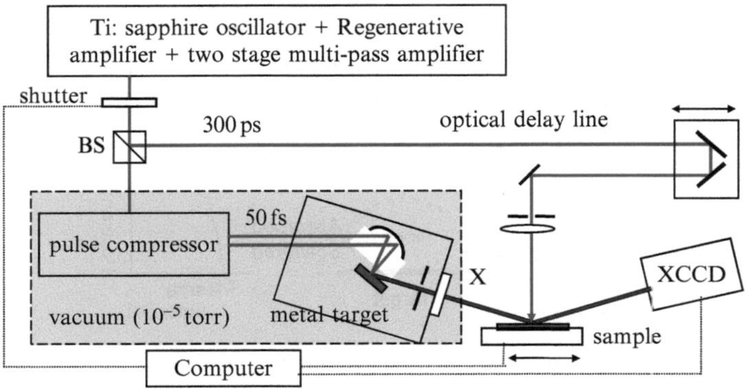

Fig. 8.14 Time-resolved X-ray diffraction from laser-irradiated Si (111)

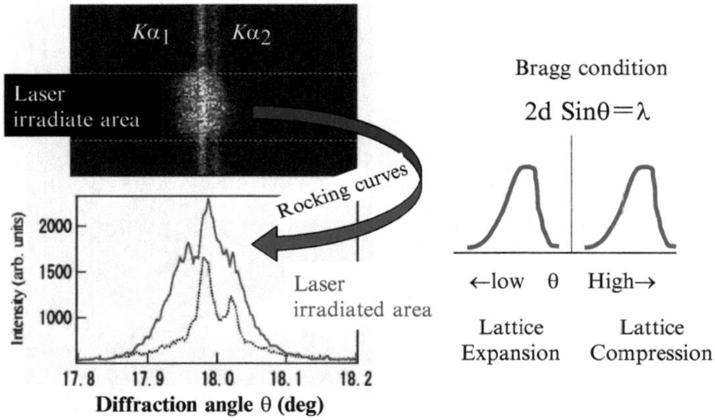

Fig. 8.15 X-ray diffraction: single-shot experiment

is directed towards the delay line and focused onto the Si sample to get excitation phenomena.

A typical example of an X-ray CCD camera image is shown in Fig. 8.15. Only part of the sample is laser-irradiated. The two observed lines correspond to Cu-$K_{\alpha 1}$ and $K_{\alpha 2}$ lines. In the laser-irradiated area some deformation can be seen in the X-ray diffraction patterns. These are rocking curves. There are some changes in the angle component. In the symmetric Bragg diffraction geometry, $2d \sin\theta = \lambda$, where θ is the diffraction angle and λ is the wavelength of the X-rays. So the changes to lower- and higher-θ angle components correspond to lattice expansion and compression, respectively.

We performed two types of X-ray diffraction experiments on the laser-irradiated Si (111) crystal: single-shot and the other multiple-shot. In the single-shot experiment, seen in Fig. 8.16, the laser beam was irradiated on the silicon sample for one shot and the X-rays were directed onto that sample at the same point. Then, in order to avoid damage to the sample, we moved it for the next laser shot, and this point

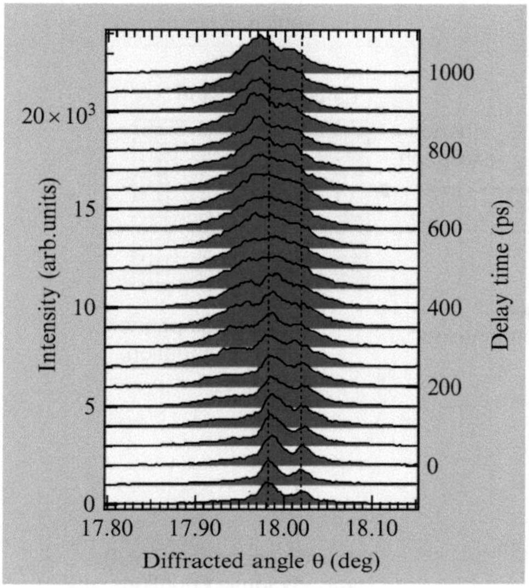

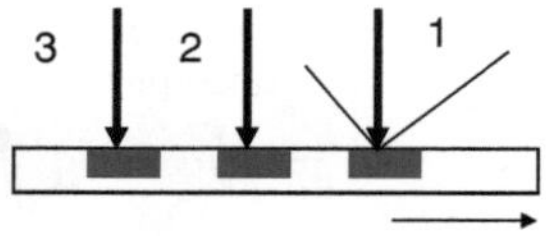

Picosecond time-resolved X-ray diffraction from Si(111) crystal under 300 ps laser irradiation at 1J/cm^2. Time step is 50 ps and accumulation of 600 data

Fig. 8.16 Single-shot experiment. Picosecond time-resolved X-ray diffraction from Si (111) crystal under 300-ps laser irradiation at 1 J/cm^2. Time step is 50 ps and accumulation of 600 data [5]

was irradiated once with the X-rays. We moved the sample 600 times to accumulate sufficient X-ray diffraction data.

The time-resolved X-ray diffraction was carried out every 50 ps after the laser irradiation. When the laser beam was irradiated, the lower-angle component increased. After 200 ps, three lines were observed and they merged into a broad single line at 600 ps. The change in the spectrum continued and two lines were observed which were shifted towards lower angles compared to the original positions. This means that lattice expansion due to laser heating takes place, and transient lattice deformation in laser heating process is directly observed.

We can explain the lattice expansion as follows using Fig. 8.17. The energy of the 800-nm laser light is approximately 1.5 eV. This value is higher than the band gap of Si (1.15 eV), and then electron-hole pairs are generated due to real photo-absorption. Energy of the photo-excited carriers is transferred to kinetic energy of other electrons by Auger recombination process. Then energy in the electronic system is transferred to the lattice system, the temperature of the sample increases, and the crystal lattice expands from the top surface.

These phenomena can be expressed by using the rate and diffusion equations for carriers and hydrodynamic equation for lattice motion:

$$\frac{\partial N}{\partial t} + \frac{N}{\tau_R} = D\frac{\partial^2 N}{\partial x^2} + \frac{\alpha(1-R)}{E} I(x,t) + \frac{\beta(1-R)^2}{2E} I(x,t)^2, \tag{8.2}$$

$$\frac{\partial}{\partial t} I(x,t) = -\left[\alpha + \Theta N + \beta I(x,t)\right] I(x,t), \tag{8.3}$$

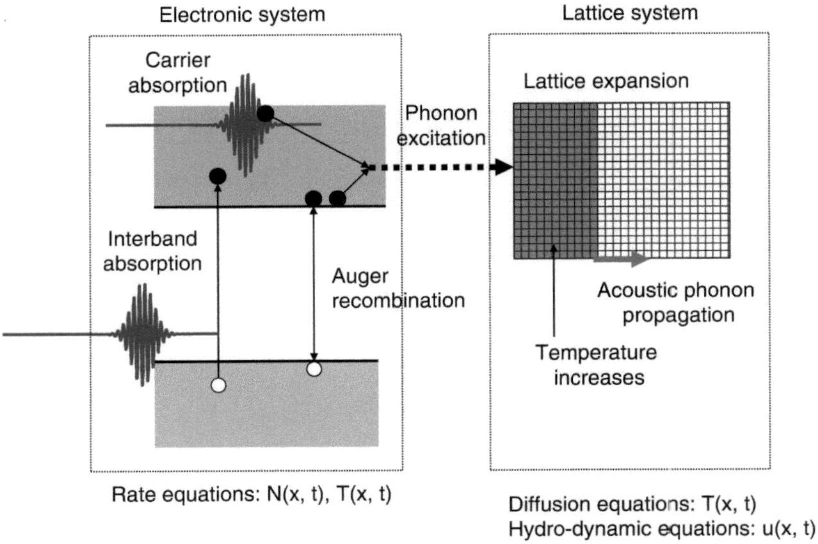

Fig. 8.17 Schematic of energy transfer

$$\tau_R = \frac{1}{\gamma N^2}, \tag{8.4}$$

$$\frac{\partial T}{\partial t} = D_T \frac{\partial^2 T}{\partial x^2} + \frac{E_g N}{C_v \tau_R} + \left(\frac{E - E_g}{C_v}\right) \left[\left(\frac{\alpha(1-R)}{E} + \Theta N\right) + \frac{\beta(1-R)^2}{2E} I(x,t)\right] I(x,t), \tag{8.5}$$

$$\rho \frac{\partial^2 u}{\partial t^2} = \rho v^2 \frac{\partial^2 u}{\partial z^2} + B \frac{\partial E_g}{\partial P} \frac{\partial N}{\partial z} + 3B\beta \frac{\partial T}{\partial z}. \tag{8.6}$$

From these equations, we can calculate the carrier density and the temperature at the topmost surface.

When we irradiate the laser beam, the carrier density increases as the laser intensity increases. Using the 300-ps laser beam, at a delay time of about 200 ps, Auger recombination occurs and the temperature increases to 1,600 K at maximum. By using the temperature distribution, we can calculate the strain profile inside the materials (Fig. 8.18). Figure 8.19 shows the strain profile as a function of depth. Depth zero corresponds to the topmost surface. The topmost surface expands to 0.3 and 0.8% at delay times of 50 and 550 ps, respectively. There is a boundary between expansion and compression at approximately 4 μm from the topmost surface. This boundary travels inside the materials with the velocity of sound; in this case, it is 9 km/s.

After a very long delay, the strain profile looks like a single exponential curve, which corresponds to the linear absorption of the laser light. By using this strain

Fig. 8.18 Model of laser heating and photon generation

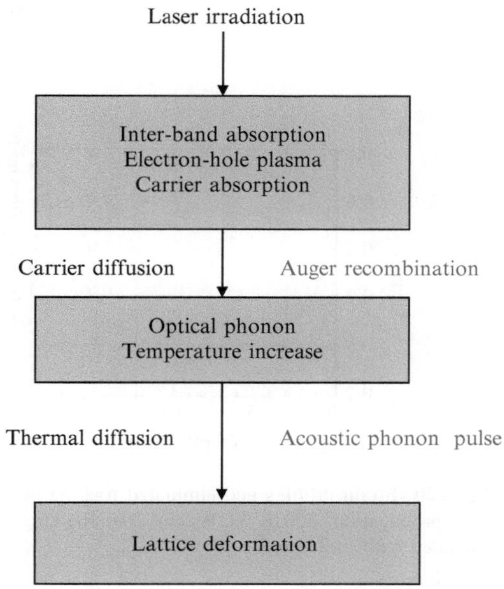

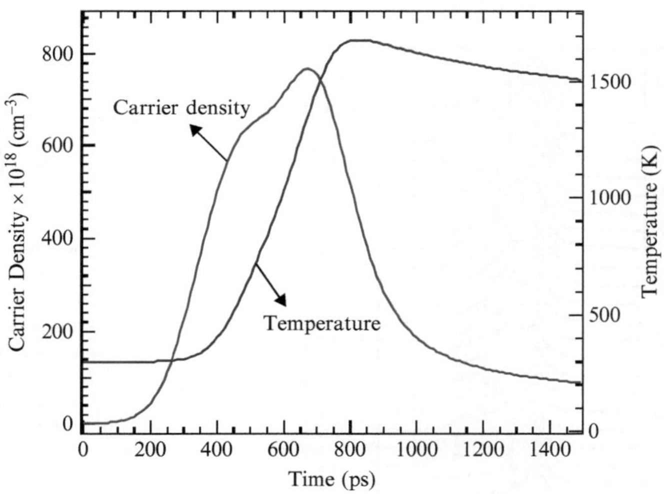

Fig. 8.19 Carrier density and temperature at surface: 300 ps pulse 3 GW/cm^2

profile (Fig. 8.20), we can simulate X-ray diffraction using the dynamic X-ray diffraction theory. The lines represent simulation results and the circles represent experimental data. The simulations well explained the experimental data. Therefore, we can directly observe the transient lattice deformation due to the acoustic phonon propagation in that material by using picosecond time-resolved X-ray diffraction.

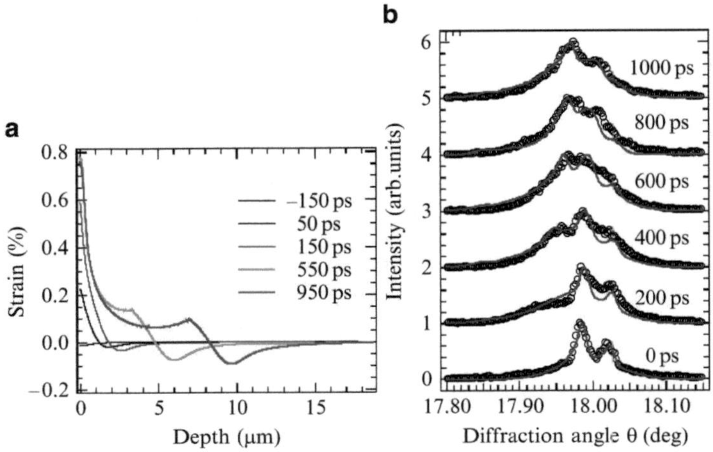

Fig. 8.20 Strain profiles and simulated rocking curves [5]. (**a**) Strain profile, Acoustic phonon pulse propagation, 300 ps 3 GW/cm^2. (**b**) Rocking curves obtained from X-ray diffraction and simulated with strain profiles

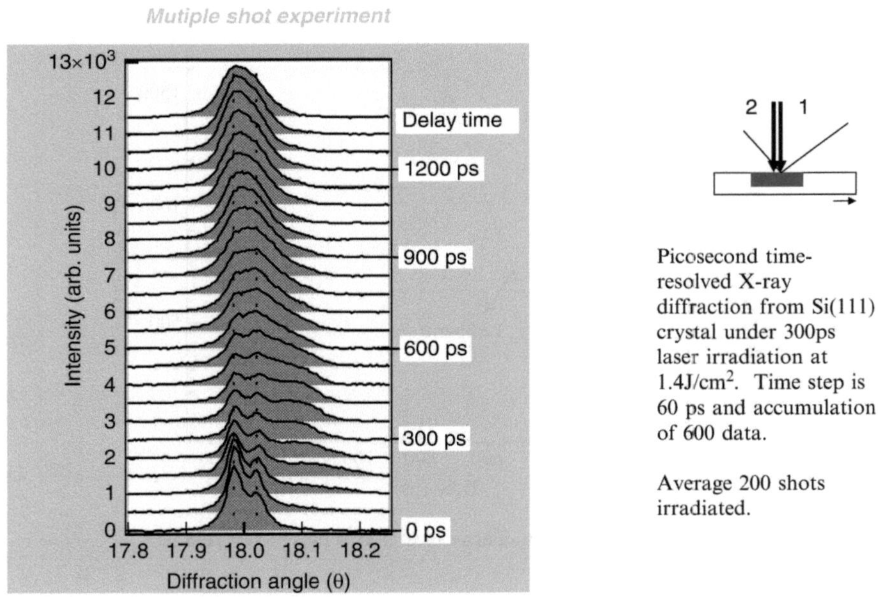

Fig. 8.21 Multiple-shot experiment. Picosecond time-resolved X-ray diffraction from Si(111) crystal under 300 ps laser irradiation at 1.4 J/cm^2. Time step is 60 ps and accumulation of 600 data. Average 200 shots irradiated

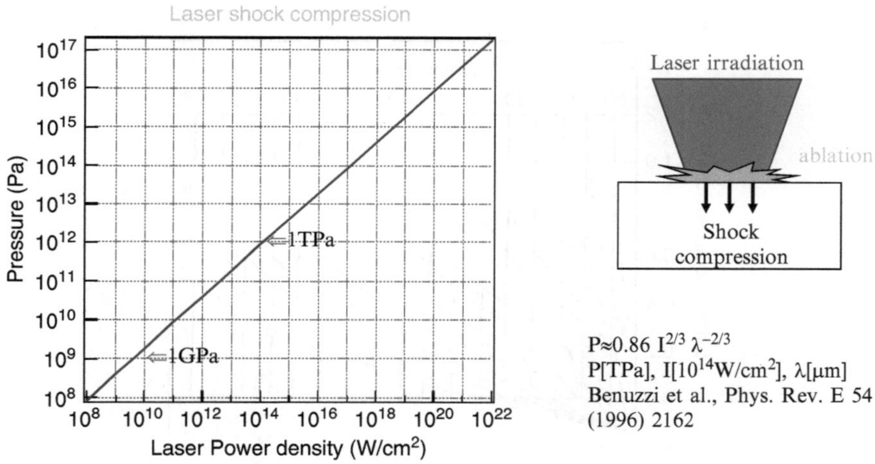

Fig. 8.22 Laser shock compression [6]

Some findings of the second type of experiment, the multiple-shot experiment, are shown in Fig. 8.21. In this case, 200 laser shots nwere irradiated at the same point on average. In the case of the single-shot experiment, an increase in the lower-θ angle component was observed. However, the multiple-shot experiment had a change in the higher-θ angle component. There was a maximum change, and then the higher-θ angle component returned to its original position. An increase in the higher-θ angle component is related to lattice compression.

The compression is due to the shock compression induced by laser ablation (Fig. 8.22). Usually, if we irradiate the pulse laser with a power density of $10^{10}\mathrm{W/cm^2}$, a high pressure (1 GPa) is induced inside the materials as a shock wave.

We also estimated the strain profile inside the material in order to represent the time-resolved X-ray diffraction data. After a delay of 180 ps, the maximum of the compression was approximately 1% and the shock wave front was located at approximately 1.8 μm from the surface. This front traveled inside the materials with the shock wave velocity of 9.4 km/s. The peak pressure corresponded to 2 GPa. Therefore, we directly measured the transient lattice deformation by the shock wave propagation (Fig. 8.23).

8.5 Summary

In summary, we have generated high-energy electrons or ions with energies higher than 1 MeV and used them to carry out time-resolved electron radiography measurements including detectionof the laser ablation process. By using picosecond

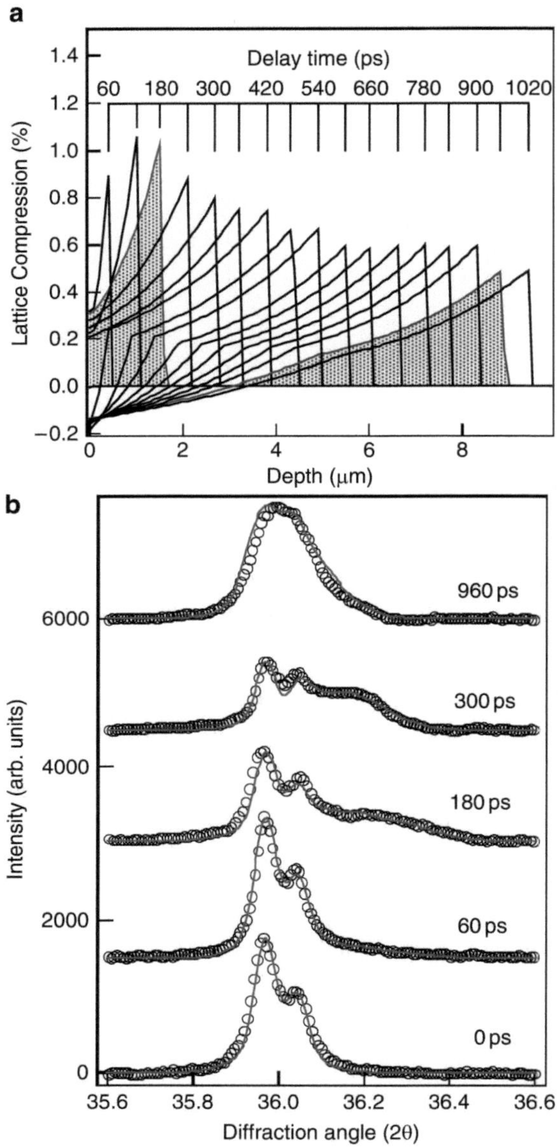

Fig. 8.23 Shock wave propagation. Peak pressure: 2.18 GPa, shock velocity 9.4 km/s. (**a**) Strain profiles. (**b**) Rocking curves obtained from X-ray diffraction and simulated with strain profiles

X-ray diffraction, we were able to demonstrate direct measurement of transient lattice deformation by shock wave propagation and acoustic phonon propagation. The quantum emissions are useful for investigation of ultrafast materials dynamics.

References

1. M. Yoshida et al., Appl. Phys. Lett. **73**, 2393 (1998)
2. Y. Fujimoto et al. Jpn J. Appl. Phys. **38**, 6754 (1999)
3. Y. Fukuda et al., Appl. Phys. Lett. **85**, 5099 (2004)
4. Y. Okano, K.G. Nakamura et al., Appl. Phys. Lett. **83**, 1536 (2003)
5. H. Kishimura et al., J. Chem. Phys. **117**, 10239 (2002)
6. A. Benuzzi et al., Phys. Rev. E **54**, 2162 (1996)

Chapter 9
Filamentation Nonlinear Optics

See Leang Chin

Abstract The physics and the characteristics of filamentation are described. This discussion is followed by presenting a few applications that make use of the unique properties of the filament core. These include nonlinear optics inside the filament (third harmonic generation and four-wave mixing), the feasibility of remotely detect chemical and biological agents in air based upon the characteristic clean fluorescence. Finally, the excitation of super-excited states of molecules is briefly described.

9.1 Introduction

Filamentation is actually the same concept as self-focusing, which has been known since the invention of the laser beam. However, with the advent of femtosecond laser pulse systems, the concept of filamentation appears to have been modified. In fact, the concepts of physics are always the same; it is just that the phenomenon appears to change into something that was totally unexpected. (For recent comprehensive reviews, see [1–4].)

9.2 Self-Focusing and Filamentation Physics

In the concept of filamentation (Fig. 9.1), we assume that a plane wavefront is present where there is an intensity distribution that is Gaussian so that at the central region, the intensity is high. If this wavefront propagates in any optical medium – solid, liquid, or gas – that is transparent to this wavefront, then the index of refraction will be $n = n_0 + n_2 I$ at the center where I is the intensity. This is nonlinear optics; i.e., the nonlinear part of the index of refraction, $n_2 I$, plays an important role. The

S.L. Chin
Laval University, Quebec, Canada

K. Yamanouchi (ed.), *Lectures on Ultrafast Intense Laser Science 1*, Springer Series in Chemical Physics 94, DOI 10.1007/978-3-540-95944-1_9,

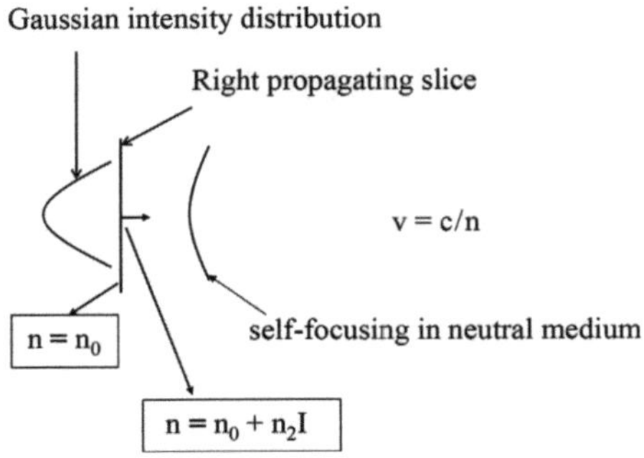

Fig. 9.1 The beginning of self-focusing of a wavefront of an assumed Gaussian beam. There is no guarantee yet for filamentation if linear diffraction is stronger than weak self-focusing, giving slowly diverging wave than normal diffraction

velocity of propagation is c/n; the central part has a higher index of refraction when n_2 is positive. At the wings of the right propagating slice, the intensity is low; then $n = n_0$, the linear index of refraction. Hence, when this wavefront propagates in any optical material, it will curve forward, the central part propagating slower than the wing. The consequence is a concave (focusing) spherical wavefront if we assume that the wavefront is part of a Gaussian beam. This is called self-focusing. We shall analyze the consequence of this self-focusing by considering different intensities I and peak powers.

If the intensity I is not high, filamentation cannot be generated immediately because the wave will diverge itself by normal diffraction. In other words, linear diffraction will compete with the self-focusing process. Therefore, in this self-focusing process, if I is not very strong then it will be weaker than diffraction in such a way that the object will slowly diverge and no filamentation occurs. Divergence of this wavefront is slow and considerably slower than the normal diffraction.

When we increase the intensity such that the linear diffraction and self-focusing will balance each other in their diverging and converging processes, then we can obtain the critical power P_c for self-focusing by solving the Maxwell equation for a nonparaxial Gaussian beam as in [5]:

$$P_{\mathrm{c}} = \frac{3.77\lambda^2}{8\pi n_2 n_0}. \tag{9.1}$$

It should be noted here that P_c is independent of intensity. This was achieved in the 1970s. Here, it should be noted that this critical power for self-focusing means that the power of the laser itself is dependent on n_2, which is the nonlinear index. The linear index n_0 does not depend on the intensity.

Recently, we have measured the critical power, which depends on the pulse duration [6] (Fig. 9.2). If the pulse is very short, then the interaction is basically with the electrons. Thus, a rapid change of the index through the electronic polarization can be observed. This gives a higher critical power for self-focusing. This means that n_2 is very small. However, when nuclear motion is also involved, a combination of electronic (fast) and nuclear (slower) motions contributing to the critical power for self-focusing is observed. This implies that n_2 changes with pulse duration.

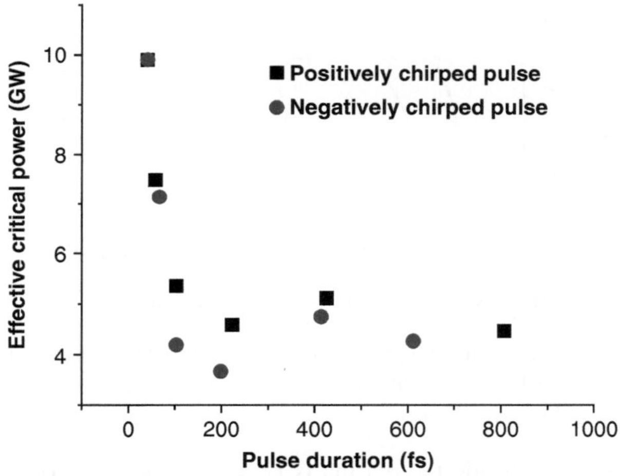

Fig. 9.2 Experimental effective critical power for self-focsuing vs. pulse duration [7]

If the laser pulse is short, there are a lot of frequency components and each component would propagate at different speeds. This is called group velocity dispersion (GVD). Hence, pulse lengthening is attained after a certain distance of propagation; thus, a lower peak power is obtained. If the peak power is low, it may fall below the critical power for self-focusing and no more self-focusing takes place. However, if the peak power of the pulse is slightly higher than the critical power, then it would be slightly self-focusing. As mentioned before, sooner or later because of this GVD, the peak power will become lower than P_c and it would not self-focus anymore and the pulse would diffract like a linear pulse.

Therefore, no filamentation has occurred yet for the situations just described above. Under all these conditions, the GVD and linear diffraction should be overcome in such a way that it will result in real self-focusing, everything else remaining the same. If we have a femtosecond laser pulse, its peak power should be higher than P_c in order to have filamentation. In the case of condensed matter, a peak power a few tens of percent higher than P_c is sufficient, as we found out many years ago in our work [8]. This would immediately result in self-focusing.

After self-focusing starts, the process will continue until the pulse becomes a very small spot. There is high intensity at this "hot" spot and any material confronting such high intensity would sooner or later be ionized, thereby generating free electrons.

Once free electrons are generated, the change of index $(\Delta n)_p$,

$$(\Delta n)_p = \frac{N_e(I(t))}{2N_{crit}} \cong -\frac{4\pi e^2 N_e(t)}{2m_e\omega_0^2}, \tag{9.2}$$

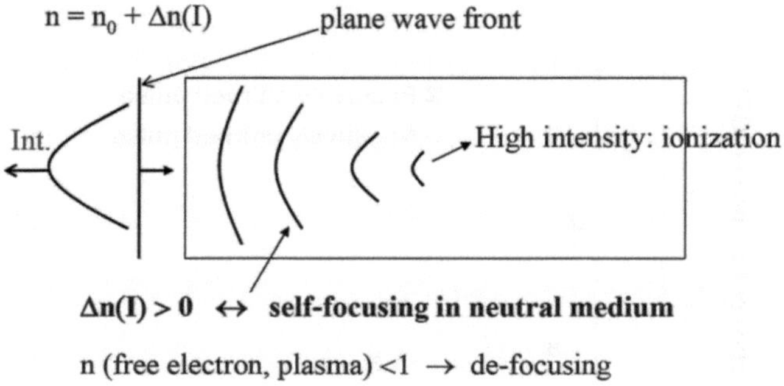

Fig. 9.3 An illustration of the interplay between self-focusing and plasma defocusing through the consideration of their contributions to the indices of refraction

for the free electrons inside the plasma is negative (Fig. 9.3). This means that if plasma is generated, the index of refraction is less than 1 and some defocusing effect takes place. $(\Delta n)_{\mathrm{p}}$ is obtained by dividing the electron density by the critical density for the plasma. In this case, if N_{crit} (from plasma physics) is substituted in the equation, we obtain a result where N_{e} is the electron density, which is created by the high-intensity point indicated in the drawing. The total index n is given by

$$n \cong n_0 + n_2 I - \frac{4\pi e^2 N_{\mathrm{e}}(t)}{2m_{\mathrm{e}}\omega_0^2}. \tag{9.3}$$

One can see that there are two nonlinear terms acting against each other: the self-focusing term and the plasma term, both depend on the intensity.

As an example, we consider air. In air the reaction will be unimolecular because the time is very short. Thus, either tunnel ionization or multiphoton ionization takes place; whichever occurs, the number of electrons generated is proportional to I^m:

$$N_{\mathrm{e}}(t) \propto I^m, \tag{9.4}$$

where m is of the order of 8. This is not the multiphoton law, but an empirical relationship from experiments and normally with femtosecond laser pulses, it is tunnel ionization [9]. Therefore, self-focusing and self-defocusing compete with each other. Initially, the focusing wavefront is obtained; however, because of the negative sign in the equation:

$$n = n_0 + n_2 I - \frac{4\pi e^2}{2m_{\mathrm{e}}\omega_0^2} k I^m, \tag{9.5}$$

this wavefront will spread out. That is, the wavefront is not curved so much anymore; it becomes flatter (Fig. 9.4) in such a way that when I is increased further in the propagation, the second and third terms in the right-hand side of (9.5) will be equal, that is,

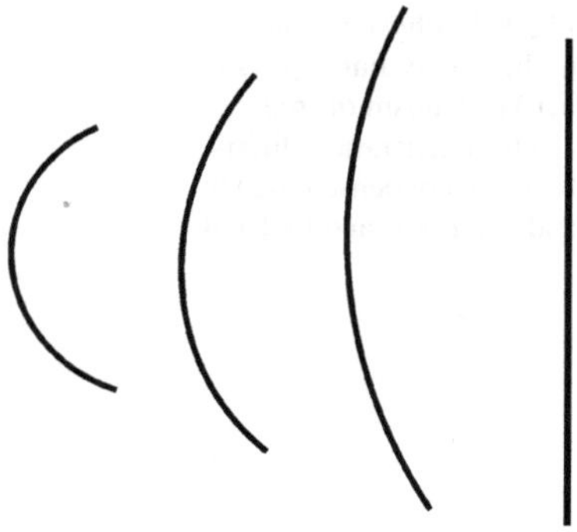

Fig. 9.4 An illustration of the balance between self-focusing by the neutrals and the self-defocusing by the free electrons in the plasma, resulting in the intensity clamping

$$n_2 I = \frac{4\pi e^2}{2m_e\omega_0^2} k I^m. \tag{9.6}$$

When they are equal, we are in the regime of balancing between self-focusing and self-defocusing by the self-created plasma. Hence, in this case, we can obtain a plane wavefront that diverges later.

This balance defines intensity clamping, which is one of the important phenomena associated with filamentation. When intensity clamping takes place, it means that at a particular position, the intensity is a maximum and later, the wavefront diverges out and is less intense. Hence, whenever self-focusing takes place, intensity increases first and later it does not increase anymore. Even if more energy is added, only the size increases [10].

Intensity clamping depends on I^m and m depends on the material. If gases are used then m depends on the ionization potential. In the case of solids or liquids, the value of m depends on the band gap between the valence band and the conduction band [11, 12].

When $\Delta n(I)$ of the neutral atom is equal to $\Delta n(I)$ of the free electron, that is, when

$$\Delta n_{\text{neutral}}(I) = \Delta n_{\text{free e}}(I), \tag{9.7}$$

there is a balance of self-focusing and de-focusing [13]. We can calculate the critical intensity or clamped intensity in the air by

$$n_2 I_{\text{crit}} \cong N_e(I)/2N_{\text{crit}}, \tag{9.8}$$

and it is of the order of 4–5 $\times 10^{13}$ Watts/cm^2 [13]. This result was based upon our experimental value $N_e(I)$ for the generation of free electrons:

$$N_{\text{crit}} = \varepsilon_0 m\omega^2/e^2 = 1.7 \times 10^{21}\text{cm}^{-3}(\text{at } 800\text{ nm}). \tag{9.9}$$

The term ionization (Fig. 9.5) refers to multiphoton absorption or tunnel ionization. In gases with femtosecond laser pulses, this type of transition into the ionization continuum is observed in which multiphoton absorption or tunneling takes place. In condensed matter, very often it is proven that ionization is due to the transition from the valence to conduction bands and following this transition, there would be some

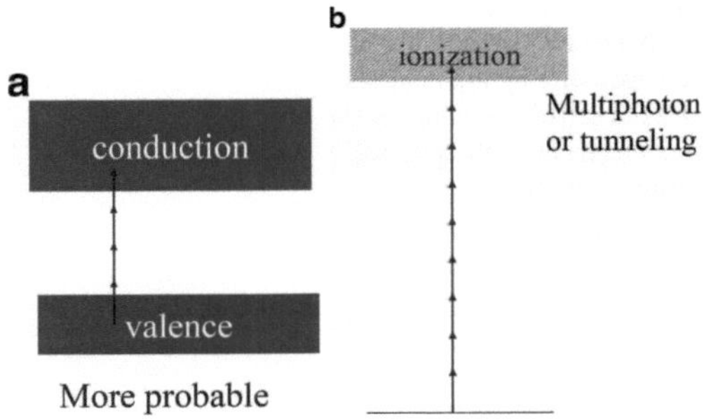

Fig. 9.5 Physical concept of multiphoton/tunnel physics: (**a**) Mechanism of free electron generation differs between condensed matters and (**b**) atoms and molecules

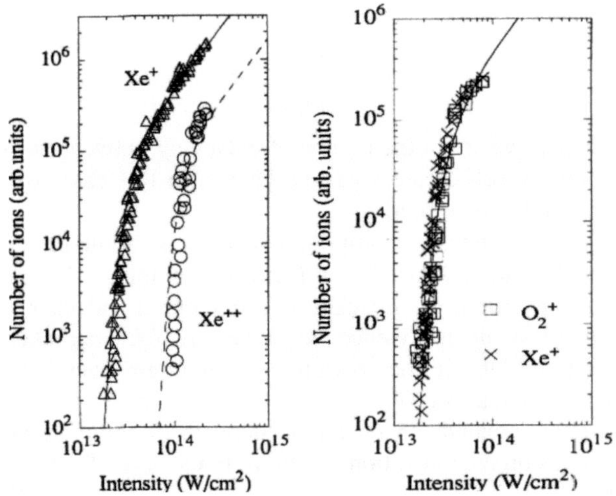

Fig. 9.6 Some results on tunnel ionization of xenon and oxygen using a CO_2 laser [14, 15]. First experimental observation of tunnel ionization using a CO_2 laser ($\lambda = 10.6\,\mu m$) showing the "Perfect" agreement with ADK theory

collisional ionization via inverse Bremsstrahlung. However, avalanche ionization might not be attained because there is not enough time for a sufficient number of collisions.

Today, tunnel ionization is popularly accepted as a physical process when using femtosecond laser pulses. However, it was not so in the beginning of the laser era when laser pulses are long (nanosecond). To prove the existence of tunnel ionization with long nanosecond type pulses, it is necessary to use long wavelength pulses.

In 1985, we were the first to observe tunnel ionization by a CO_2 laser pulse (Fig. 9.6), the wavelength of which is in the infrared, $10.6\,\mu m$ [14], but it had been proposed theoretically in the 1960s by Keldysh [16]. In the 1960s and 1970s, tunnel ionization was not accepted by the research community of multiphoton physics. Hence, the KFR theory was not used unfortunately, because at that time nobody believed that tunnel ionization could take place. Twenty-five years later, we have found that the theory is working. We were the first to prove this convincingly [14]. A discussion on multiphoton and tunnel ionization can be found in [9].

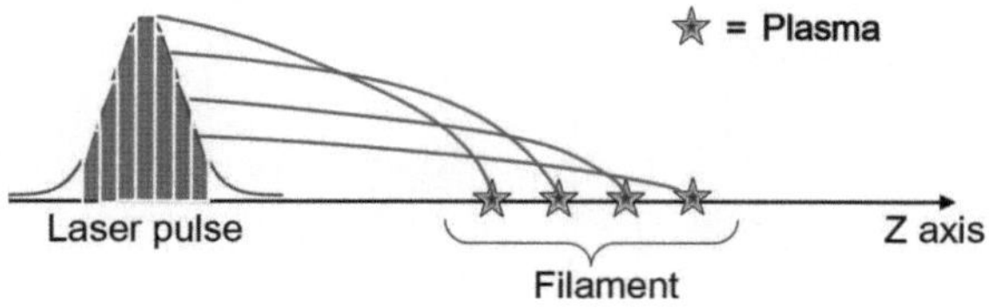

Fig. 9.7 Slice-by-slice self-focusing and intensity clamping. Critical intensity for clamping in air is $\approx 5 \times 10^{13}\,\mathrm{W/cm^2}$ [13]

Now let us come back to filamentation. When a slice of the laser pulse self-focuses to a certain point generating ionization, this position is called the self-focusing distance z_f,

$$z_f = \frac{0.367ka_0^2}{\left\{\left[\left(P/P_c\right)^{1/2} - 0.852\right]^2 - 0.0219\right\}^{1/2}}, \tag{9.10}$$

and it has been calculated already [5]. This distance is explicitly dependent on the peak power of the laser pulse of that slice. In this equation, a_0 is the radius of the laser beam and k is the wave vector of the electromagnetic field. Note that a_0 is defined as the radius at $1/e$ level of the peak intensity. That is to say, z_f is implicitly dependent on the intensity.

We assume that a pulse propagates in the direction z (Fig. 9.7). If we consider one central slice similar to a wavefront, the pulse will self-focus at some point and the higher the peak power, the shorter will be this distance in the direction z [see (9.10)]. Other slices at the front part of the pulse will self-focus further and further away in the direction z. Therefore, as long as all the slices have peak powers higher than the critical power for self-focusing, they will self-focus. Each slice will stop self-focusing when plasma is generated that defocuses the slice back into the main body of the pulse (reservoir). The series of continuous plasma dots constitute a perception of a thin line by us. This is thus called the filament. The back part of the pulse undergoes the same type of self-focusing, except when a slice self-focuses, it interacts with the plasma on its way before reaching its own self-focal point. When it interacts with the plasma, it becomes a diverging object. However, it should be noted that the complete pulse undergoes self-focusing and not only individual slices.

The plasma density in the filament is not very high; it is roughly of the order of $10^{15}\,\mathrm{cm}^3$ [17].

9.3 Theoretical Model and Simulation

A lot of calculations have been done by solving the Maxwell equation by inserting the nonlinear polarization and electron generation terms into it.

In the nonlinear optical model of femtosecond pulse interaction with air,

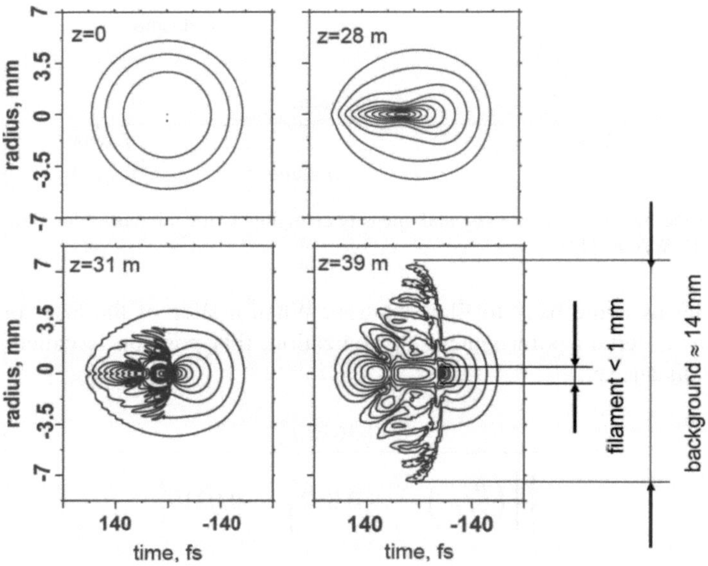

Fig. 9.8 Numerical solution of the filamentation nonlinear equation. The figures are intensity contours of the filamenting pulse at different positions of propagation in air [18]

$$2\mathbf{ik}\left(\frac{\partial}{\partial \mathbf{z}}+\frac{1}{\mathbf{v_g}}\frac{\partial}{\partial \mathbf{t}}\right)\mathbf{E}=\underbrace{\Delta\bot\mathbf{E}}-\mathbf{kk}''_{\omega}\frac{\partial^2\mathbf{E}}{\partial \mathbf{t}^2}+\mathbf{i}\left(\frac{1}{\mathbf{v_g}}\mathbf{k}''_{\omega}+\frac{1}{3}\mathbf{kk}'''_{\omega}\right)\frac{\partial^3\mathbf{E}}{\partial \mathbf{t}^3}$$
$$+\frac{2\mathbf{k}^2}{\mathbf{n}_0}\Delta\mathbf{n_k}(\mathbf{E})\mathbf{E}+\frac{2\mathbf{k}^2}{\mathbf{n}_0}\Delta\mathbf{n_p}(\mathbf{E})\mathbf{E}-\underbrace{\mathbf{ik}\alpha\mathbf{E}}, \qquad (9.11)$$

the material dispersion term is a nonlinear term. There are also terms for Kerr nonlinearity, plasma nonlinearity, and ionization losses [18].

Initially, when $z = 0$, the laser pulse has a uniform distribution that is Gaussian (Fig. 9.8). After a certain time and propagation distance, self-focusing can be observed. The back part of the outer pulse interacts with the plasma. At any instant, there is only one spot that is the self-focus and the pulse always changes its shape [18].

When the pulse changes it shape, it will always have only one hot spot. This has been demonstrated in calculations (Figs. 9.9 and 9.10).

Initially, a Gaussian pulse is observed and then the front part of the pulse becomes narrower. This means that the self-focusing moves towards the front which is the propagation direction and the peak intensity is almost the same over time as Fig. 9.10 shows.

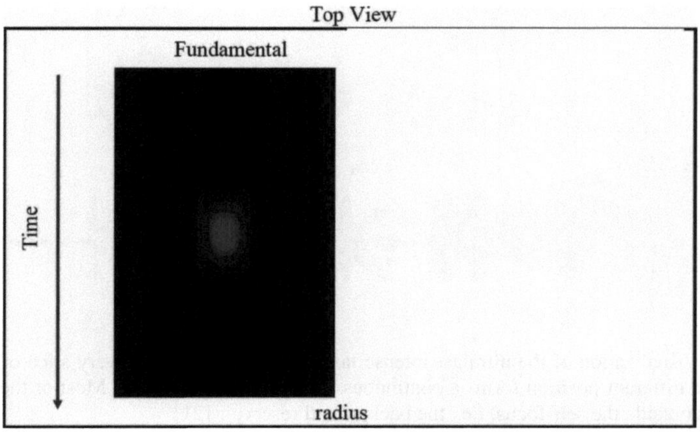

Fig. 9.9 2D grayscale representation of the initial pulse's intensity distribution [19]

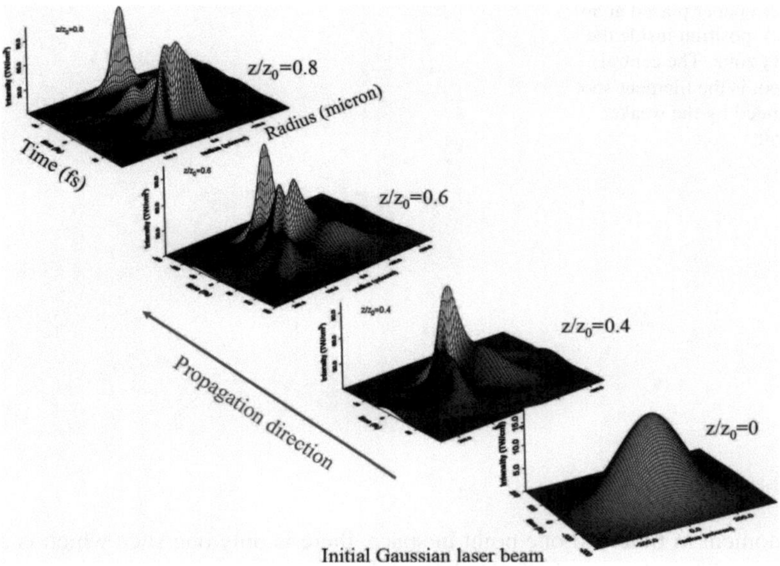

Fig. 9.10 Simulation of filamentation similar to Fig. 9.9 but in 3D-plots. The vertical axis is the intensity [20]: $I_0 = 10^{13}\,\mathrm{W/cm^2}$, $\tau_0 = 150\,\mathrm{fs}$ (FWHM), $w_0 = 250\,\mu\mathrm{m}(1/e^2)$, $z_0 = \pi w_0^2/\lambda_0$

9.4 Background or Energy Reservoir

The concept of background reservoir is very important. Figure 9.11 shows this schematically. The self-foci are generated because of the self-focusing of the slices from the outside. This means that one slice comes after the other and all the outside energy that comes into focus makes up the background reservoir. At any

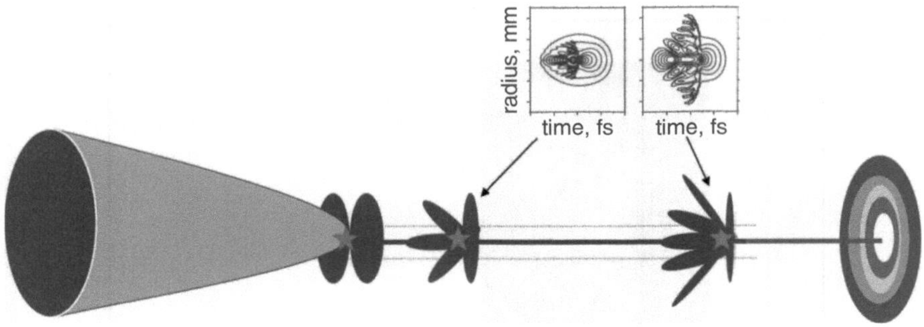

Fig. 9.11 Visualization of the ultrafast intense laser propagation in air. Every slice of light self-focuses at a different position forms a continuous series of foci (filament). Most of the energy of the pulse is outside the self-focus, i.e., the background reservoir [21]

Fig. 9.12 A real burn pattern on a burn paper placed at an arbitrary position inside the filament zone. The central dark spot is the filament spot surrounded by the weaker reservoir

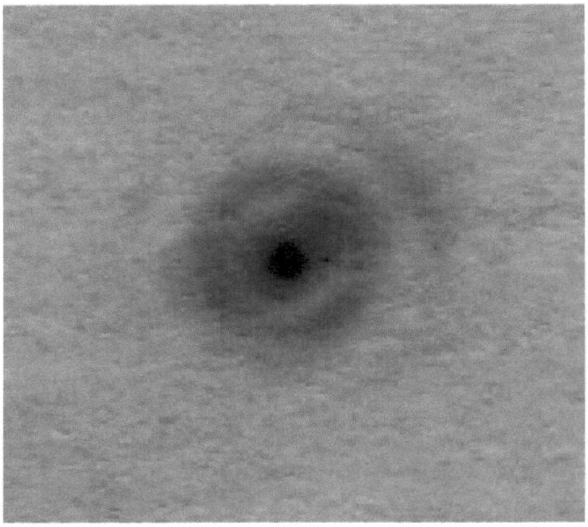

one moment in time and one point in space, there is only one slice which is at the self-focus. The others are in the background reservoir. Figure 9.12 shows a real burn pattern on a burn paper placed at an arbitrary position inside the filament zone. The central dark spot is the filament spot surrounded by the weaker reservoir. If we carry out the Fourier transform of the field distribution in Fig. 9.8 or Fig. 9.10, a white light chirped laser pulse is obtained with a very broad frequency spectrum.

We conducted a simple experiment to show the white light laser pulse (Figs. 9.13 and 9.14). The laser was shot into a long corridor (26 m length), which had a moveable white screen. A camera was placed so that it is focused on the screen. The camera and screen were moved along the pulse propagation axis and an image was captured using the camera every 0.5 m. The obtained images are shown at four different positions of the propagation [2].

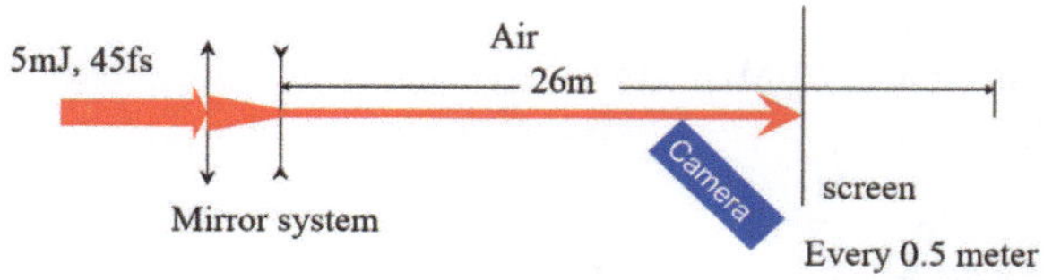

Fig. 9.13 An experiment showing the evolution of the transverse pattern of an intense fs laser pulse propagating in air

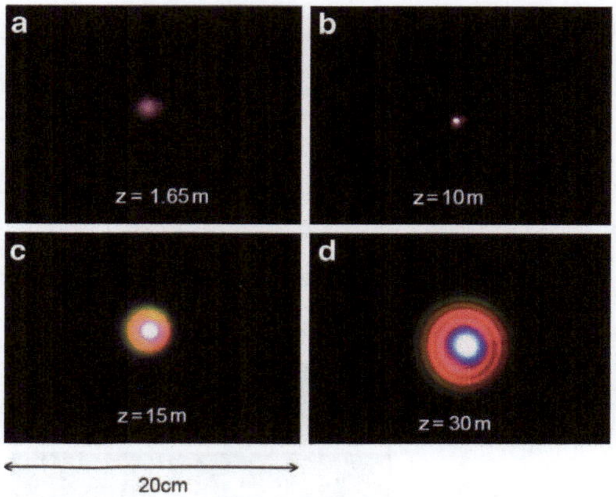

Fig. 9.14 Single filamentation [2]

Essentially, the background reservoir can be blocked using a pinhole. Thus, the concept of filamentation cannot be explained as a filament (a sharp needle) that can pierce through the material. That is to say, if the background reservoir is responsible for the hotspots, there will be no more energy at the output of the pinhole when the background is blocked (Fig. 9.14). Liu et al. [22, 23] carried out the experiment to investigate this problem (Fig. 9.15). A laser beam was passed through a pinhole. They could capture images of the filament using an ICCD camera. If the pinhole is not used, a filament will be obtained (Fig. 9.16). This is because there is a large amount of energy outside which is the background energy. When they changed the size of the pinhole, the length of the filament behind the pinhole varied depending on the size of the pinhole. When the pinhole is too small, there will be no filament behind the pinhole.

The experimental images agreed with the simulation results as seen in Fig. 9.16.

This experimental verification of the background reservoir was important as the reservoir is a very important concept with regard to filamentation. This background reservoir and self-focusing can be found if a slice-by-slice self-focusing model is used. This idea of background energy reservoir was theoretically proposed first by Moloney's group at the University of Arizona [7, 24] and was expanded later by Kandidov's group at Moscow State University [18].

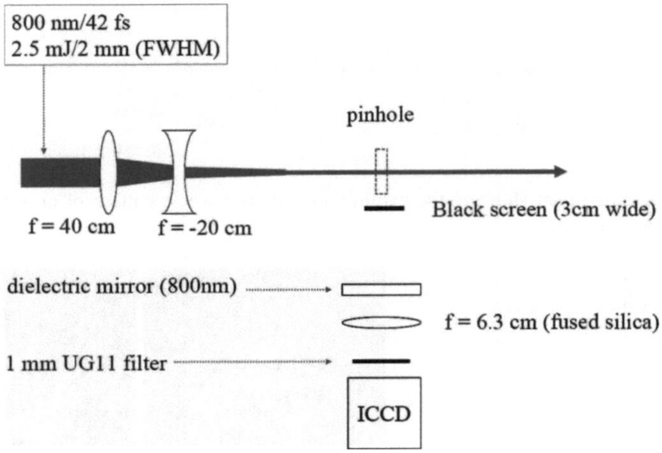

Fig. 9.15 Experimental setup to study the concept of energy reservoir [22, 23]

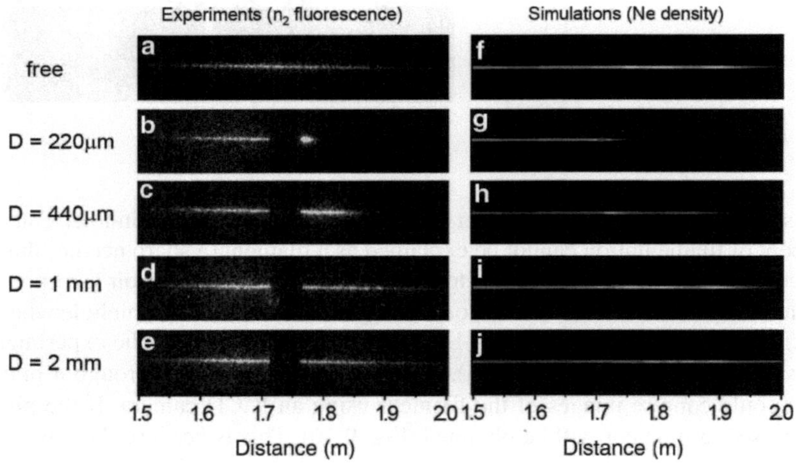

Fig. 9.16 Comparison of experimental and simulation results on the concept of reservoir. *D* is the diameter of the pinhole [23]

9.5 Extraordinary Properties of Filaments

We can summarize the fundamental understanding of the filamentation process. A filament in an optical medium has some extraordinary optical properties. The first is the ability to cause self-remote projections at high intensity into the atmosphere. The second is self-avoidance of breakdown in air, which implies that clean fluorescence can be obtained. The third and fourth are self-stabilization at high intensity which refers to intensity clamping and self-transformation into the white light short laser pulse implying a supercontinuum. The fifth property is self-pulse compression down to a single cycle level. The sixth is self-spatial filtering which implies that a single spatial mode can be obtained in the filament core. The last is self-group phase locking of other pulses inside a filament which is due to the high clamp intensity, particularly in air. This high intensity would influence cross-phased modulation or cross-phased interaction with any other laser pulse or electromagnetic pulse inside the filament and would lead to dramatic changes.

9.6 Long-Distance Propagation in Air

One of the most recent developments in self-remote projection is to shoot the laser pulse into air using the Teramobile, which is a joint system owned by scientists from Germany and France (Fig. 9.17). In one application, they have obtained images of scattering from a cloud layer approximately 1–2 km above the earth. Self-avoidance of breakdown in air to obtain clean fluorescence is very important in such an application.

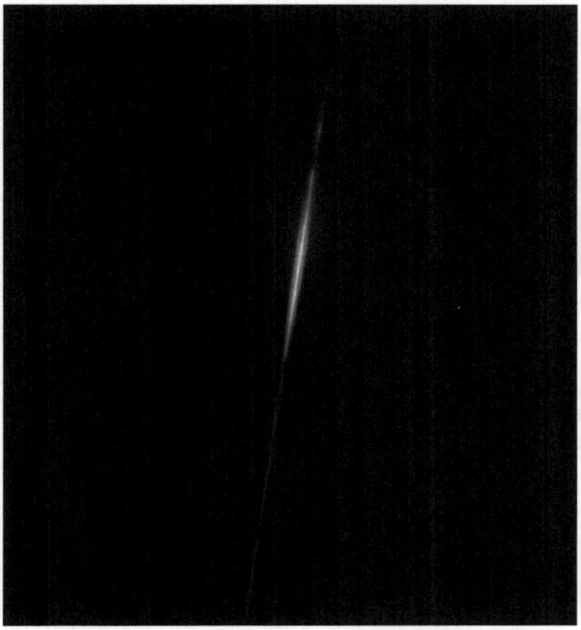

Fig. 9.17 An image of the scattering from a cloud layer approximately 1–2 km above the earth when a 5TW femtosecond laser pulse was shot into air early in the morning in Jena, Germany. Courtesy of Roland Sauerbrey Teramobile group (France, Germany)

9.7 Clean Fluorescence

We wanted to conduct some experiments inside the laboratory rather than in a long corridor, so we used a lens of long focal length (5 m) for the focusing. The filament can be seen in the image of Fig. 9.18. There are some dark sections in this long filament, which is an indication of refocusing. Finally, this is not breakdown, but the fluorescence coming from nitrogen.

The Moscow State University group (Figs. 9.19 and 9.20) has done calculations showing the same phenomenon. That is, when a transform-limited pulse is propagated through air a long fluence distribution can be obtained and if the pulse is short, then longer sections can be obtained.

In laser-induced breakdown, a spark is obtained; even if an energy of 700 mJ is used in 10-ns pulses, pure white scattered light will be generated from the plasma (Fig. 9.21).

Figure 9.22 reproduces two spectra obtained with different pulse lengths. A distinct difference can be observed between them. There are atomic lines from neutral

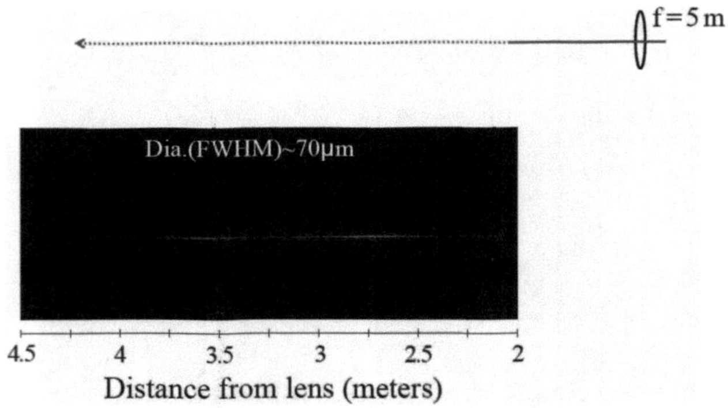

Fig. 9.18 A filament in air with refocusing. Fluorescence from N_2 and N_2^+ is observed. Ti-sapphire laser, 800 nm, 45 fs, 10 Hz, 13 mJ/pulse, UG11 filter, ICCD gain = 200, 10 ns gate width, 100 shots averaged

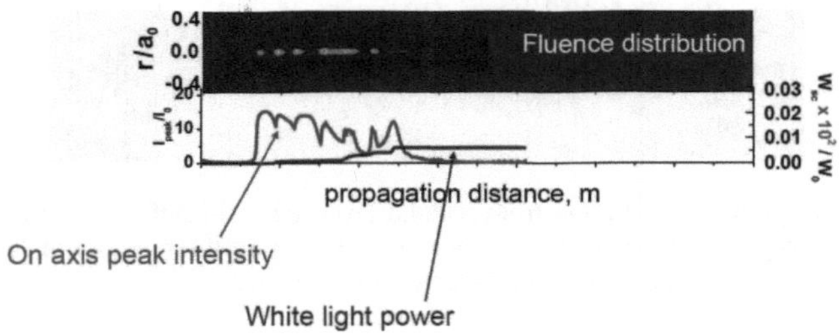

Fig. 9.19 Simulated fluence distribution during filamentation of a 42 fs transform-limited laser pulse in air. Pules energy: Ti-S laser, 60 mJ, 42 fs, $I_0 \sim 10^{13}\,\mathrm{W/cm^2}$ (transform-limited pulse) [25]

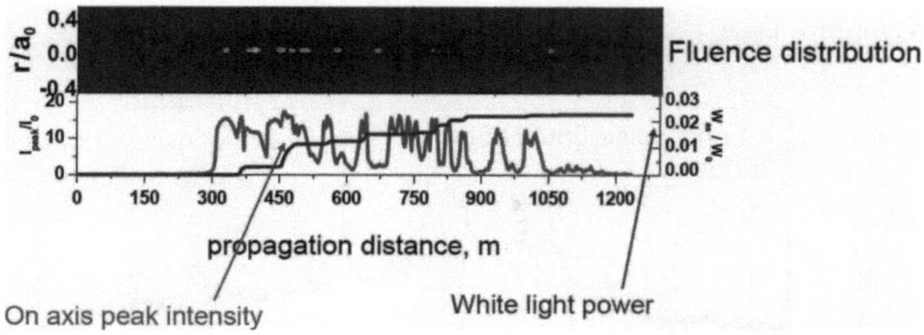

Fig. 9.20 Same as Fig. 9.19 using a 1,200 fs negatively chirped pulse. Pulse energy: Ti-S laser, 60 mJ, 1,200 fs, negative chirp, $I_0 = 10^{13}\,\mathrm{W/cm^2}$ [25]. Note the longer filament than in Fig. 9.19

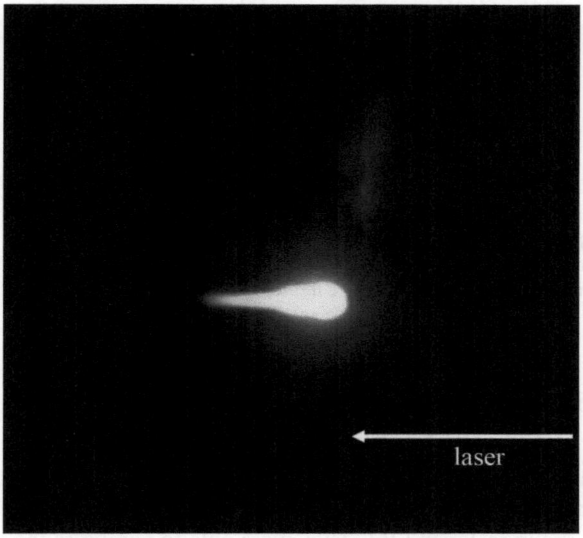

Fig. 9.21 A breakdown spark in air induced by a focused 10-ns, 700-mJ pulse. $\lambda = 532\,\mathrm{nm}$, $700\,\mathrm{mJ}/10\,\mathrm{ns}$, $f = 50\,\mathrm{cm}$

atoms, N+, and O+ and the white light plasma from air when a long pulse is used. But for the short pulse, molecular lines are obtained from N_2^+ and N_2. In this latter case, since there is no measurable plasma white light, we call this "clean fluorescence," which means that if we generate any filament in air, the intensity will be very high and it will ionize and dissociate any molecule inside the filament core. Some of them will undergo or emit clean fluorescence. No breakdown can occur because the pulse is too short to allow any collisional absorption of the light through inverse Bremsstrahlung.

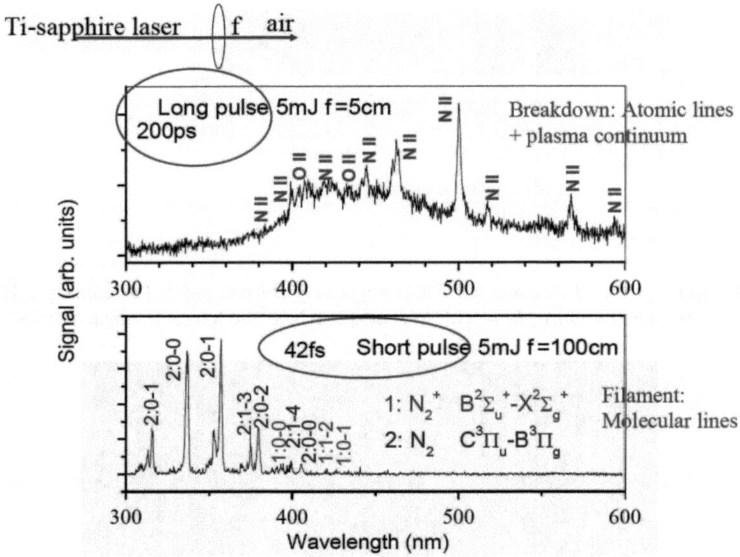

Fig. 9.22 Comparison of the spectrum of the breakdown in air induced by a long (200-ps) Ti-sapphire laser pulse and the spectrum of the filament induced by a short (42 fs) pulse. A clean fluorescence emission spectrum in the lower figure shows that no breakdown occurs in air with fs laser pulses

9.8 Self-Pulse Compression

During filamentation, the pulse becomes narrower and narrower in time; i.e., self-compression occurs. Initially, the self-compression of the pulse was carried out by Midorikawa's group in Japan [26]. They were able to compress the pulse up to ten times. Recently, in some studies, the pulse has been shortened to a few cycles of about 6 fs [27].

Shortening the pulses to even single cycle pulses is the current challenge for experimentalists so that these types of pulses can be used to create high-order harmonics. These high-order harmonics could be used to create single intense attosecond pulses (Fig. 9.23).

Researchers have attempted to obtain a single short pulse as noted earlier in the discussion with Fig. 9.10, by using a diaphragm or a pinhole. But some amount of background will be associated with it, which cannot be eliminated. So a way to clean up the pulse is needed.

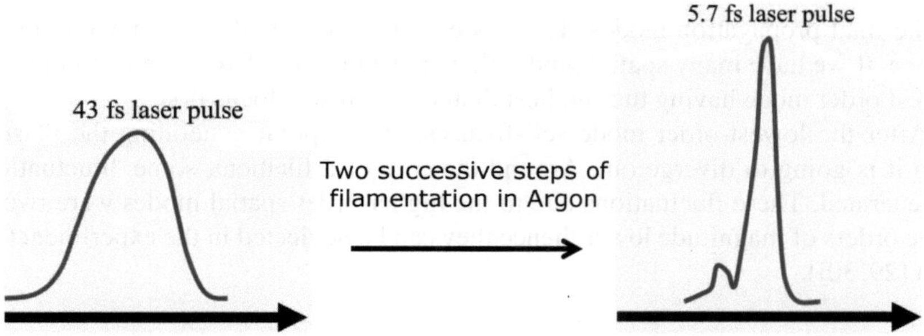

Fig. 9.23 Self-pulse compression during filamentation. By the compression, it became possible to shorten the pulse down to 6 fs (few cycle laser pulses) from 43 fs by two successive steps of filamentation in Ar [27]. The first detailed experiment carried out by Midorikawa's group in 2000 demonstrated 30-fs pulses were obtained by the ten time compression [26]

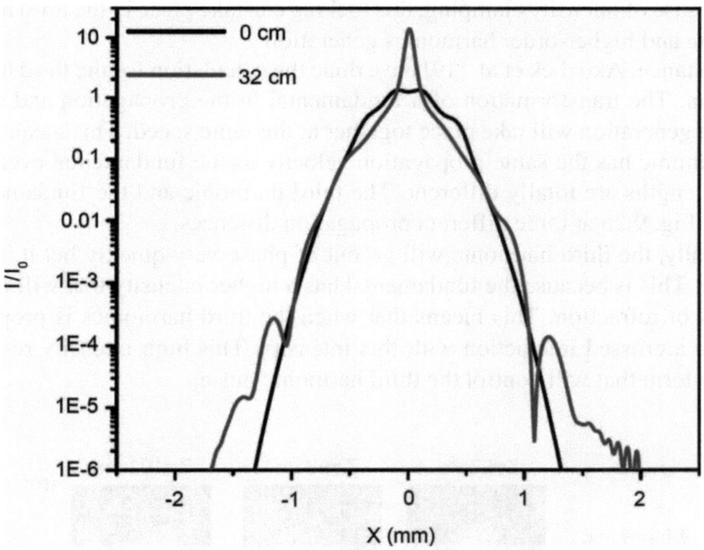

Fig. 9.24 A simulation of self-spatial filtering obtained by solving a nonlinear Schroedinger equation and cw propagation in air: Gaussian (diam. $= 1\,\text{mm}\ (1/e^2)) + 10$ random perturbations (200 μm diam.); self-focus at 32 cm, $I = 5 \times 10^{13}\,\text{W/cm}^2$, clamped intensity in air)

9.9 Self-Spatial Filtering

During filamentation, a single mode can be obtained in the core; the reason for this was analyzed by Liu in simulations and confirmed through an experiment [28]. For a simulation (Fig. 9.24), it can be assumed that there is a wave front with several special modes that are propagating in air. The wavefront is allowed to propagate till the self-focusing distance of 32 cm. The simulation result shows the distribution of the final propagation modes. The mode at the center is the fundamental mode. Hence, if we have many spatial modes that are trying to self-focus at one point, the lowest order mode having the smallest diameter will self-focus first.

After the lowest-order mode self-focuses, at one point generating the plasma, then it is going to diverge out. Around the core of a filament, some "fluctuation" is generated. These fluctuations due to the higher-order spatial modes were two or three orders of magnitude lower; hence they can be neglected in the experiment (see also [29, 30]).

9.10 Self-Group Phase Locking

In another experiment, the last optical property, self-group phase locking, was examined. Because of intensity clamping, this locking can take place in the third harmonic generation and higher-order harmonics generation.

For instance, Aközbek et al. [19] have done the calculation for the third harmonic generation. The transformation of a fundamental in the propagation and the third harmonic generation will take place together at the same speed. This means that the third harmonic has the same propagation velocity as the fundamental even though the wavelengths are totally different. The third harmonic and the fundamental are shown in Fig. 9.25 at three different propagation distances.

Normally, the third harmonic will go out of phase very quickly but it is not the case here. This is because the fundamental has a higher intensity that will influence the index of refraction. This means that when the third harmonics is propagating, it can see a crossed interaction with this intensity. This high-intensity results in a nonlinear term that will control the third harmonic pulse.

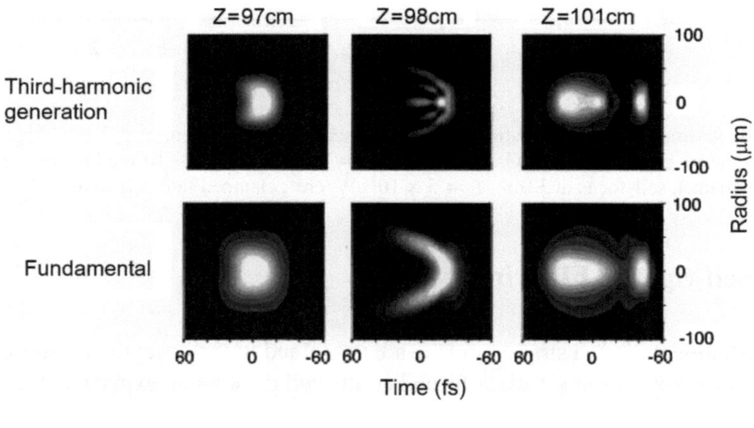

Fig. 9.25 Self-group phase locking: the fundamental and the third harmonic pulses propagate together at the same speed [19]

The calculation by Aközbek et al. [19] is summarized as follows: Qualitatively, this is applicable to all parametric processes as long as there is cross field influence of the pump. The pump field is considerably stronger than the new field. Under typical conditions, $N_{\text{neutral air}} \sim 3 \times 10^{19}\,\text{cm}^{-3}$ and $N_\text{e} \sim 10^{15}\,\text{cm}^{-3}$, Plasma effect, and the following equations are obtained:

$$\begin{aligned} k_{\text{eff}}^{\omega} &= k_\omega + \Delta k_\omega + (n_{\omega 0} + \Delta n_\omega)k_0 = n_{\text{eff}}^{\omega} k_0 \\ &= \left(n_{\omega 0} + n_2\left(A_\omega^2 - A_\omega A_{3\omega} + 2A_{3\omega}^2\right) - \tfrac{2\pi e^2}{k_0^2 mc^2} N_\text{e}\right) k_0 \\ &\cong \left(n_{\omega 0} + n_2\left(A_\omega^2\right)\right) k_0 \text{ (plasma effect weak, } A_\omega \gg A_{3\omega}), \end{aligned} \tag{9.12}$$

$$\therefore n_\omega \cong n_{\omega 0} + n_2\left(A_\omega^2\right) = n_{\omega 0} + n_2 I_\omega, \tag{9.13}$$

$$\begin{aligned} k_{\text{eff}}^{3\omega} &= k_\omega + \Delta k_\omega = 3(n_{3\omega 0} + \Delta n_{3\omega})k_0 = 3k_0 n_{\text{eff}}^{3\omega} = \\ &= \left(n_{3\omega 0} + n_2\left(2A_\omega^2 - \frac{1}{3}\frac{A_\omega^3}{A_{3\omega}} + A_{3\omega}^2\right) - \frac{2\pi e^2}{9k_0^2 mc^2} N_\text{e}\right) 3k_0 \\ &\cong \left(n_{3\omega 0} + n_2\left(2A_\omega^2 - \frac{1}{3}\frac{A_\omega^3}{A_{3\omega}} + A_{3\omega}^2\right)\right) 3k_0 \quad \text{(plasma effect weak)}, \end{aligned} \tag{9.14}$$

$$\begin{aligned} \therefore n_{3\omega} &\cong n_{3\omega 0} + n_2\left(2A_\omega^2 - \frac{1}{3}\frac{A_\omega^3}{A_{3\omega}} + A_{3\omega}^2\right) \\ &= n_{3\omega 0} + n_2\left(2I_\omega - \frac{1}{3}\frac{A_\omega^3}{A_{3\omega}} + I_{3\omega}\right). \end{aligned} \tag{9.15}$$

When the refractive indices are represented as:

$$n_\omega = n_{0\omega} + \Delta n_\omega, \tag{9.16}$$

$$n_{3\omega} = n_{0.3\omega} + \Delta n_{3\omega}, \tag{9.17}$$

the plasma effect in the change in the refractive index Δn_ω,

$$\Delta n_\omega = \Delta n_\omega(\text{netural air}) + \Delta n_\omega(\text{weak plasma in filament}), \tag{9.18}$$

can be neglected, and

$$\Delta n_\omega \cong \Delta n_\omega(\text{natural air}) \cong n_2 I_\omega(\text{clamped}) \tag{9.19}$$

is obtained. By assuming that $I_\omega(\text{clamped}) \sim 5 \times 10^{13}\,\text{W/cm}^2$ and $I_{3\omega} \sim 10^{12}\,\text{W/cm}^2$, the calculation and analysis were done in such a way that n_ω in air, including the nonlinear term, was approximately 1.00029. The value of $n_{3\omega}$ was also 1.00029.

$$n_\omega = n_{0\omega} + \Delta n_\omega(\text{nonlinear}) = \ldots. \cong 1.00029004 \tag{9.20}$$

$$n_{3\omega} = n_{0,3\omega} + \Delta n_{3\omega}(\text{nonlinear}) = \ldots. \cong 1.00029174. \tag{9.21}$$

This means that during the self-group phase locking, the indices of refraction of the fundamental and the third harmonic are almost the same; i.e., $n_\omega = n_{3\omega}$, hence, the speeds of propagation are the same.

This is because there is a nonlinear term in the expression for $n_{3\omega}$ which involves the amplitude A_ω of the pump pulse. This term is considered to control everything. This is self group phase locking. Everything else will be similar if any pulse other than the third harmonic is propagating inside the filament. In such a case, this type of phase locking would still be valid.

9.11 Nonlinear Optics Inside the Filament Core

The above experiments and calculations have demonstrated the unique optical properties of the filament, including self-projection very far into the atmosphere by self-avoidance of breakdown in air resulting in clean fluorescence; self-stabilization of high intensity; self-compression into almost a single cycle; self-spatial filtering to obtain a single mode in the core; and self-group phase locking of other pulses. It should be noted that all these processes take place inside the core of the filament and not outside.

In order to take advantage of the filament core, some interaction should be carried out inside it. But if a linear interaction is carried out with the filament core, signals with high quality are not obtained because interaction takes place with the reservoir also. Therefore, in order to obtain good quality in the central core, some nonlinear sampling has to be carried out. Luckily, such nonlinear sampling can be achieved easily because the intensity inside the central core is much higher than that in the reservoir. This nonlinear sampling results in a stable signal because of intensity clamping. In intensity clamping, a chemical or physical reaction has to be carried out inside the filament core. Because of the stable intensity, the interaction rate will be constant. This means that a stable output signal is obtained. Stable self-spatial filtering renders excellent beam quality in terms of the third harmonic or four-wave mixing because of the high clamping intensity.

Normally, nonlinear and nonresonant processes are unstable, fluctuating, and inefficient but these do not happen inside the filament core. We call this filamentation nonlinear optics [29].

9.12 Four-Wave Mixing Inside the Filament Core

An example of this filamentation nonlinear optics has been studied by Théberge et al. [30] (Fig. 9.26). Four-wave mixing was conducted inside the filament in air and argon. A beam splitter was used to separate the pulse into two parts; one part propagated freely and the other was used to pump an OPA, thereby resulting in infrared pulses tunable from 1.2 to 2.4 μm. This latter was continuously tunable.

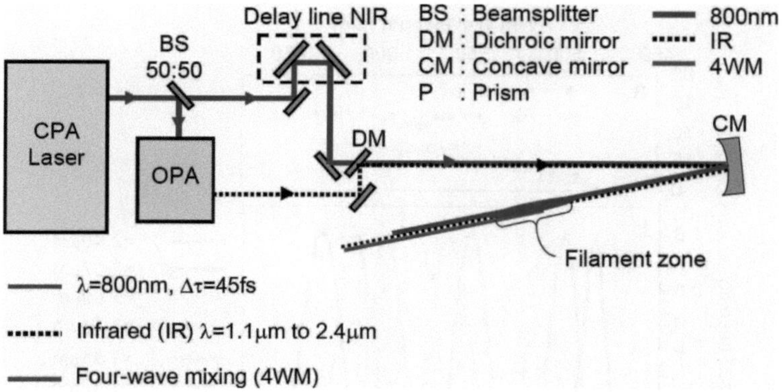

Fig. 9.26 Experimental setup for the four-wave mixing occurring inside the filament core, resulting in tunable short visible pulses [30]

The pump pulse duration was 45 fs. Here, the OPA pulse duration was approximately 30 fs. These were mixed together and passed through the dichroic mirror. The beam was then focused by the concave mirror into the filament zone. One pulse was superimposed onto the other and then four-wave mixing was done as $\omega_{4\,\mathrm{WM}} = 2\omega_{800\,\mathrm{nm}} - \omega_{\mathrm{IR}}$. The visible pulse obtained was passed through the prism compressor and then auto correlated and the spectrometer was used to measure the spectra.

With our experimental setup, spectra obtained were continuously tunable from approximately 475 to 650 nm (Fig. 9.27). The conversion efficiency was defined as the energy of the four-wave mixing (4WM) pulse divided by the infrared energy and could be as high as ~30% in air and ~60% in argon (1.5 atm). This is very efficient; if we estimate the conversion efficiency without the filament, it is only in the range of 10^{-5}.

We have also done both experiments and calculations regarding the self-compression property and obtained good agreement between them (Fig. 9.28).

Figure 9.29 shows our results to confirm the self-stability of the laser pulse in the filament core generated with our setup. We obtained a root mean square fluctuation of 4.8% theoretically and 5.2% experimentally of the 4WM pulse when there is no filamentation:

$$\mathrm{RMS}_{4\mathrm{WM}} \approx 2 \times \mathrm{RMS}_{800\mathrm{nm}} + \mathrm{RMS}_{\mathrm{IR}} = 4.8\%. \tag{9.22}$$

This is in very good agreement with the prediction of statistics. However, when there is filamentation, the root mean square fluctuation of the 4WM pulse $\mathrm{RMS}_{4\mathrm{WM}}$ becomes 1.8%:

$$\mathrm{RMS}_{4\mathrm{WM}} \approx 2 \times \mathrm{RMS}_{800\mathrm{nm}} + \mathrm{RMS}_{\mathrm{IR}} = 1.8\%. \tag{9.23}$$

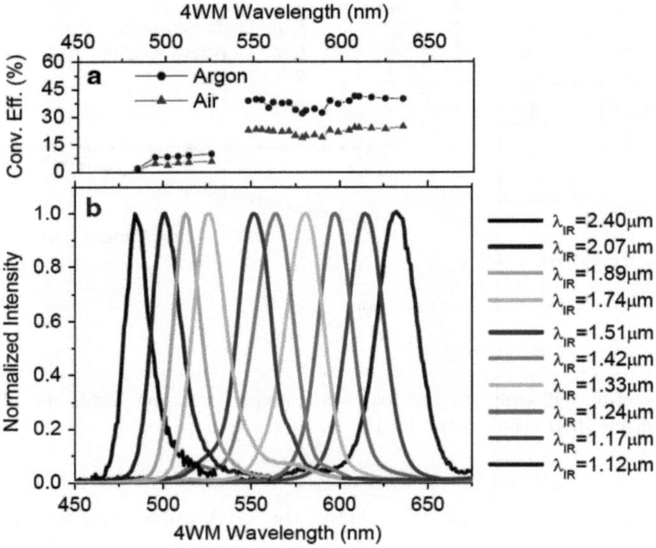

Fig. 9.27 Tunable output in the visible region as a result of four-wave mixing inside the filament core in air and in argon [31]

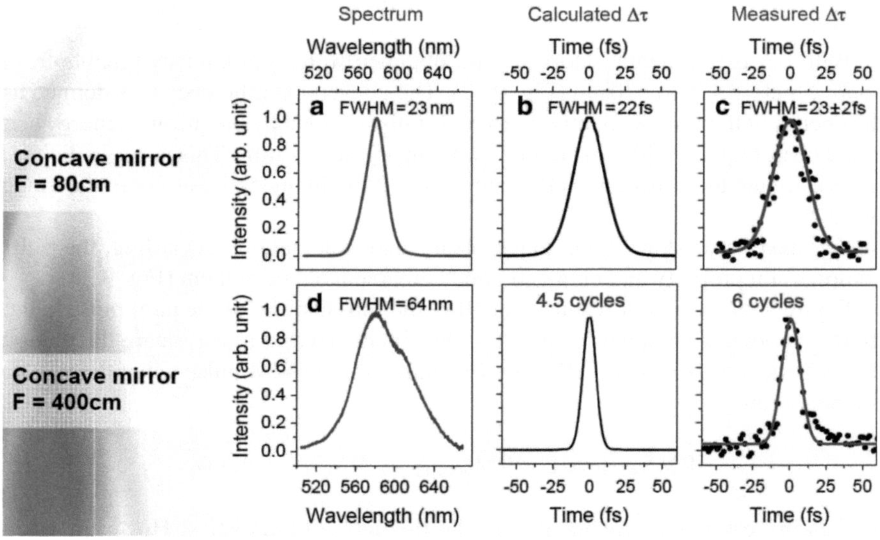

Fig. 9.28 Self-compression of the pump pulse leading to compression of the 4WM pulse [17]. Comparison between experimental and theoretical results

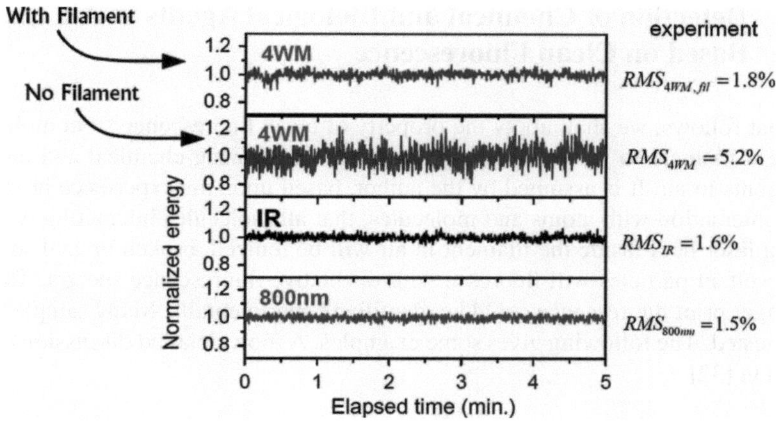

Fig. 9.29 The self-stabilization of the pump pulse inside the filament core due to intensity clamping [17]

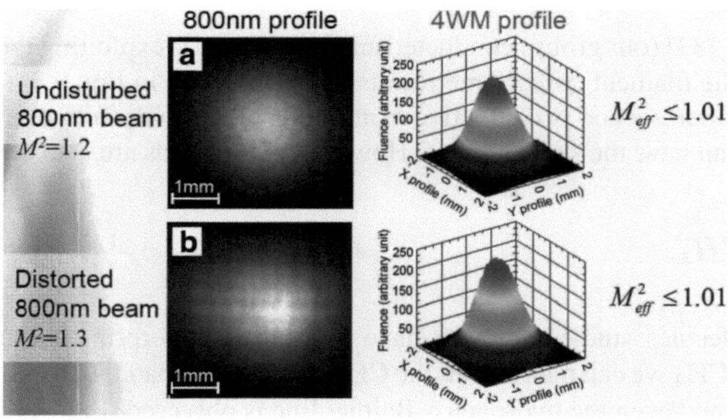

Fig. 9.30 Self-spatial filtering of the 4WM pulse [17]

A calculation using $RMS_{IR} = 1.6\%$ and $RMS_{4WM} = 1.8\%$ shows that the root mean square fluctuation of the filament core of fundamental pulse is less than 0.1%, i.e., self-stabilization. This is because of intensity clamping.

We also have observed self-spatial-filtering during 4WM (Fig. 9.30). We started with a laser pulse distribution, where $M^2 = 1.2$. We generated the four-wave mixing pulse. We modified this spatial distribution by inserting a square opening and generating an interference pattern so that $M^2 = 1.3$. Both of the cases result in a 4WM pulse with M^2 less than 1.01.

We concluded that generating pulses at high efficiency with wavelength tunability has been achieved. We confirmed that it was possible to generate pulses of a few cycles and that there was self-stabilization of energy. Finally, we confirmed excellent beam quality for the 4WM pulse because of the self-spatial-filtering property.

9.13 Detection of Chemical and Biological Agents in Air Based on Clean Fluorescence

In what follows, we shall apply the property of clean fluorescence from molecules inside a filament to study the feasibility of remote sensing chemical and biological agents in air. It is assumed by the author, based upon his experience in intense laser interaction with atoms and molecules, that all molecules interacting with the strong laser field inside the filament in air will be ionized, broken up and some of the resultant particles will fluoresce with distinctive fluorescence spectra. The latter finger print fluorescence could be identified experimentally. Many samples have been tested. The following gives some examples. A more detailed discussion can be found in [32].

9.13.1 Halocarbons

Gravel et al. [33] (our group) conducted an experiment by exploiting the high intensity inside the filament core. Some results are reproduced in Fig. 9.31.This kind of fingerprint fluorescence is distinctive from molecule to molecule, even though the molecules can have the same spectra. However, other bands are also observed.

9.13.2 CH_4

All the molecules studied thus far have distinctive fingerprint fluorescence. For instance, in CH_4 we can measure all the CH bands. The C-band, B-band, and A-band can be seen and even the hydrogen α Balmer line is observed (Fig. 9.32).

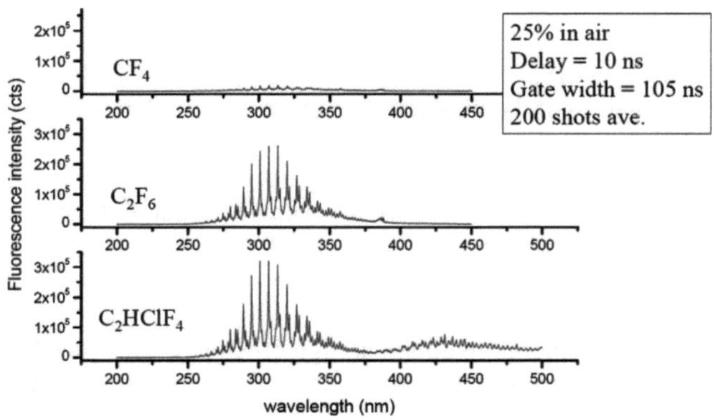

Fig. 9.31 Fluorescence spectra from some halocarbons inside the filament in air. Only CF_2 Fluorescence was detected by temporary gating out N_2 fluorescence [33]

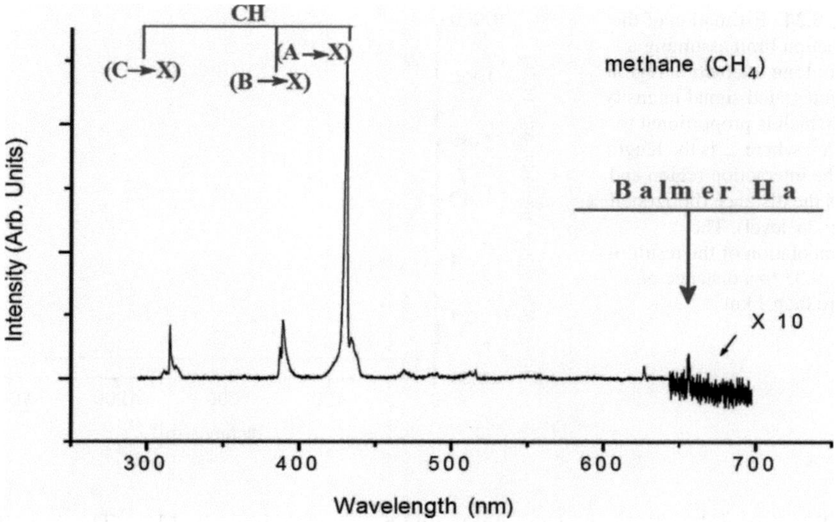

Fig. 9.32 Finger-print fluorescence from CH_4. The spectrum was obtained with a Ti:sapphire laser (800 nm, 42 fs). The laser intensity is $2 \times 10^{14}\,\mathrm{W/cm^2}$ and the gaseous pressure is 3 torr

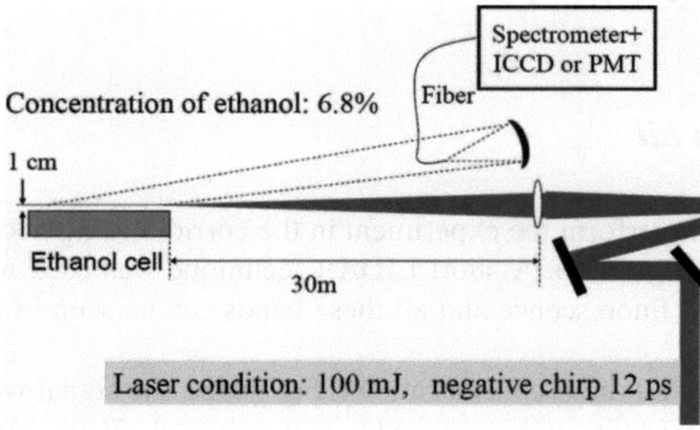

Fig. 9.33 Measurement of the fluorescence of ethanol molecules inside a filament in air from a distance of 30m using the technique of LIDAR. Back-scattered fluorescence is detected [34]

9.13.3 Ethanol Vapor

Using ethanol, we attempted a long-distance experiment. The laser (Fig. 9.33) was shot using a focusing telescope, which controlled the filament intensity. The filament was 1 cm above the ethanol cell inside the corridor. A large amount of vapor was generated. We estimated the concentration to be 6.8%, and we used a LIDAR technique to collect all the generated fluorescence that was directed into the spectrometer from a distance of 30 m.

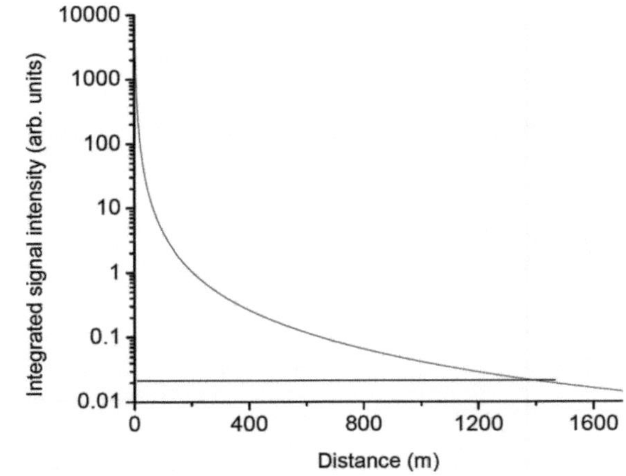

Fig. 9.34 Estimation of the detection limit assuming a 20-m long interaction region by integrated signal intensity I_S, which is proportional to L/R^2, where L is the length of the interaction region and R is the distance (horizontal line: 3σ level). The extrapolation of the result in Fig. 9.33 to a distance of more than 1 km

We extrapolated our measurements down to approximately 1 km. The σ is the standard deviation of the background noise level in our experiment. This type of signal can be detected at approximately 1.3 km, where the signal intensities are the same as the 3σ noise level (Fig. 9.34).

9.13.4 CH_4 in air

When we could not perform the experiment in the corridor, we performed it inside a pipe in the lab (Fig. 9.35). A short LIDAR technique was used to measure the maximum scattered fluorescence and all these bands can be seen in Figs. 9.36 and 9.37.

By extrapolating all these concentrations (Fig. 9.38), the signal was obtained as a function of concentration and it reduced to approximately 50 ppm inside the lab.

9.13.5 Bio-agents: Egg White and Yeast Powders

Figure 9.39 is the setup we used for measurements of a biological model of egg white and yeast. The laser is shot onto the powdered surface, and breakdown takes place inside the material; this is called femtosecond breakdown and the phenomenon is femtosecond laser-induced breakdown spectroscopy, or FIBS [31, 35]. This type of breakdown is very different from the nanosecond-induced breakdown. The former is not very strong; we can distinguish the signal soon after the laser pulse is gone, whereas when a nanosecond laser pulse is used, to distinguish the signal, microsecond or millisecond delays have to be used before distinctive signals are observed.

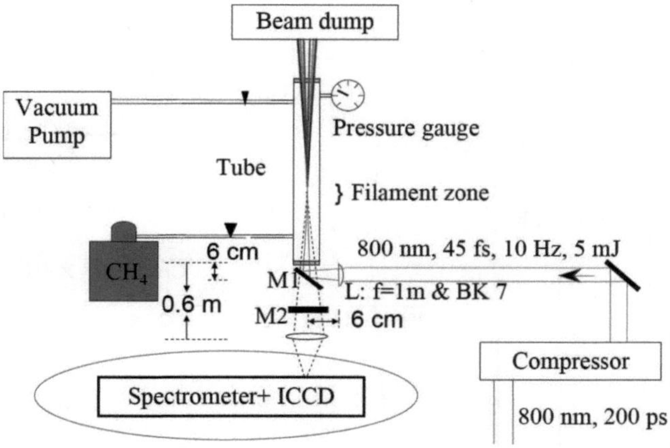

Fig. 9.35 Measurement of CH_4 fluorescence mixed with air inside a pipe. Experimental setup for the back-scattered fluorescence detection of CH_4: M1, M2:dielectric mirrors (diam. 76.2 mm), with high reflectivity at 800 nm and high transmission for UV light [36]

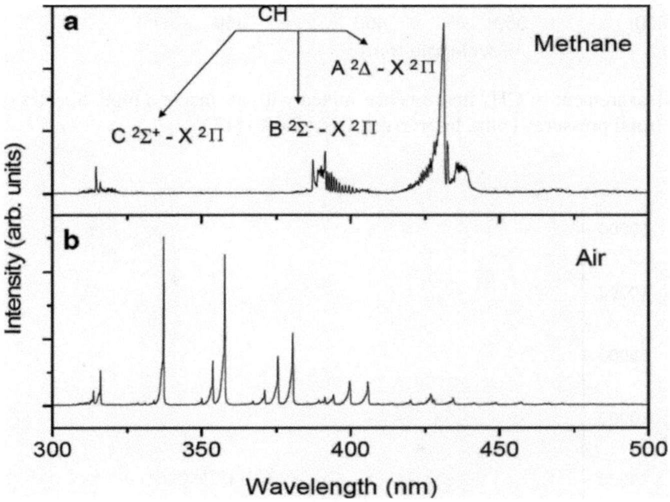

Fig. 9.36 Measurement of CH_4 fluorescence mixed with air inside a pipe. Fluorescence spectra (**a**) Methane at 20 Torr. (**b**) Air at atmospheric pressure. Laser: 45 fs and 5 mJ [36]

We placed the filament onto the sample, which then undergoes an explosion; the detailed theory behind this explosion is not known yet.

Fluorescence is emitted (Fig. 9.40) and the fluorescence signal is obtained by integrating all the light that returns. In this case, no discernable signal can be found because of the strong plasma light.

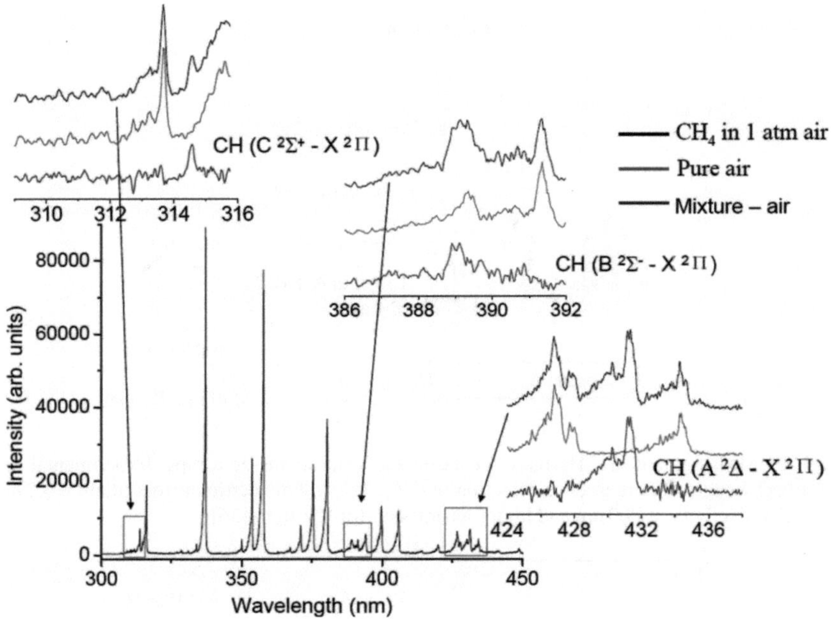

Fig. 9.37 Measurement of CH_4 fluorescence mixed with air inside a pipe. Spectra of mixture of CH_4 and air (total pressure: 1 atm, laser: 5 mJ, CH_4: 2.6%) [13]

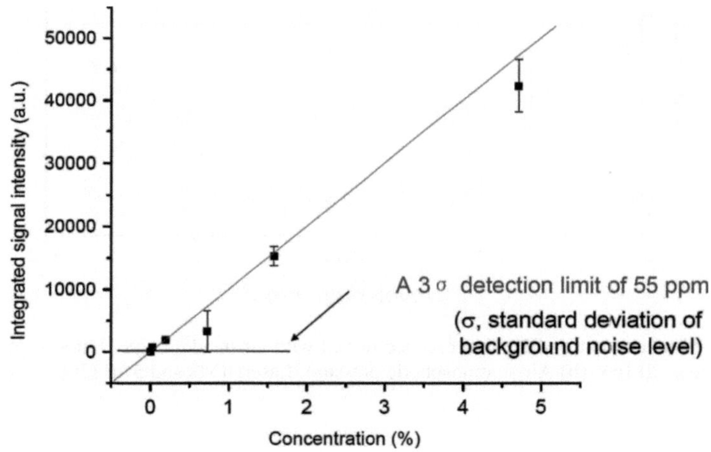

Fig. 9.38 The signal intensities as a function of concentration of CH_4 obtained by subtracting air background. Extrapolation of the result in Figs. 9.34–9.36 to obtain the detection limit CH_4 concentration in air

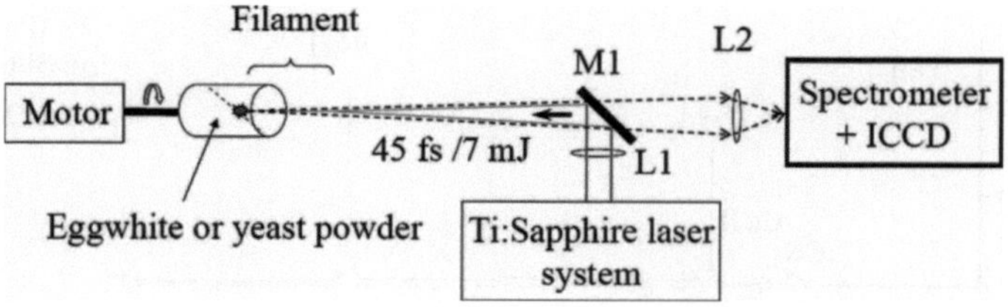

Fig. 9.39 Detection of bio-agents from a distance [37]

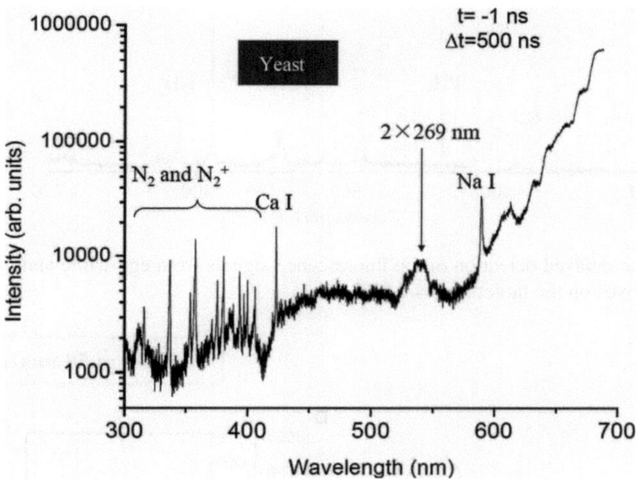

Fig. 9.40 All the fluorescence signals were recorded giving rise to a huge continuum [37]

However, using the known delayed detection technique, we could see the distinctive fluorescence spectra. For example, if we use $t = 10$ ns, i.e., capturing the signal 10 ns after the laser pulse has passed, we can observe the signal that returns. Egg white and yeast have distinctively different clean spectra (Fig. 9.41).

We can also observe the molecular CN spectrum and distinctive signals from yeast powder from even a distance of 50 m away (Fig. 9.42).

With this type of experiment, which is a remote sensing experiment, we can extrapolate to all the signals that are 1 km away.

9.13.6 Summary of Remote Sensing Feasibility Using Only One Laser

In fingerprint fluorescence, there is no measurable plasma signal and it is clean. Some other features are summarized as follows:

1. Finger print fluorescence;
2. "Clean" (in gases, almost free of plasma continuum because of self-avoidance of breakdown);

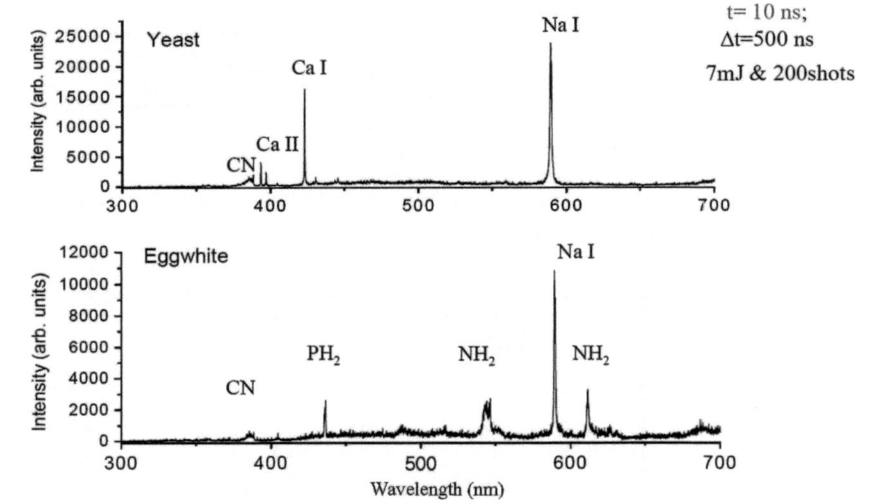

Fig. 9.41 Time-delayed detection of the fluorescence signals from egg white and yeast powders. Laser pulse arrives on the target at $t = 0$ [37]

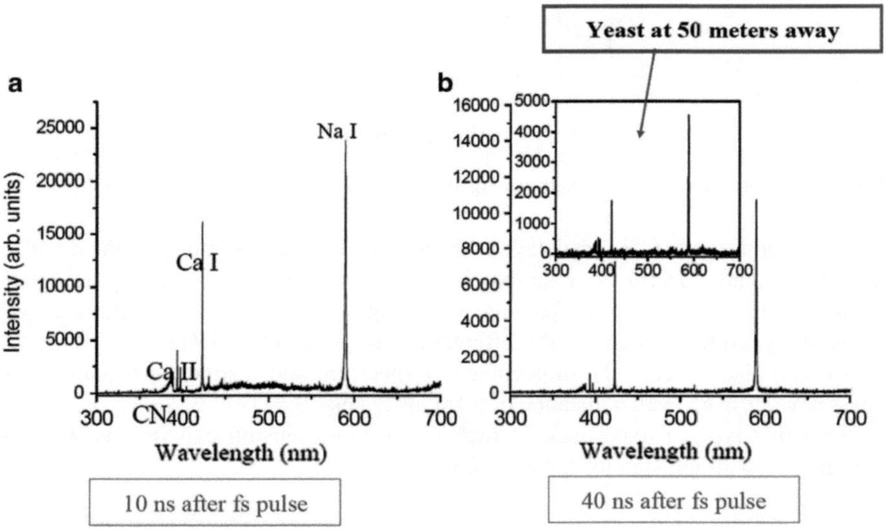

Fig. 9.42 Temporal evolution of delayed fluorescence from yeast powder up to a distance of 50 m [37]

3. Remotely detectable from km range (extrapolation of lab results);
4. Applicable to gases, vapors, bio-target, metals, etc.;
5. fs LIBS in cases of solids

In remote sensing, the filamentation and the ionization in the filament lead to clean fingerprint fluorescence of chemical and biological agents. So this is a valuable

Table 9.1 Comparison of the fluorescence spectra of CH_4 excited by synchrotron radiation and by intense femtosecond laser pulses [38]

Molecule	Synchrotron radiation			Intense laser	
	Products	Excitation energy	Ref	Products	Ref
CH_4	CH (A, B → X) H (Balmer-α, β, γ, δ)	14.5–29 eV	*a*	CH (A, B → X) H (Balmer-α)	*b*

property. We could conclude that, in principle, one laser can detect all pollutants from a distance.

9.14 Super-Excited States of Molecules Inside a Filament

Kong [38] has proposed that the excited molecular states of gaseous molecules giving rise to clean fluorescence are super-excited states. His group observed this with methane molecule in a synchrotron radiation experiment (Table 9.1). We also made a similar observation (cf. Fig. 9.32).

We have attempted to define the super-excited state using the diagram in Fig. 9.43. The super-excited state is an electronic excited state of the molecule that is embedded in the continuum and it is very similar to the auto ionizing state of an atom. This definition is valid for big molecules because it can be assumed that many of the electrons can be simultaneously excited into certain quantum states in such a way that the total energy state is inside the continuum. Once they are inside the continuum, results of studies conducted in synchrotron radiation have suggested that the UV or XUV photon can be used to excite the molecules to this super-excited state. When excited, they have the probability of dissociating into neutral products and some of them may generate fluorescence. We call this phenomenon neutral explosion, but it is also known as neutral fragmentation.

This kind of fluorescence is similar to that generated when using a strong field laser (Fig. 9.44). If we are using the laser, the states undergo excitation by absorbing many photons by multiphoton absorption or tunneling. If these excited states are attained, neutral fragmentation is obtained.

From the molecular analysis, it was found that the super-excited CH_4 was roughly situated in all these positions shown in Fig. 9.45. Some of them decayed into the hydrogen line, the others decayed into the CH, C-, B-, and A-bands.

The MPQ group, particularly Schroeder, has proposed this same idea to explain multiphoton ionization and tunneling ionization experiments (Fig. 9.46). They generated a correspondence with similar excited states by using a particular wavelength or photon energy. By making a statistical calculation of all combinations using the absorption cross section of the synchrotron radiation, they found that these combinations corresponded exactly to the cross section of the multiphoton and tunnel ionizations.

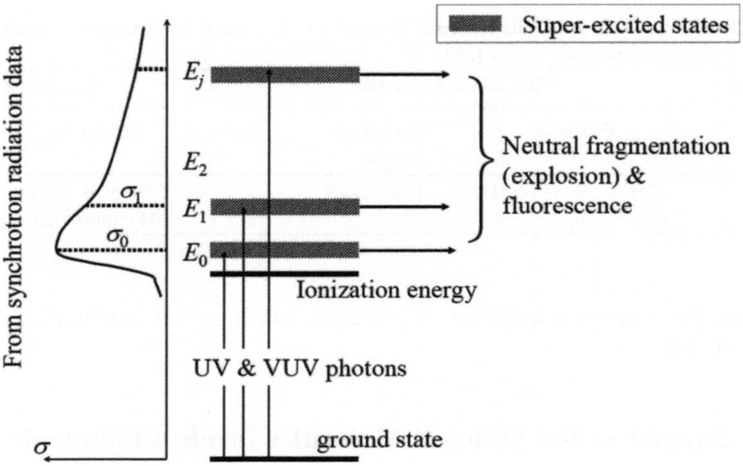

Fig. 9.43 Diagrammatic definition of super-excited states through interaction with XUV light from synchrotron radiation [39]

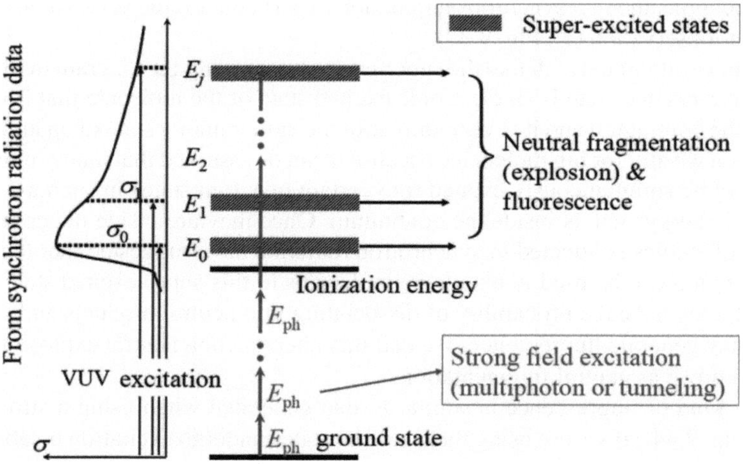

Fig. 9.44 Excitation of super-excited states by an intense femtoseocnd laser pulse. Strong field excitation of super-excited states: equivalent to multiple VUV photons excitation

9.15 Looking Ahead and Conclusion

There are other possibilities that can be explored. We can consider the superposition of one or more pulses with a filamenting pulse and self-phase locking high-quality sources that can be produced from the UV to radio frequencies. If we consider four-wave mixing, the infrared source can be changed into any other tunable source and terahertz pulses can be obtained. Already, tetrahertz pulses have been used in other

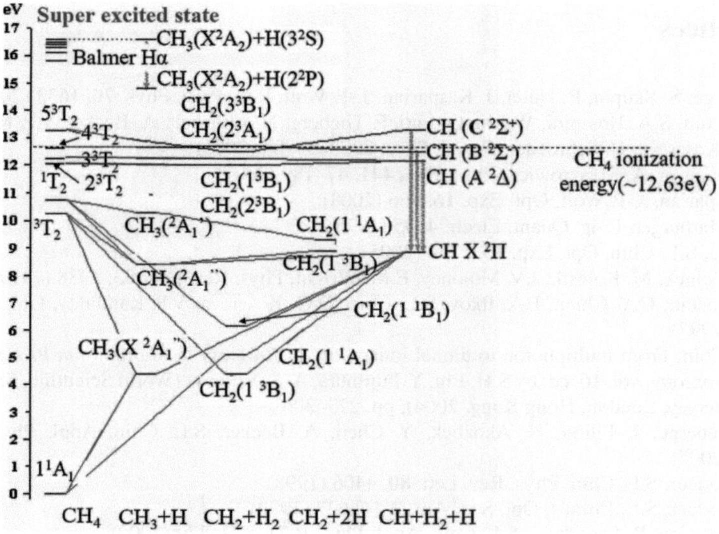

Fig. 9.45 Energy diagram illustrating the pathways of excitation of CH_4 into various super-excited states [38]

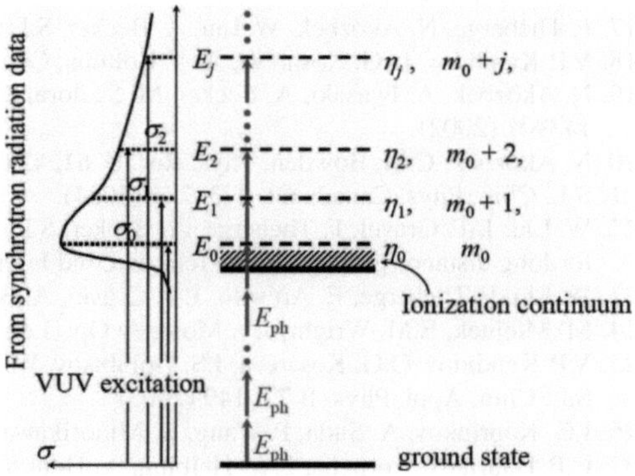

Fig. 9.46 The correspondence between XUV ionization and multiphoton and tunnel ionization [40]

studies; for example, second harmonic and 800-nm pulses were mixed together to generate very efficient tetrahertz pulses.

The interaction rate should remain constant during the stable interaction of the molecules inside the filament with or without the superposition of other pulses due to intensity clamping. In the case of high harmonic generation, for instance, the intensity-dependent phase is present. If clamped intensity is present, then the phase will be constant inside the filament for high harmonics generation.

As a general conclusion, we can say that filamentation nonlinear optics is a new method to study high-quality ultrafast nonlinear optical physics.

References

1. L. Bergé, S. Skupin, R. Nuter, J. Kasparian, J.-P. Wolf, Rep. Prog. Phys. **70**, 1633 (2007)
2. S.L. Chin, S.A. Hosseini, W Liu, Q. Luo, F. Théberg, N. Aközbek, A. Becker, V.P. Kandidov, O.G. Kosareva, H. Schroeder, Can. J. Phys. **83**, 863–905 (2005)
3. A. Couairon, A. Mysyrowicz, Phys. Rep. **441**, 47–189 (2007)
4. J. Kasparian, J.-P. Wolf, Opt. Exp. **16**, 466 (2008)
5. J.H. Marburger, Prog. Quant. Electr. **4**, 35 (1975)
6. W. Liu, S.L. Chin, Opt. Exp. **13**, 5750 (2005)
7. M. Mlejnek, M. Kolesik, J.V. Moloney, E.M. Wright, Phys. Rev. Lett. **83**, 2938 (1999)
8. A. Brodeur, C.Y. Chien, F.A. Ilkov, S.L. Chin, O.G. Kosareva, V.P. Kandidov, Opt. Lett. **22**, 304 (1997)
9. S.L. Chin, From multiphoton to tunnel ionization, in *Progress in Multiphoton Processes and Spectroscopy*, vol. 16, ed. by S.H. Lin, Y. Fujimura, A.A. Villaeys (World Scientific, Singapore, New Jersey, London, Hong Kong, 2004), pp. 273–306
10. F. Théberge, J. Filion, N. Akozbek, Y. Chen, A. Becker, S.L. Chin, Appl. Phys. B **87**, 207 (2007)
11. A. Brodeur, S.L. Chin, Phys. Rev. Lett. **80**, 4406 (1998)
12. A. Brodeur, S.L. Chin, J. Opt. Soc. Am. B **16**, 637 (1999)
13. J. Kasparian, R. Sauerbrey, S.L. Chin, Appl. Phys. B **71**, 877–879 (2000)
14. S.L. Chin, Y. Yergeau, P. Lavigne, J. Phys. B **18**, L213 (1985)
15. T.D.G. Walsh, F.A. Ilkov, S.L. Chin, J. Phys. B.: At. Mol. Opt. Phys. **27**, 3767 (1994) © IOP Publishing
16. L.V. Keldysh, Sov. Phys. JETP **20**, 1307 (1965) (Russian original: Keldysh, LV (1964) Zh Eksp Teor Fiz 47:1945.)
17. F. Théberge, N. Aközbek, W. Liu, A. Becker, S.L.Chin, Phys. Rev. Lett. **97**, 023904 (2006)
18. V.P. Kandidov, O.G. Kosareva, A.A. Koltuna, Quantum Electron. **33**, 69–75 (2003)
19. N. Aközbek, A. Iwasaki, A. Becker, M. Scalora, S.L. Chin, C.M. Bowden, Phys. Rev. Lett. **89**, 143901 (2002)
20. N. Aközbek, C.M. Bowden, Phys. Rev. E. **61**, 4540–4549 (2000)
21. S.L. Chin, Phys. Canada **60**, 273–281 (2004)
22. W. Liu, J.-F. Gravel, F. Théberge, A. Becker, S.L. Chin, Background reservoir: its crucial role for long distance propagation of femtosecond laser pulses in air. Appl. Phys. B**80**, 857 (2005)
23. W. Liu, F. Théberge, E. Arévalo, J.-F. Gravel, A. Becker, S.L. Chin, Opt. Lett. **30**, 2602 (2005)
24. M. Mlejnek, E.M. Wright, J.V. Moloney Opt. Lett. **23**, 382 (1998)
25. V.P. Kandidov, O.G. Kosareva, I.S. Golubtsov, W. Liu, A. Becker, N. Akozbek, C.M. Bowden, S.L. Chin, Appl. Phys. B **77**, 149 (2003)
26. I.G. Koprinkov, A. Suda, P. Wang, K. Midorikawa, Phys. Rev. Lett. **84**, 3847–3850 (2000)
27. C.P. Hauri, W. Kornelis, F.W. Helbing, A. Heinrich, A. Couairon, A. Mysyrowicz, J. Biegert, U. Keller, Appl. Phys. B **79**, 673–677 (2004)
28. W. Liu, S.L. Chin, Abnormal wavelength dependence of the self-cleaning phenomenon during femtosecond-laser-pulse filamentation. Phys. Rev. A **76**, 013826 (2007)
29. S.L. Chin, F. Théberge, W. Liu, Filamentation nonlinear optics. Appl. Phys. B **86**, 477 (2007)
30. F. Théberge, W. Liu, P.Tr. Simard, A. Becke, S.L. Chin, Phys. Rev. E **74**, 036406 (2006)
31. Ph. Rohwetter, K. Stelmaszczyk, L. Wöste, R. Ackermann, G. Méjean, E. Salmon, J. Kasparian, J. Yu, J.-P. Wolf, Spectrochimica Acta Part B **60**, 1025–1033 (2005)
32. S.L. Chin, H.L. Xu, Q. Luo, F. Théberge, W. Liu, J.-F. Daigle, Y. Kamali, P.T. Simard, J. Bernhardt, S.A. Hosseini, M. Sharifi, G. Méjean, A. Azarm, C. Marceau, O. Kosareva, V.P. Kandidov, N. Aközbek, A. Becker, G. Roy, P. Mathieu, J.-R. Simard, M. Châteauneuf, J. Dubois, Filamentation 'remote' sensing of chemical and biological agents/pollutants using only one femtosecond laser source. Appl. Phys. B **95**, 1 (2009)
33. J.-F. Gravel, Q. Luo, D. Boudreau, X.P. Tang, S.L. Chin, Analytical Chem. **76**, 4799–4805 (2004)

34. W. Liu, F. Théberge, J.-F. Daigle, P.T. Simard, S.M. Sarifi, Y. Kamali, H.L. Xu, S.L. Chin, Appl. Phys. B **85**, 55–58 (2006)
35. K. Stelmaszczyk, Ph. Rohwetter, Appl. Phys. Lett. **85**, 3977 (2004)
36. H.L. Xu, J.F. Daigle, Q. Luo, S.L. Chin, Appl. Phys. B **82**, 655 (2006)
37. H.L. Xu, W. Liu, S.L. Chin, Opt. Lett. **31**, 1540 (2006)
38. F. Kong, Q. Luo, H. Xu, M. Sharifi, D. Song, S.L. Chin, J. Chem. Phys. **125**, 133320 (2006)
39. Y. Hatano, Interaction of vacuum ultraviolet photons with molecules. Formation and dissociation dynamics of molecular superexcited states. Phys. Rep. **313**, 109 (1999)
40. C.J.G.J. Uiterwaal, C.R. Gebhardt, H. Schröder, K.-L. Kompa, Eur. Phys. J. D. **30**, 379–392 (2004)

Chapter 10
Diagnosing Intense and Ultra-intense Laser–Matter Interactions: Status and Future Requirements

Leonida A. Gizzi

Abstract In this chapter we describe experimental techniques for investigating laser–matter interactions that have been developed in recent years within the framework of international collaborations. We introduce phenomena observed in the wide range of interaction regimes and plasma conditions, including long-scale-length plasmas, laser–plasma coupling in inertial fusion studies, and electron acceleration driven by ultrashort pluses. We focus on laser–matter interactions, particularly laser–solid interactions and laser–gas jet interactions. We also discuss in details the temporal and spatial features of laser pulses amplified by the chirped pulse amplification (CPA) technique (Strickland and Mourou, Opt. Commun. 56:212, 1986) with special attention to the properties that play an important role in laser–solid interactions, including pulse duration, contrast, prepulses and focusing configurations.

We investigate plasma formation and plasma density evolution by optical interferometry techniques. We also discuss techniques for recovering information on plasma density evolution when optical interferometry with ultrashort pulses is adopted. Further, we introduce time-resolved X-ray spectroscopy of plasmas generated by ultrashort laser pulses, then we describe recent measurements of fast electron transport in solids conducted by using a novel monochromatic imaging technique. Finally, we briefly analyze some future instrumentation requirements for investigating laser–matter interaction when femtosecond laser pulses are adopted.

10.1 Introduction on Ultra-Intense Laser–Matter Interactions

When an intense, short laser pulse is focused on a solid target, several features of the pulse have to be considered because there are many mechanisms that can take place, and the dominating ones need to be identified. Among others, consideration needs

L.A. Gizzi
Intense Laser Irradiation Laboratory, Istituto Nazionale di Ottica, Consiglio Nazionale Delle Ricerche, Area della Ricerca CNR, Via G. Moruzzi, 1 – 56124 Pisa, Italy
and
INFN, Pisa, Italy

K. Yamanouchi (ed.), *Lectures on Ultrafast Intense Laser Science 1*, Springer Series in Chemical Physics 94, DOI 10.1007/978-3-540-95944-1_10,

to be given to the prepulse features, long pulse to short pulse contrast, prepulses, and so on. As discussed later in the chapter, one of the key applications of intense, short laser pulses is the *fast ignition* (FI) approach [1, 2] to *inertial confinement fusion* (ICF) where ultrahigh intensity interaction is used to generate energetic (MeV) electrons that propagate in the compressed core and initiate the nuclear fusion reactions. In these circumstances, there is a great interest in understanding how fast electrons are produced, how they propagate in the target substrate, how they transfer their energy to the compressed material, what is the production efficiency, and what fraction of their energy can be dumped in the target. All these features depend strongly on the conditions that the laser pulse finds on the target surface; and after generation of the electrons begins, their transport will be affected by several mechanisms, including the electron beam parameters and the resistivity of the material.
This leads us to take a close look at CPA lasers. In typical CPA systems based upon Ti:Sa, a nJ-level femtosecond, ultrashort laser pulse, generated by an oscillator, is stretched in a dispersive medium (e.g., optical fiber, gratings) to reach the sub-ns pulse duration. As shown in Fig. 10.1 for the case of the ILIL laser system at INO-CNR, in these systems a stretched seed pulse is preamplified by a high-gain preamplifier, typically a so-called *regenerative amplifier*. The mJ level pulse thus generated is further amplified in a multipass configuration and then optically compressed, typically using a grating compressor. A view of the multipass amplifier (right) and of the vacuum compressor of the femtosecond laser system at ILIL taken during test operations using a remotely controlled camera is shown in Fig. 10.2.

In general, such lasers are characterized by the so-called *pedestal*, due to *amplified spontaneous emission* (ASE), which is characterized by a nanosecond temporal duration. Then, there are some features of the pulse that arise from limitations in the optical compression of the stretched, amplified laser pulse due to grating imperfections. Finally there are low energy pre- and postpulses and, of course, the main pulse.

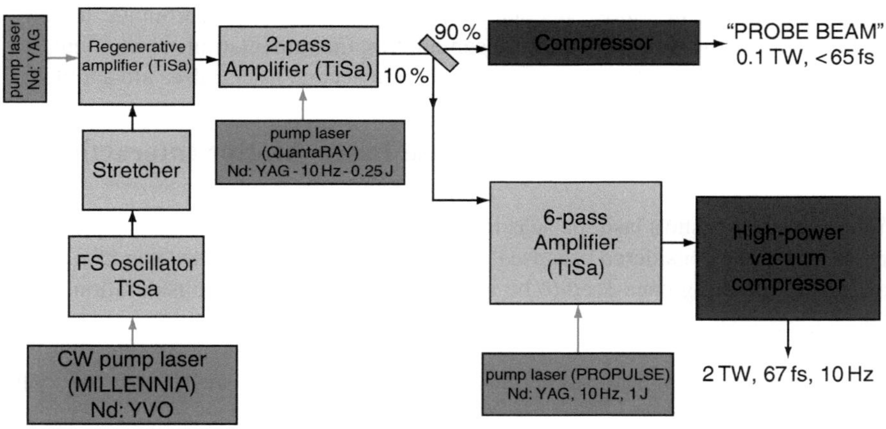

Fig. 10.1 Schematic layout of the FIXER femtosecond laser system at ILIL, showing the front-end and the power amplifier

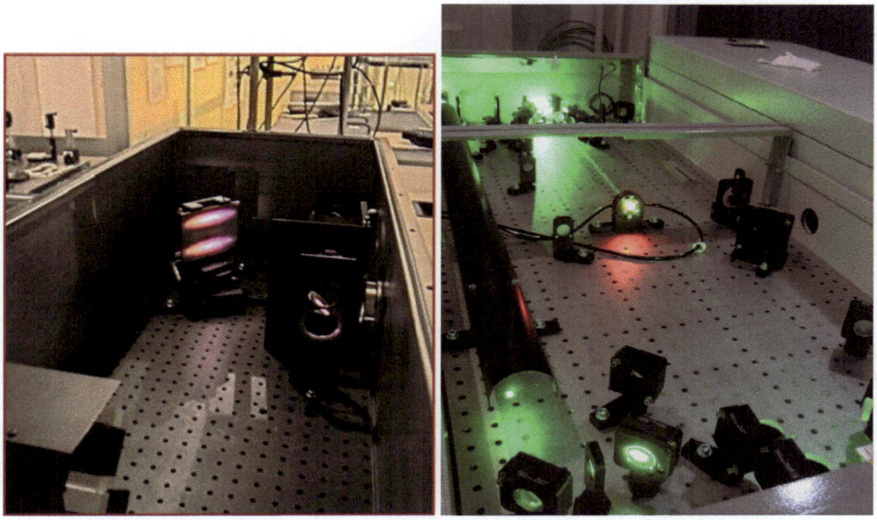

Fig. 10.2 Images of Ti:Sa femtosecond FIXER laser system at the ILIL lab at INO-CNR, Pisa. On the *right* is a view of the 6-pass amplifier showing the pumping beams (*green*) and the fluorescence of the Ti:Sa crystal (*red*). On the *left* is a view of the inside of the vacuum compressor showing one of the gratings with the patterns of the forward- and backward-propagating 800 nm beam

The most important feature to be considered in CPA laser–target interactions is indeed the ASE that can deliver a significant amount of energy on target before arrival of the main short pulse, at intensity that can lead to premature plasma formation [3]. The temporal profile of a typical CPA pulse appears as seen in Fig. 10.3 with the ASE contrast, i.e., the ratio of the peak intensity to the ASE intensity ranging from 10^5 to a best performance of 10^{10} in recent, contrast enhanced systems. For high focused intensities of up to $10^{20}\,\mathrm{W/cm^2}$, the ASE level will range approximately from a minimum of 10^{11} to a maximum of $10^{15}\,\mathrm{W/cm^2}$. In the case of interactions with solid targets, these intensities generate a precursor plasma on the target. If no measures are taken to prevent this, the CPA pulse will interact with the preformed plasma generated by the ASE rather than with the solid target surface.

Recently, effective techniques have been developed to suppress precursor radiation and prepulses to increase the contrast. With regard to ASE, one of these techniques is the plasma mirror, the effectiveness of which has been established by recent experiments with ultrathin foil targets [4]. Presently, the only drawback of this technique is the limited repetition rate because the surface on which the plasma mirror is generated has to be replaced at every shot.

Figure 10.4 shows an example of a more realistic temporal profile obtained from a third-order cross-correlation measurements of the front end (10 mJ) output of the FIXER (Femtosecond Interaction X-ray Emitter) laser system. In this case the pulse is compressed after a two-stage amplifier including the regenerative amplifier and

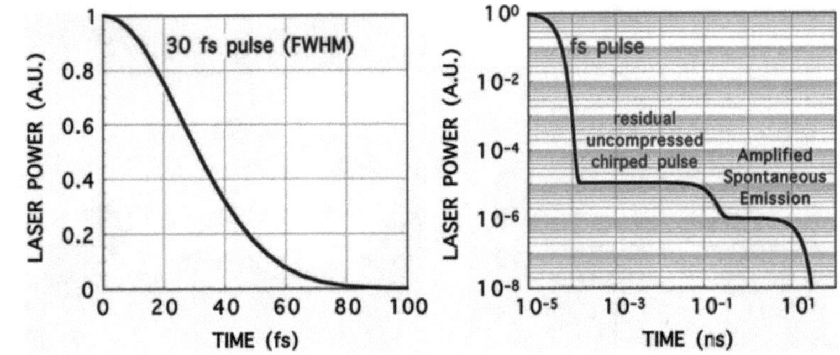

Fig. 10.3 Schematic profile of laser pulse power on a linear scale (*left*). The same laser pulse temporal profile plotted on a log scale (*right*) shows low power features due to amplified spontaneous emission in the amplifier chain and spurious light due to imperfections in the optical compression. These components, depending on the experimental configuration, can give rise to premature plasma formation on the target with a consequent change of interaction conditions

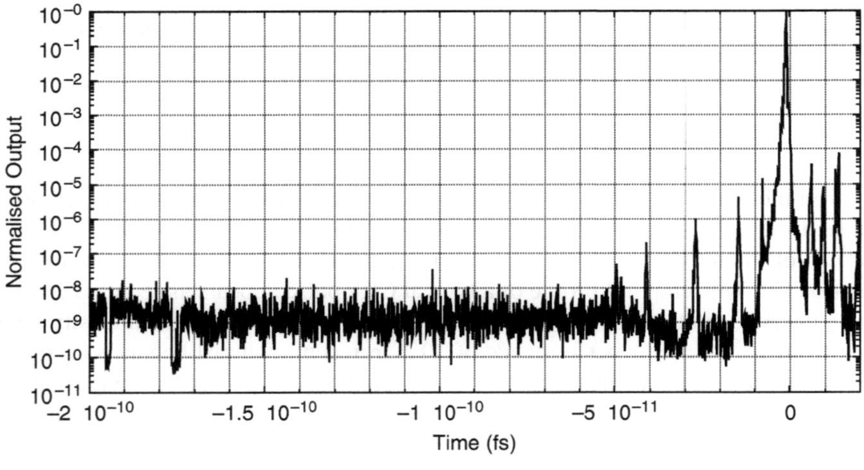

Fig. 10.4 Laser pulse contrast measurement obtained from third-order cross-correlation trace of the ILIL FIXER laser pulse after a two-stage amplification (10 mJ). Measurement by Amplitude Tech using a *SEQUOIA* cross-correlator

a two-pass amplifier that works in a relatively low-gain (3X) regime. According to this profile, the FIXER laser system has a contrast between 10^8 and 10^9.

The cross-correlation trace also shows low-intensity short pulses preceding the main pulse. These pulses are unlikely to play a crucial role in the interaction, even in the case of solid targets, because of their short duration that, in the worst case, will lead to generation of a very small, cold plasma. Clearly visible in the plot is the main pulse appearing at about time zero, followed by post pulses.

To study the effect of such complex structure of femtosecond pulse profiles, experiments have been performed using thin plastic foils as targets. A range of experiments performed in the past two decades used this technique for generating test plasmas with long-scale-length plasmas. In fact, when the laser intensity on a foil target is higher than the damage threshold of the foil, the foil explodes and is then heated by the laser radiation. This method was used to create conditions that are similar [5] to ICF coronal plasmas where the fuel *pellet* is heated and a plasma is created all around. The use of exploding-foil targets was found to be effective to achieve coronal-like plasmas for laser–plasma coupling studies. In the case of irradiation by long pulses, a simple analytical model [6] can be used to obtain the expected values of density and temperature scale lengths of these plasmas in a self-similar expansion model. Alternatively, numerical codes [7, 8] can be used to obtain full description of the expected hydrodynamics of the plasma produced by laser irradiation of such foils.

In the case of irradiation with femtosecond CPA pulses, thin foil plastic targets have peculiar properties owing to their higher damage threshold (in comparison with metallic targets) and remain therefore undamaged up to relatively high intensities, typically between 10^{11} and 10^{12} $\mathrm{W/cm^2}$.

Since thin plastic foils are optically transparent, if the laser intensity is increased starting from well below the damage threshold, the electric field of the electromagnetic (e.m.) wave will be allowed inside the target. Thus, interaction with the volume of the target (rather than with the surface) will take place. In general, if the intensity of ASE on target is below the damage threshold for the thin foil, one may have a condition in which no precursor plasma is generated and interaction of the main pulse with a steep density gradient on the target may be achieved. When preplasma free conditions have been achieved at very high peak intensity, anomalous propagation effects have been observed showing unexpectedly high transmissivity and large ionization blue shift [9]. Absorption of the laser energy by the target will occur mainly at the critical density due to resonance absorption or the Brunel effect [10] and fast electrons will be generated directly or by the collision-less dumping of the resonantly excited electron plasma waves. As discussed below, these circumstances can be verified by investigating the polarization dependence of secondary radiation like X-rays and/or laser harmonics [11].

In contrast, at sufficiently high ASE levels, the thin foil will be exploded by the ASE prior to the arrival of the main pulse. In these circumstances, the above-cited analytical and numerical models can be used to estimate the conditions of the precursor plasma generated by the ASE, so that CPA pulses interact with known plasma conditions. A similar scheme has been used to successfully perform laser–plasma electron acceleration experiments [12].

In the following we present a few examples of the use of thin foils and how to diagnose the interactions. The first is a simple experiment in which a femtosecond laser pulse is focused on the target and X-ray and optical emission is detected to characterize the interaction in different linear polarization conditions. In fact, if the interaction does not suffer from excessive precursor plasma, then the pulse can reach the critical density. In this case, the absorption mechanisms, which are expected to

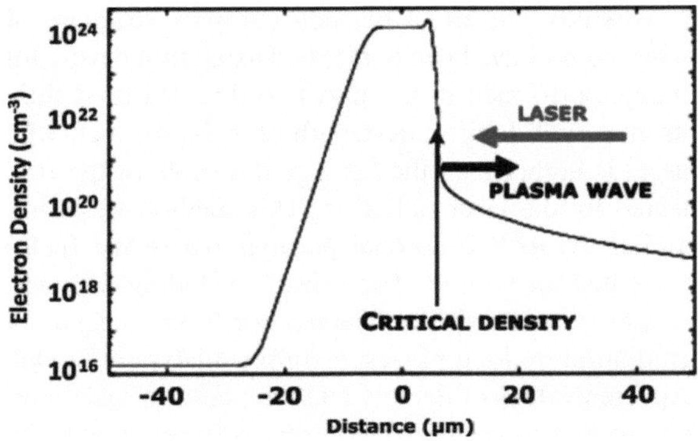

Fig. 10.5 Typical density profile of a plasma produced by interaction of an intense, short pulse with a solid target, for a relatively low level of ASE precursor radiation. A key role is played by the critical density layer where laser absorption takes place

take place, will be polarization dependent. Indeed, the experiment [11] shows that a strong dependence on the polarization is observed by looking at both the X-ray emission and the second harmonic emission from the specular reflection direction. These measurements show that polarization studies provide an effective method to identify the interaction mechanisms. On the other hand, when femtosecond laser target interactions are used as X-ray sources [13] this method can possibly be used to control the intensity of X-rays.

A possible scenario for the observed results is that energy absorption is dominated by resonant [14] or not-so resonant absorption [10] as schematically described in Fig. 10.5. Here we assume a relatively steep profile because of the expected low level of ASE. Therefore, the plasma is generated either by the CPA pulse itself or by picosecond-level prepulses shown in Fig. 10.3 (right). In either case, the gradient will have a scale length of the order of a micrometer or less. If the incident laser light is p-polarized, i.e., with the electric field in the plane of incidence, then at the turning point the electric field points in the direction of the density gradient in the target. In this way the laser field resonantly excite electron plasma waves at the laser frequency [14]. Then the plasma waves undergo noncollisional damping processes, thereby generating the so-called *fast electrons*, i.e., a population of electrons with an energy significantly higher than the energy of the electrons in the rest of the plasma. These *fast* electrons were observed many years ago in quite different experimental conditions as described in [15] (and references therein) and are now gaining importance, due to their potential role in the ICF FI scheme [1].

If metal targets are used instead of plastic (dielectric) targets, than even low levels of ASE can generate a large preplasma that can change the interaction conditions, making collisional absorption more effective [14].

However, at sufficiently high intensities, collisional absorption is suppressed due to the high velocity of electrons quivering in the laser electric field [14]. In these circumstances, even though there is a large plasma in front of the target, little energy of the laser is transferred to the plasma until the laser radiation reaches the critical density. Hence, the critical density is reached and the dependence on the polarization can still be observed. This is not the case when a long pulse or a low-intensity short pulse with the precursor plasma are used.

Relatively thick targets are instead used for the generation of Kα radiation as an X-ray sources for applications [13] and for the study of fast electron transport [16]. The basic idea here is to focus intense ultrashort laser pulses on a solid targets to generate fast electrons as described above. These fast electrons propagate through the cold target substrate and generate fluorescence Kα radiation that escapes from the target. For an efficient production of Kα radiation, the target thickness is large compared to the interaction processes, but is comparable with or smaller than the range of the fast electrons. In fact, the requirement for an efficient, short pulse emission of Kα radiation is that electrons release the maximum possible energy in the target and at the same time Kα radiation can escape from the target.

Also, in order to keep the duration of the Kα emission as short as possible, the extension of the Kα emitting region should be small to limit collection of radiation from regions where multiple scattering of electrons takes place, which would increase the duration of X-ray emission. Hence, the thickness of the target should be carefully matched to preserve the quality of the output X-ray pulse. Since the early works on Kα sources [13], an extensive cross-disciplinary literature is available demonstrating that laser-driven Kα sources are now commonly used for applications in which short and ultrashort pulses are needed [17–21].

As discussed above, laser radiation may still reach the critical density provided its intensity exceeds $10^{17}\,\mathrm{W/cm^2}$ for one micron wavelength, which is the limit for suppression of collisional absorption due to nonlinear effects. Under these conditions, half-integer harmonic emission occurs as one of the signature features of this regime. This emission originates from propagation of the laser light in the long-scale-length plasma at the quarter critical density region where either two-plasmon decay [22] or Raman instability [23] can develop generating longitudinal electron plasma waves at half of the laser frequency. Nonlinear coupling of these plasma oscillations with the e.m. wave of the laser at the fundamental laser frequency generates the 3/2 harmonic emission. If the critical density layer is reached, second harmonic emission can still take place, mainly in the direction of specular reflection, provided that the critical density surface is not significantly distorted. From an experimental viewpoint, the scattering of optical radiation in the specular direction can be used to monitor the interaction in order to locate the regions where the most effective laser–plasma coupling takes place.

An example of optical scattering for the $(3/2)\omega$ emission is shown in Fig. 10.6. This data was taken using the femtosecond laser system of the Laboratoire d'Optique Appliquée (LOA) (in Palaiseau, France). Here, the optical spectrum was obtained by collecting the radiation emitted in the specular direction, and by spatially resolving it in the direction perpendicular to the spectral dispersion axis. Space resolved spectra like this can be used to identify the quarter critical density layer, provided that electron plasma waves generated by SRS or TPD couple with the laser field before significant propagation takes place [24]. An important conclusion of the entire experiment in which this measurement was taken was that the emission at $(3/2)\omega$ was strongly correlated with the presence of a large preplasma. In other

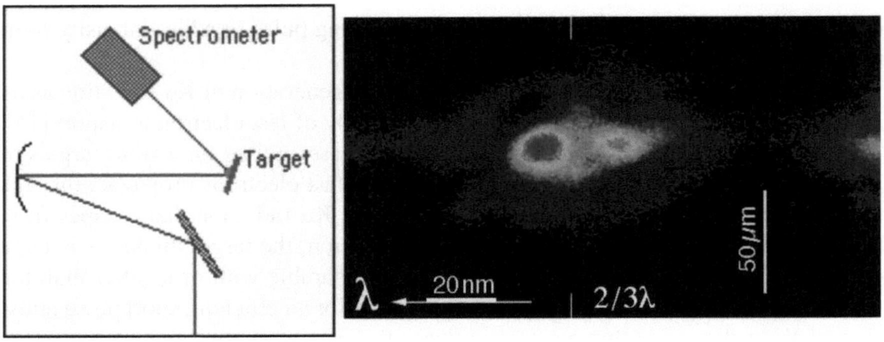

Fig. 10.6 Space resolved spectrum of specular $(3/2)\omega$ emission from the interaction of a 35 fs laser pulse at an intensity of 10^{19} W/cm^2 with a 1-μm thick plastic target. The intensity of the ASE was 10^{12} W/cm^2, i.e., sufficiently high to lead to premature plasma formation. The spectrum shows a very broad emission that is spatially localized on the position of the laser spot on target. Both *red* and *blue* components are present

words, this emission could be used as a marker of premature plasma formation by the ASE.

In the case of 2ω, a similar situation takes place. In this case, the emission originates from the critical density layer and the structure of the emission can be used to diagnose the interaction and to detect where the interaction is generated and what are the main mechanism for the energy exchange between the laser and the plasma wave.

An example of emission at 2ω is shown in Fig. 10.7, where a space-resolved spectrum of specular emission is displayed. The spectrum shows complementary blue and red shifted components where the shift is likely to originate from a combination of propagation and ionization effects. We also observe that the pattern shown in Fig. 10.7 is not reproducible from shot to shot, indicating that the regions from which the emission originates change. This variability may also depends on the laser beam stability and quality. In fact, if hot spots are present in the laser focal spot, some of these hot spots may activate emission.

In the general case of laser interaction with solids, when second harmonic emission occurs, in general it is generated by interaction at the critical density region. Another example of this emission is given in the plot of Fig. 10.8. It shows the emission from the interaction of the JETI laser system at the Institute of Optics and Quantum Electronics (IOQ) (Jena, Germany) with a 12- μm thick titanium foil. The laser pulse in this case was 60 fs FWHM and the intensity was 10^{19} W/cm^2. Visible in the spectrum of Fig. 10.8 are both the 2ω and the $(3/2)\omega$ emission indicating nonlinear coupling of the laser light at the critical density layer as well as with electron plasma waves at $\omega/2$.

A much similar situation is found also when significantly longer laser pulses are used. The plot of Fig. 10.9 shows the spectrum obtained from interaction of a 500-fs laser pulse using the TITAN laser at the Lawrence Livermore National Laboratory.

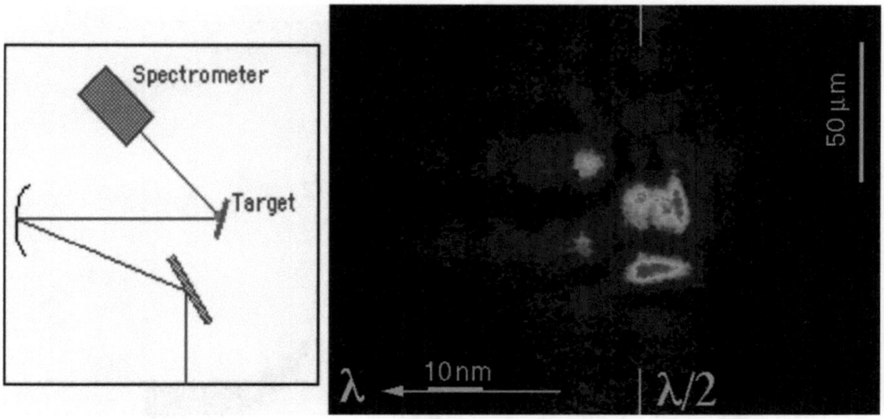

Fig. 10.7 Space-resolved spectrum of specular 2ω emission from the interaction of a 35 fs laser pulse at an intensity of $5 \times 10^{18}\ \mathrm{W/cm^2}$ with a 1-μm thick plastic target. The intensity of the ASE was $10^{12}\ \mathrm{W/cm^2}$, i.e., sufficiently high to lead to premature plasma formation. The spectrum shows a very broad (spectrally) and large (spatially) emission with complementary red- and blue-shifted components

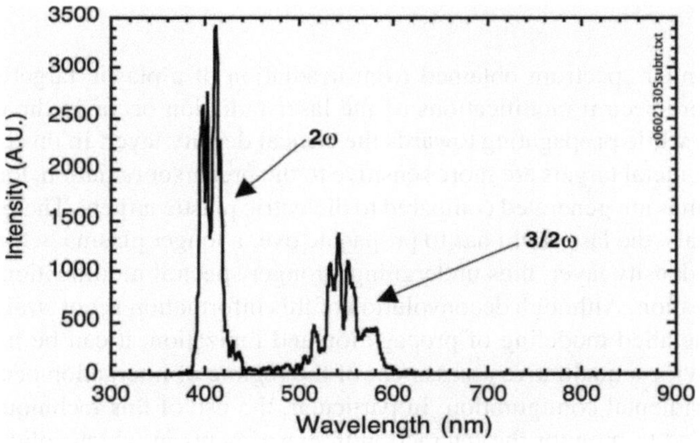

Fig. 10.8 Spectrum of optical scattering from laser interaction of a 60 fs laser pulse at an intensity of $10^{19}\ \mathrm{W/cm^2}$ with a 12-μm thick solid target. The spectrum was taken with a spectrometer capable of resolving the entire visible range, showing simultaneously the 2ω and the $(3/2)\omega$ emission components

In this case, a solid steel target was irradiated at an intensity exceeding $10^{19}\ \mathrm{W/cm^2}$. Clearly visible in the spectra are the second harmonic and $(3/2)\omega$ emissions. In this case, the $5/2\omega$ emission is also visible, indicating a stronger nonlinear coupling of the laser e.m. field with the plasma waves.

Additional measurements performed with the same experimental also show that the second harmonic emission spectrum obtained from the steel target is broader

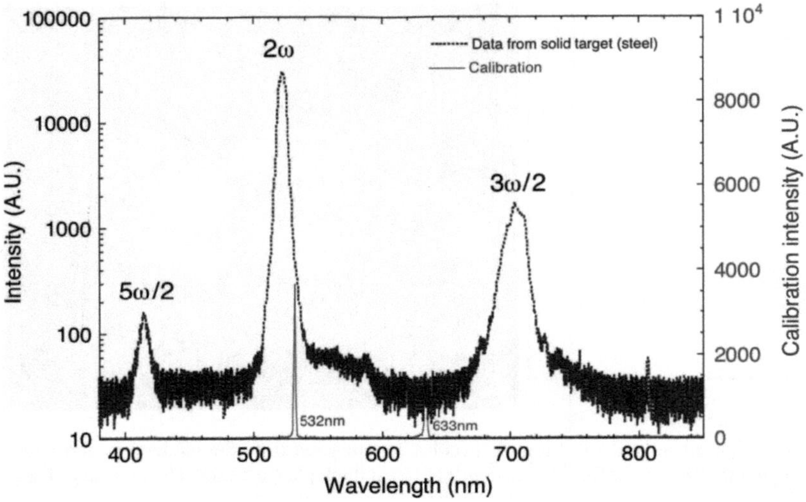

Fig. 10.9 Spectrum of the radiation emitted from interaction of a 500-fs laser pulse at 10^{19} W/cm^2 with a solid (steel) target. The spectrometer was set to detect the radiation emitted in the specular direction

than the similar spectrum obtained from irradiation of a plastic target suggesting that stronger spectral modifications of the laser radiation occur in the case of the steel target, while propagating towards the critical density layer. In this case, due to the fact that metal targets are more sensitive to the precursor radiation, longer-scale-length plasmas are generated compared to dielectric plastic targets. Therefore, in the case of metals, the laser light has to propagate over a longer plasma before reaching the critical density layer, thus undergoing stronger spectral modifications, possibly due to ionization. Although deconvolution of this information is not straightforward without a detailed modeling of propagation and ionization, it can be immediately used to provide a qualitative assessment of the regime of interaction occurring in a given experimental configuration. In particular, the use of this technique is highly recommended to monitor the interaction in experiments in which solid targets are used to activate Kα generation [25,26]. In these circumstances, optical spectroscopy of the scattered radiation from the target (e.g., in the reflection direction) can be used as an monitor for the control of laser–target interaction for Kα sources, to stabilize the source intensity.

An example of a possible drift of long term behavior of the second harmonic and $(3/2)\omega$ emissions in laser–target interaction for Kα generation is shown in Fig. 10.10 where the spectra obtained from a sequence of 800 shots obtained in similar conditions as in the spectrum of Fig. 10.8 is shown. The image consists of a sequence of individual spectra displayed as 1D false color images placed next to each other to form a single image. The sequence was sorted from left to right according to the increasing order of the corresponding yield of X-ray emission, including Kα fluorescence, as measured from a separate CCD detector. The plot

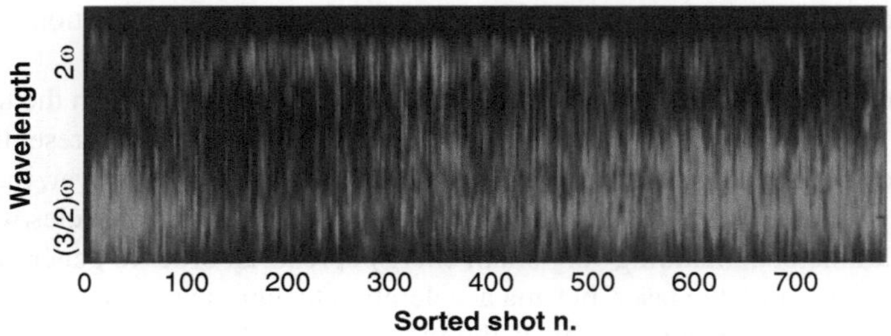

Fig. 10.10 Multi-shot spectra of optical emission from laser interaction of a 60 fs laser pulse at an intensity of 10^{19} W/cm^2 with a 12-μm thick solid target. The image consists of a sequence of spectra as that of Fig. 10.9, displayed as a false color image. The sequence is sorted from left to right according to increasing yield of X-rays as measured from a separate detector

shows that there is a region of high $(3/2)\omega$ emission corresponding to low X-ray emission yield. As the X-ray yield increases, optical emission is characterized by a weaker $(3/2)\omega$ emission and a brighter 2ω emission. Finally, at higher X-ray yield, lower 2ω emission and brighter $(3/2)\omega$ emission occur, with the $(3/2)\omega$ showing a broadened and spectrally modulated emission.

A possible explanation of the behavior at higher X-ray yield is that a strong interaction occurs in the lower-density plasma region, reducing the laser energy that reaches the critical density. Hence, efficient coupling occurs at the quarter critical density, likely producing higher-energy electrons as discussed above. These high-energy electrons escape and pass through the target producing *Bremsstrahlung* emission with little Kα emission. In contrast, the high second harmonic emission occurring for the shot number range from 150 to 500 suggests that more efficient coupling occurs at the critical density region, and the condition becomes more relevant to efficient generation of Kα X-rays. In summary, these measurements show simple spectroscopic measurements can be used to monitor, and, possibly, control laser–plasma X-ray sources.

10.2 Optical Interferometry and Propagation Issues

If the temporal evolution and detailed spatial features of the plasma have to be investigated during propagation, additional techniques like plasma probing should be used. In fact, there are several reasons for studying the propagation of intense laser pulses in gases and most of them are related to the laser acceleration of electrons in gas-jets [27] or in capillaries for which energy up to the GeV range has already been demonstrated [28]. However, there are still significant concerns for the stability and reproducibility of the main parameters of the accelerated electron bunches. Recent experiments based upon counter-propagating laser pulses have shown that some control of the electron bunch energy can be achieved [29]. However, it is clear that reproducibility and stability are strictly related to laser propagation issues and a lot remains to be investigated.

Presently, a significant effort is being made worldwide to establish the technique from the viewpoint of the high-energy physics (HEP) community, presently facing difficulties due to the gigantism of conventional accelerators. However, the need of the HEP community is focused on reliable ways of accelerating particles with high-quality parameters, including very small energy spread. In order to generate interest from that viewpoint, the laser–plasma acceleration technique must establish itself in terms of quality and stability. This is the focus of the effort of many groups worldwide. In Italy, the laser–plasma acceleration community and the HEP community are now entering this arena with an entirely new project [30–32] based upon the FLAME laser [27], a 20-fs, 300-TW Ti:Sa 10-Hz system specifically designed to investigate the full potential of laser-driven acceleration.

To illustrate the basic issues in laser–plasma acceleration, in the following we explore a series of experiments aimed at studying propagation of an ultra-intense laser pulse in a gas-jet using optical probing and interferometry. The techniques are the same used in the past to diagnose plasmas produced by long laser pulses. However, special attention is dedicated here to highlight new effects arising from the use of ultrashort pulses to probe plasmas. In fact, several issues arise including ionization front smoothing due to finite probe pulse transit time [33] and loss of fringe visibility [34].

The basic experimental setup for the study of laser propagation in a gas is shown in Fig. 10.11. In this set up the laser pulse is focused on a target consisting of a gas-jet capable of producing a slab of gas with steep edges and peak density exceeding 10^{19} el/cm^3. The probe pulse propagates along the direction perpendicular to the propagation direction of the main laser pulse to detect changes of refractive index induced by ionization.

Figure 10.11 also shows the phase shift map of a He gas-jet. The image shows a positive phase shift due to the small refractive index of neutral He atoms. As

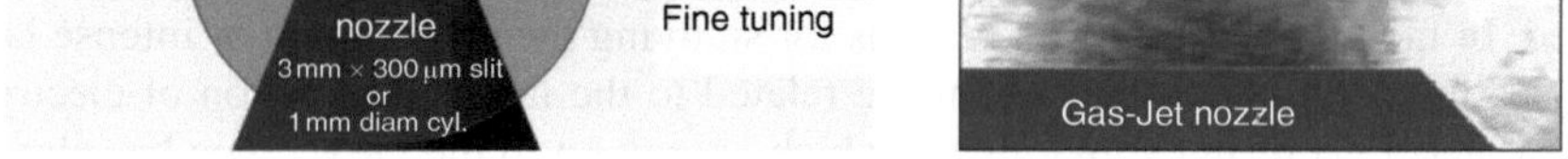

Fig. 10.11 Typical schematic experimental setup for study of propagation of an intense ultrashort laser pulse in a gas-jet target. The focal spot position can be moved relative to the gas-jet and the probe beam (*gray circle*) is set to propagate perpendicular to the main laser pulse. Also visible on the right is the map of the phase shift induced by neutral helium gas on the probe pulse as measured by the interferometer

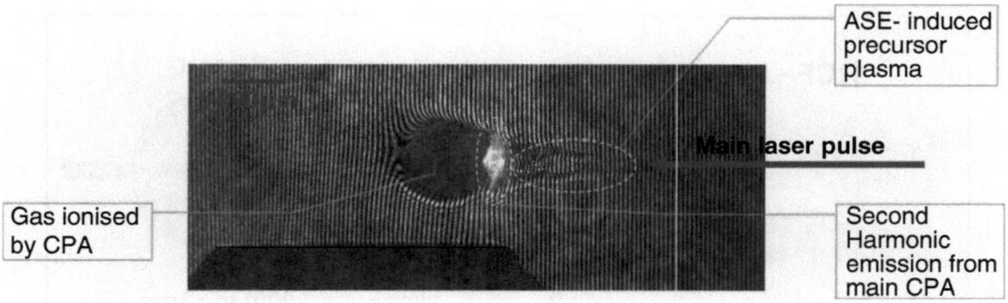

Fig. 10.12 Interferogram of the laser–plasma interaction region in a gas-jet experiment as that schematically shown in Fig. 10.11. The interferogram was taken with a short pulse probe timed to reach the interaction region 5 ps after propagation of the main pulse

discussed in [5], numerical techniques are used for the fringe pattern analysis to extract the small phase-shift information that was of the order of a small fraction of wavelength. The pattern is noisy because of the numerical noise, however, the shape of the gas-jet is clearly visible.

Figure 10.12 shows an interferogram taken 5 ps after the arrival of the main laser pulse on the gas-jet. Clearly visible is the effect of a short pulse propagating in the gas. It also shows the effect of the ASE that, in this case, gives rise to premature gas breakdown with a shockwave propagating radially as visible on the right-hand side of the image. This premature plasma formation is the reason why ASE is regarded as one of the sources of nonreproducibility of the interaction due to its strong level fluctuations from shot to shot. Because of this, several techniques are being implemented to reduce the level of ASE. Also visible in the image is second harmonic emission. This emission is due to a different mechanism [35] than the specular second harmonic emission from the critical density layer discussed above since the former is being observed at 90°. Finally, in the left-hand side of the image of Fig. 10.12, a third region is visible in which the laser has ionized the medium and this plasma region is opaque to the probe radiation, hence, no fringes are visible.

Under these conditions, electrons are still accelerated as shown in Fig. 10.13, but acceleration conditions are strongly affected by the presence of the ASE precursor plasma that prevents the main pulse to be properly focused due to uncontrolled refraction and defocusing effect. Measurements of accelerated electrons were performed using a technique [36] developed to detect the electrons and measure their energy. The technique uses a stack of layers of radiochromic films to measure the dose released by the electrons in the layers. By using some known stopping materials in between the layers, the energy of the incoming beam can be reconstructed. The angular distribution of these electrons can also be determined. In this case, a particular sequence of aluminum layers and radiochromic films was used as shown in the layout of Fig. 10.13. Actual raw data are reproduced in the figure, enhanced in terms of contrast for the purpose of displaying the shot-to-shot change. As one can see, strong fluctuations from shot to shot occur with the electron beam being produced by approximately only 30% of the shots. The initial plasma conditions

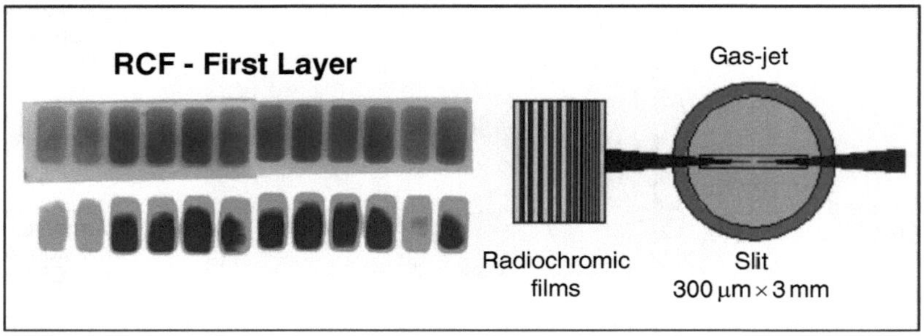

Fig. 10.13 Energetic electron production from a sequence of 12 laser interaction events with a gas-jet, in the presence of precursor plasma due to ASE. Measurement of angular and energy distribution using stack of radio chromic films show strong fluctuations from shot to shot

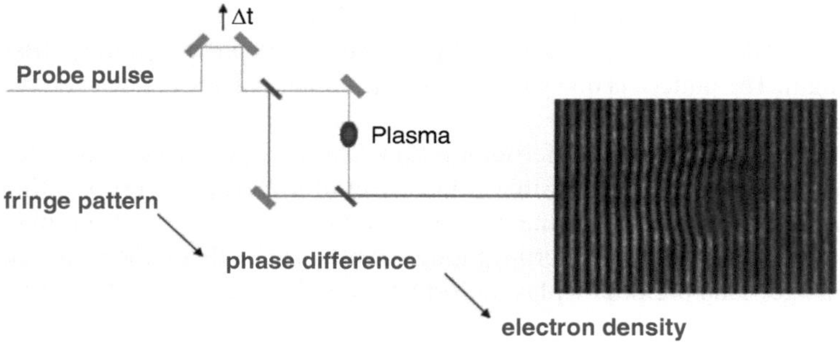

Fig. 10.14 Conceptual layout of a laser plasma interferometer showing the sequence of steps required to retrieve the map of electron density distribution from the detection of fringe shift induced by the plasma via its refractive index

(at the time of arrival of the main short pulse) are most likely the reasons for this behavior as displayed by plasma probing via interferometry.

As anticipated above, special attention requires the use of very short probe pulses in laser plasma interferometry. In view of this discussion, it is useful to recall the basic principles of plasma interferometry. Figure 10.14 shows a typical Michelson-type interferometer. The beam is split and the resulting two beams are made to interfere, after one of them has propagated through the plasma. Once the fringes are obtained, the phase difference can be retrieved and, by the deconvolution of the phase difference assuming a cylindrical symmetry, the electron density is finally obtained. A number of papers have been published on laser plasma interferometry and deconvolution techniques as discussed in [5], with the latest advances being in the use of wavelet transform, where a different technique is used to deconvolute the fringe pattern [37] as an alternative to the standard FFT technique. Special algorithms can also be used to account for a small degree of asymmetry in deconvolution [38].

If short pulses are used, then there are several limitations that should be taken into account when deconvolution is done. One is the probe transit time that occurs when the probe pulse is very short in comparison to the time taken by the probe-pulse to cross the evolving interaction region. A second limitation is related to the visibility of the fringes, which suffers from the usual losses well known [5] in the case of long pulse probing techniques, plus additional mechanisms of fringe visibility depletion due to fast evolution of the electron density due to fast ionisation.

To explain these circumstances we consider here a simple case in which the change of the refractive index is due to a glass cylinder moving in the field of view of our interferometer. This situation is somewhat similar to probing a very sharp change in the refractive index as we would expect from an intense short pulse propagating in a gas jet and generating a sharp ionization front. We consider the case of a short probe pulse with duration that is short compared to the time taken by the pulse to cross the cylinder, and the cylinder is moving while the probe pulse propagates through it. This motion limits the spatial resolution of the interferometry and, as a result, a sharp edge will be smeared out. In a similar way, if we probe the ionization front generated by an ultrashort pulse propagating in the gas, the longitudinal (along the laser propagation axis) extent of the ionization front will appear broader due to the probe transit time.

Figure 10.15 shows a typical situation in which an ionization front generated by a femtosecond laser pulse in a gas propagates from right to left. The fringes that are shifting and being displaced due to the change of refractive index in the laser ionized region. The displacement ranges from zero to the maximum. However, it is not sharp; any sharp change of the refractive index will be seen as a smooth change in the phase shift. This is a very complex problem and deconvolution is a rather difficult task. In some circumstances this analysis leads to the identification of picosecond prepulse effects as discussed in [39].

In terms of fringe visibility, the typical losses are due to absorption and refraction. Whenever a probe pulse propagates through the plasma, it will be either refracted or absorbed. Under this condition, one of the two beams that undergoes interference makes the fringe pattern disappear by decreasing its intensity and typically, the average intensity integrated over a period of the fringes is less than the average intensity of the unperturbed fringe pattern (without plasma). Fringe visibility is affected in this case and so are phase shift measurements.

There is another reason why the fringes disappear in the case of short pulses. In ultrashort interactions the changes in the refractive index are on a temporal scale that is fast compared to the probe pulse duration. Basically, if the plasma is changing while it is being probed, the instantaneous fringes will shift at the detector plane and this shift will appear as a loss of visibility due to the time integration of the detector recording the fringe pattern. An example is shown in the sequence of interferogram of Fig. 10.16 taken 1.34, 2.33, and 4.67 ps after arrival of the main pulse at the focal region, when the beam is out of the best focus region. As time goes and the main laser pulse propagates, ionization occurs and plasma is generated. The sudden ionization occurring at the leading edge of the laser pulse induces fringe motion and lower visibility. Behind this region, the density of the plasma is stationary (on the time scale of the probe pulse), and is relatively low, allowing the probe pulse

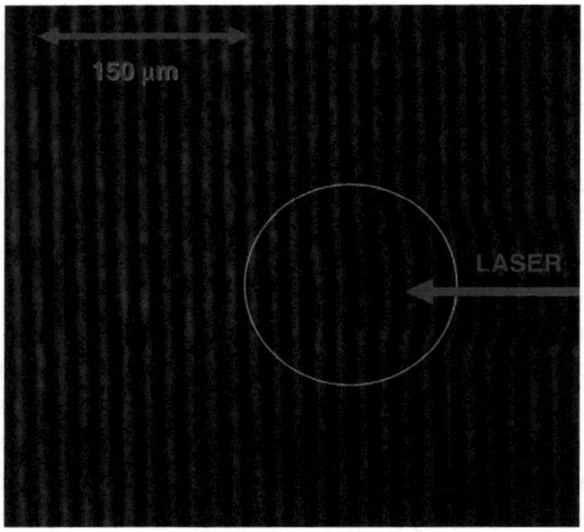

Fig. 10.15 Example of interferogram of ionization front propagating in a gas showing the smearing most likely due to the smoothing effect discussed in [39]. In this case a 30-fs laser pulse was propagating in a gas-jet at an intensity exceeding 10^{18} W/cm^2. The *circle* shows a smooth change of refractive index where a sharp transition is expected due to the ionization front

to propagate without significant absorption and the fringes are observed again. As indicated by the arrow, fringes are not visible in a region located at the leading front of the pulse.

A signature of this effect is that the average intensity of the region of the fringe pattern with low visibility becomes equal to the average intensity of the interfering pattern. This is the case for the middle interferogram of Fig. 10.16, where the intensity on the region of no fringe visibility is equal to the average intensity elsewhere. This simple estimate based upon a measurement of the average intensity allows us to distinguish between the different mechanisms of loss of fringe visibility driven either by refraction and absorption or by transient ionisation as in this case.

According to this mechanism, the speed of a fringe shift is basically given by the rate at which the density changes. A quantitative description of this has been developed [33] and according to this model, for a given density of the plasma and probe pulse duration and profile, the visibility can be estimated. For example, if there is a change of density of 2% of the critical density, a 100-fs pulse will produce a loss of fringe visibility of approximately 30%. This model has been used for a semiquantitative interpretation of experimental measurements of fringe visibility depletion [34] shown in the interferogram of Fig. 10.16.

In fact, a quantitative measurement of the loss in the fringe visibility can be obtained by using the same FFT analysis technique [40] for the extraction of the phase shift from the fringes. The whole analysis yields a complex result, the imaginary part gives the phase shift and the real part gives the fringe visibility. As a final result, the map of the fringe visibility is obtained in the range of 0–1. Figure 10.17 shows how the fringe visibility changes from frame-to-frame in a sequence obtained with a pump-and-probe like approach, i.e., by taking measurements with several shots at different probe pulse delay.

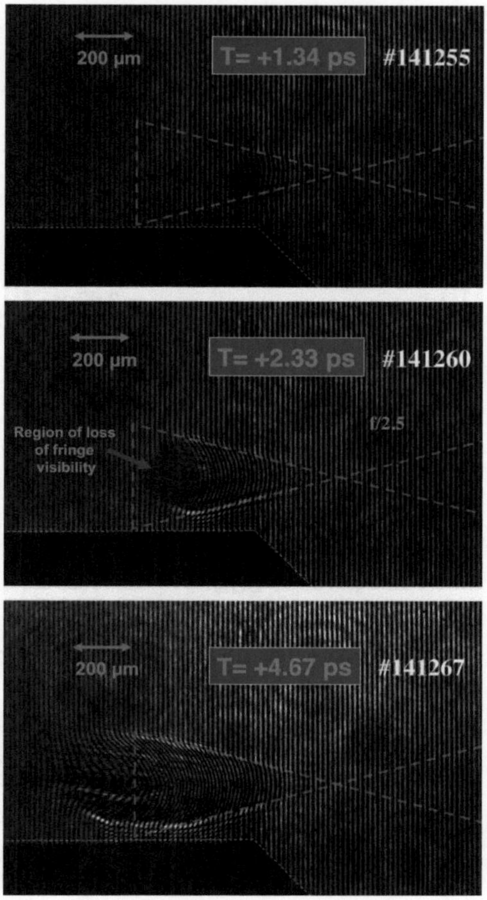

Fig. 10.16 Example of interferograms showing propagation of a laser pulse (from the *right*) in a gas. Partial loss of fringe visibility detected in the top interferogram becomes total in the middle interferogram, discussed in details in [34]. In this case a 30-fs laser pulse was used at an intensity of 5×10^{18} W/cm^2. The *arrow* shows the region where total loss of fringe visibility occurs

This result was interpreted using the model discussed above to extract information concerning the longitudinal extent of the ionization front, that was found to be shorter than 70 μm. Here the measurement is limited by the relatively long pulse duration of the probe pulse (110 fs) as discussed in detail elsewhere [34].

10.3 Time-Resolved X-Ray Spectroscopy and Imaging

In this section we discuss X-ray-based diagnostics. First we show a few examples of conventional X-ray techniques used to investigate the basic physics of laser–plasma interaction. We start discussing time-resolved X-ray spectroscopy.

Figure 10.18 shows a typical set-up for X-ray spectroscopy in which we can clearly identify the basic components, the laser beam, the target and the detection

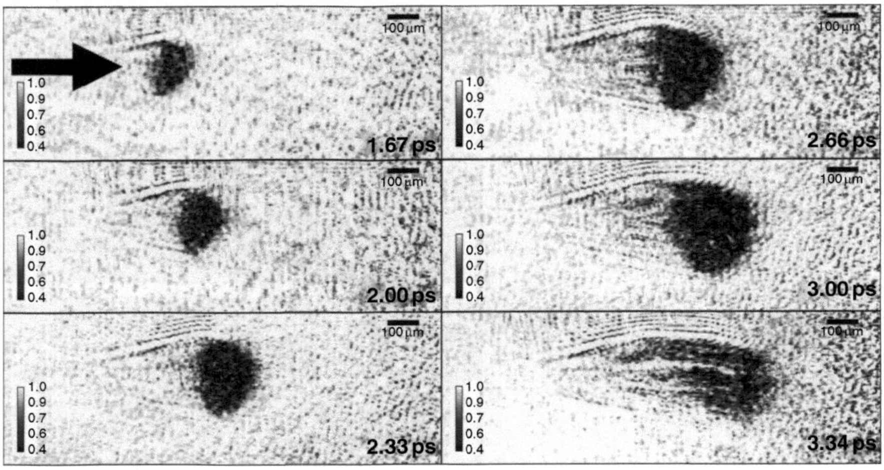

Fig. 10.17 Fringe visibility maps vs. time showing as the region where fringe visibility loss moves as the laser pulse propagates (from *left* to *right*)

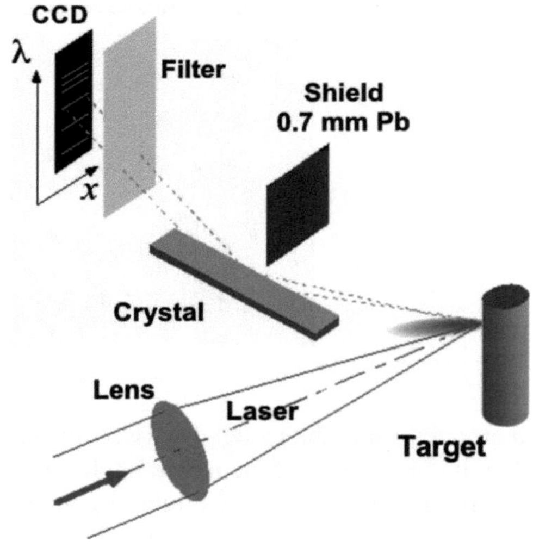

Fig. 10.18 Schematic setup for X-ray spectroscopy of laser-produced plasmas using a Bragg crystal as a dispersive element. Also shown in the image are the detector and the shield that prevents X-rays from the source to reach the detector directly. The filter is used to stop unwanted visible and UV light

system. A crystal with a known separation of the lattice planes is used to create Bragg-type diffraction. Because the source is very small, a spectrum is obtained whose resolution is limited by the source size [41].

Figure 10.19 shows a typical spectrum obtained using the setup of Fig. 10.18 from a plasma produced by laser irradiation of a solid Al target. In this case, a

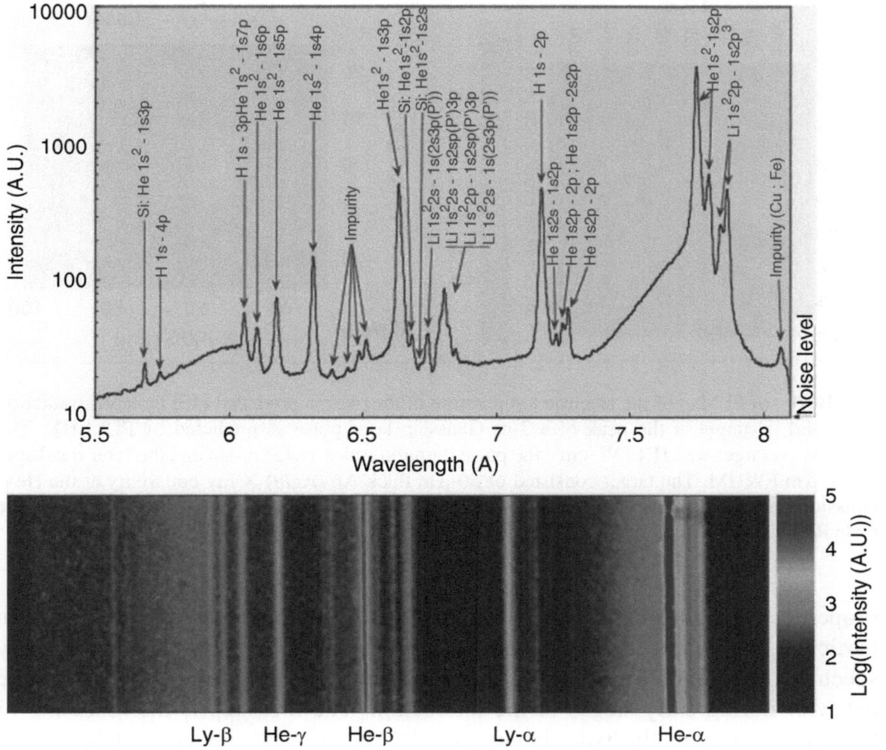

Fig. 10.19 Example of spectrum obtained using the setup of Fig. 10.18 from a plasma produced by laser irradiation of a solid Al target. Due to the small source size, the resolution is sufficiently high to allow identification of K-shell emission from helium-like and hydrogen-like aluminum

3-ns laser pulse from a Nd:YLF laser was used, focused to an intensity exceeding $10^{13}\,\mathrm{W/cm^2}$. In these experimental conditions, efficient transfer of laser energy to the plasma occurs and a hot plasma is generated leading to ionization of Aluminum up to the He-like and hydrogen-like configuration as demonstrated by the K-shell emission visible in the spectrum.

In order to unfold the temporal evolution of the plasma, temporal change in our spectra must be recorded on a very fast time scale, typically subnanosecond. The information thus obtained can then modeled using a combination of hydrodynamics and atomic physics modeling codes: the hydrodynamics provides the expected density distribution and atomic physics, the expected line intensity ratios. Then, these line intensity ratios are compared with the measured ones.

In the case of plasmas generated with long nanosecond pulses, the temporal resolution of the spectrum can be achieved by using an X-ray streak-camera to reach the temporal resolution below 10 ps. In the following, we described an experiment in which time-resolved X-ray spectroscopy is used to investigate the temporal

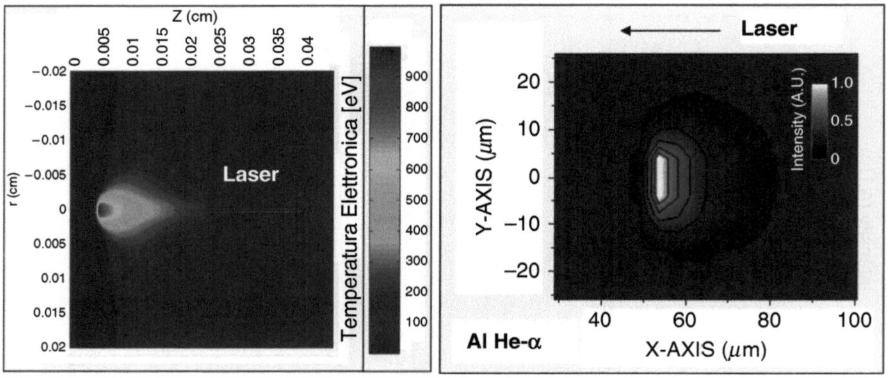

Fig. 10.20 (*left*) Map of the electron temperature of the plasma produced [43] by laser irradiation of a solid Al target at the peak of a 3-ns Gaussian laser pulse as predicted by POLLUX. The intensity on target was 1E14 W/cm^2, the pulse duration was 3-ns Gaussian and the laser focal spot was 8-μm FWHM. The target consisted of 50-μm thick Al. (*right*) X-ray emissivity at the He-α emission line calculated from electron density and temperature maps given by POLLUX using the code RATION/FLY

properties of X-ray emission in plasmas produced by nanosecond pulses [42]. In particular, the aim of the experiment was to identify, through time resolved X-ray spectroscopy, possible deviations of the plasma dynamics from the steady-state behavior during early stages of the interaction. The first step is the modeling of the interaction using the hydrodynamics code POLLUX [8] and we used the combination of RATION/FLY [42] and hydrodynamics results from POLLUX to calculate the expected emission in the spectral range of our diagnostics. The plot of Fig. 10.20 (left) shows the calculated electron temperature from the hydrodynamic calculations. This information, combined with the density map, is postprocessed using the code RATION/FLY to yield the expected emissivity at the required emission wavelengths. The plot of Fig. 10.20 (right) shows the result of this processing for a given time and a give emission wavelength in the form of a 2D map of emission.

Similar plots have been obtained [43, 44] for the entire emission time and for a set of emission lines. The ratio between α and β lines was then calculated taking into account the opacity effect that are expected to be significant for α lines. Figure 10.21 shows the final results in terms of comparison between measured and expected line intensity ratios for the steady-state and time-dependant models. The simulation predicts some changes in the nonstationary effect early in the pulse. After approximately 500 ps, the system becomes stationary. However, before that, there is a significant difference and the measurements were consistent with this separation. It should be stressed here that this effect could only be detected using temporal resolution.

In fact, since emission in the nonstationary regime only occurs early during the emission, its contribution would be largely overcome by the emission in the stationary regime and, therefore, be undetectable in time-integrated measurements.

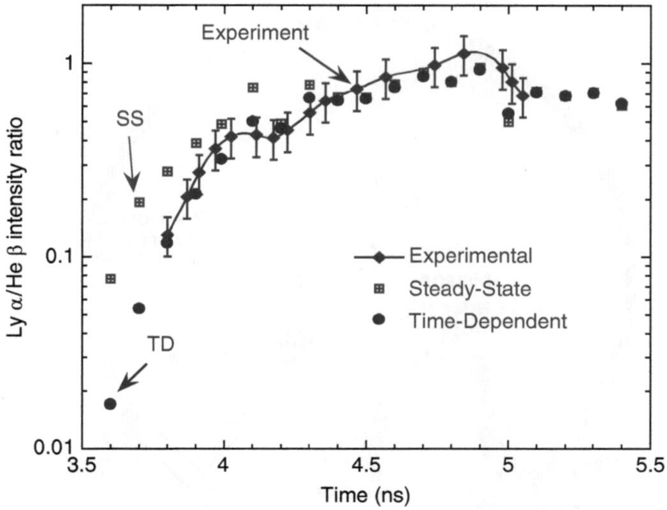

Fig. 10.21 Temporal evolution of Lyα to Heβ intensity ratio. Experiment vs. steady-state (SS) and time-dependent (TD) modeling

Figure 10.22 presents another experiment that shows the need of time-dependant X-ray spectroscopy. In this case, time-resolved spectroscopic techniques were used [45] to detect the cooling process of a plasma for different atomic species of the target.

The basic idea in this experiment was to measure the history of X-ray emission for plasmas produced by different elements and consequently infer information on radiation and cooling effects that may occur at a different rates depending on the plasma composition. Hence, it would be possible to probe the radiation cooling effects by using this X-ray spectroscopy. Figure 10.23 includes a comparison of time-resolved spectra of fluorine emissions from plasmas produced by irradiation of a range of targets containing fluorine, from the lighter LiF to the heavier SrF_2.

The analysis of the spectra of Fig. 10.23 shows that the temporal evolution of the electron temperature of this system depends on the charge state of the target. Indeed, the data show that this effect of radiation cooling plays an important role in these conditions ad discussed in details in [45].

Another interesting application of time-resolved spectroscopy concerns the study of electron transport measurements. A layered target was used and measurements were carried out to find the laser mass ablation rate, with a sufficiently short pulse that could separate the layers [46]. Targets consisting of thin layers of different materials were used to detect propagation of the heat front in the target as shown in Fig. 10.24 (left). According to hydrodynamics codes, the temperature of the first and third layers are expected to increase at different times as from the plots of Fig. 10.24 (right).

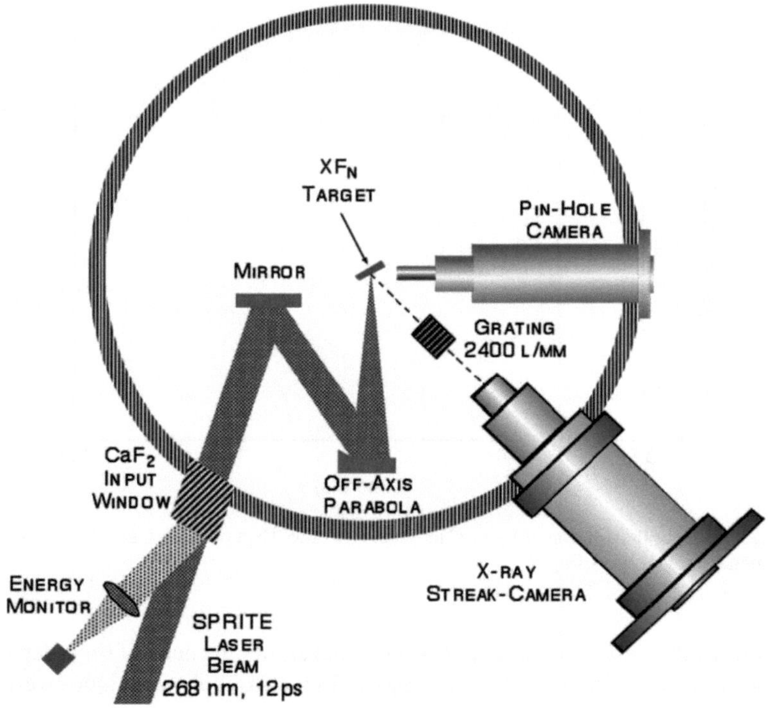

Fig. 10.22 Experimental setup of the interaction chamber for time-resolved soft-X-ray spectroscopy of picosecond plasmas [45]. The plasma was generated by the interaction of a high-contrast, 12-ps, 268-nm laser pulse with solid targets consisting of fluorine salts. The laser beam was focused on the target by means of an off-axis parabola at an intensity of $10^{15}\,\mathrm{W/cm^2}$ in a 30-μm diameter focal spot

Thus, if the X-ray emission from these layers occurs at different times, quantitative information on the transport of energy through materials is retrieved. Time-resolved measurements of X-ray emission from each of the layers was detected using characteristic line emission as shown in Fig. 10.25, which shows carbon emission from the first and third layers.

This measurement provide invaluable experimental data on the mass ablation rate that is used to investigate the flux limiter parameter in the heat propagation that takes place at a higher intensity as discussed in details in [46].

10.4 Fast Electron Production and Characterization

This section is devoted to fast electron propagation and to the description of novel experimental techniques of X-ray imaging having simultaneously micrometer special resolution and spectral resolution in the range of 1–10 keV. The motivation here

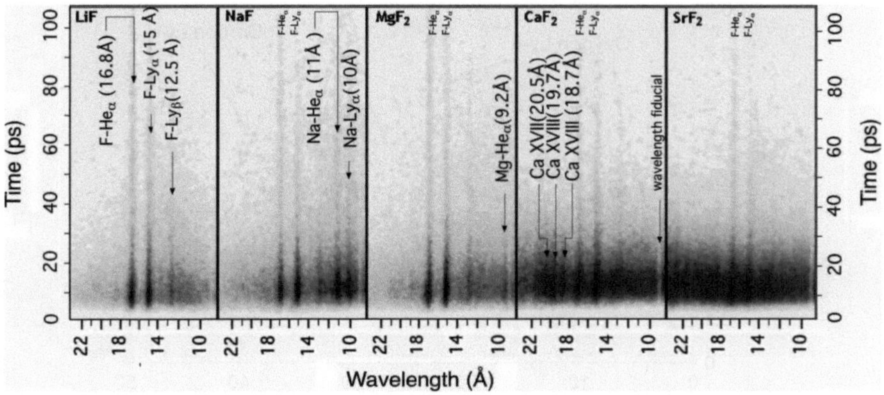

Fig. 10.23 Time-resolved spectra of the X-ray emission from LiF (*left*) to SrF2 (*right*), in the range from 10 to 22 Å, obtained with a spectrometer equipped with a flat-field grazing incidence grating (from [45]). The spectral and temporal resolution were 500 mÅ and 6 ps, respectively. Strong emission from bound–bound and free–bound transitions in H-like and He-like ions is visible. Short-lived continuum emission is also visible as a background emission due to Bremsstrahlung and recombination

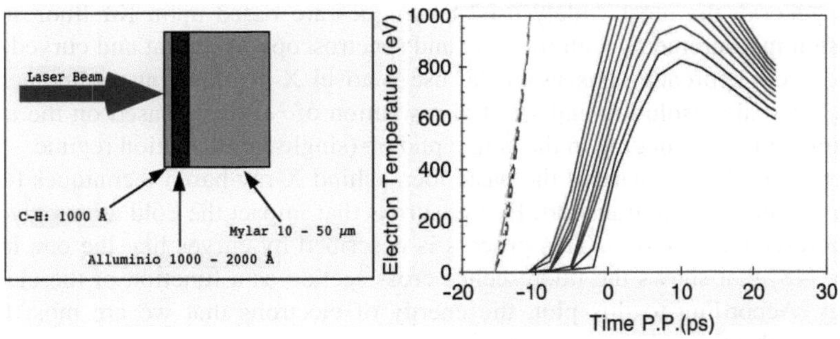

Fig. 10.24 (*left*) Target configuration for mass ablation rate measurements in intense laser–solid interactions [from [46]). The aluminum tracer layer is used to detect propagation of the ablation front. Numerical modeling (*right*) is used to calculate the expected evolution of the electron temperature to be compared with experimental measurements via time-resolved spectroscopy

is to understand how fast electrons propagate, how far they can get into the material, how much energy they have, and how much energy they release in the material. These studies are of a great relevance to the FI scheme of ICF as discussed in the introduction section.

There is no ideal experimental technique for this kind of measurements. There are instead several methods that can be used to retrieve pieces of information. Then, we need to use them together. In some cases optical probing can be used to image out the profile of the rear side of a laser irradiated foil target as described in [47]. Here measurements show the presence of a bump that is probably due to the effect of the fast electron currents.

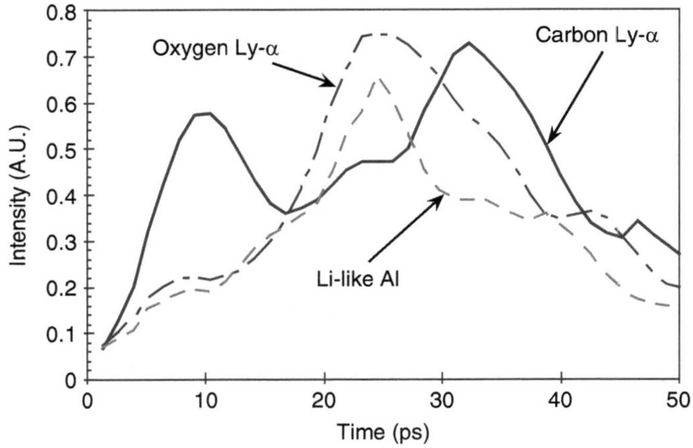

Fig. 10.25 Temporal evolution of X-UV emission line from irradiation of the target of Fig. 10.24 (*left*). Peak-to-peak delay in the H-like emission from C gives a measurement of the mass ablation rate in the Al layer. See [46]

In general, the most widely used techniques are based upon Kα fluorescence emission measurements with imaging and spectroscopy using flat and curved crystals. A new approach consists in the use a novel X-ray imaging technique with a micron-scale resolution and spectral resolution of ∼100 eV based on the use of multiple pin-hole imaging in the single photon (single-hit) detection regime.

Figure 10.26 summarizes the basic idea behind X-ray based techniques for the study of fast electron transport. Fast electrons that impact the cold atoms give rise to fluorescence emission. This process is described by curves like the one in this figure [48] that shows the fluorescence cross section as a function of the electron energy. According to this plot, the energy of electrons that we are most likely probed with fluorescence measurements ranges from a few keV to a few hundred keV. As shown in this plot, the cross-section peaks in the energy range between 10 and 100 keV. Similar behavior is found for different materials, making it possible to estimate the expected response of each target material to a given energy of the fast electrons. However, the fast electrons generated in ultra-intense laser interactions typically have a broad spectrum and fluorescence emission is generated from all electrons, up to energies of ∼10 MeV. Complementary measurements must therefore be performed to unfold the fast electron transport problem.

The use [49] of a single-hit detection technique on a *charge coupled device* (CCD) is based on a simple idea that was proposed around 15 years ago in X-ray astronomy and later applied to ultra-intense laser–plasma interactions [11]. Each X-ray photon impinging on the chip of a CCD produces a charge that is proportional to the energy of the photon. Provided the X-ray flux is sufficiently low, the charge induced by each photon can be measured and detected. The charge might be confined on a single pixel or might be spreading on several pixels depending on the energy of photons as well as on the depth of the CCD where the photon is absorbed.

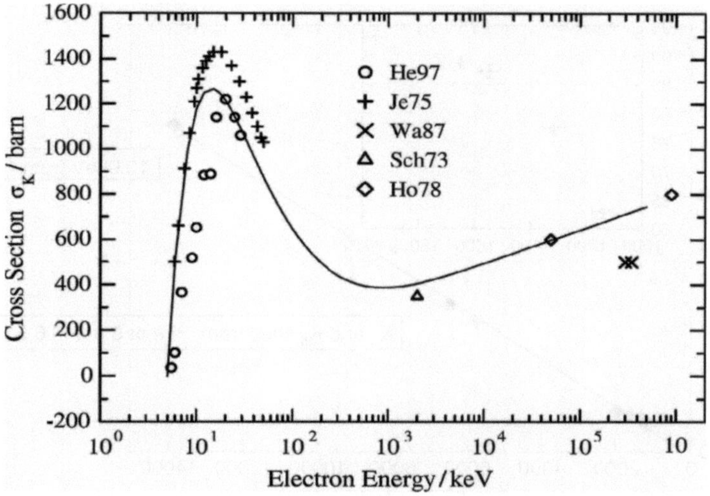

Fig. 10.26 Cross section of the fluorescence emission process in titanium (after [48]). The cross section has a pronounced peak between 10 and 100 keV. Above this value, the cross section decreases rapidly, reaching a minimum around 1 MeV and increases logarithmically for relativistic energies

For a given photon there are two different populations of "events" generated on the CCD: one is a single pixel and the other is a multipixel. Custom image processing techniques were developed to decouple the information and obtain the spectrum of the radiation [50]. Figure 10.27 shows a typical curve that gives the charge as a function of the energy of the photon incident on the CCD.

In the following we show the results of experiments in which this technique was applied to the measurement of the X-ray emission spectra from either single-layer of multilayer targets. In these experiments, single-hit measurement were carried out in a very noisy environment in which, typically, many photons and electrons scatter all around the area inside the vacuum chamber, generating a strong radiation background. To minimize the noise, several measures must be taken including the use of magnets to stop the electrons from streaming onto the CCD. In some circumstances it was also necessary to place the CCD outside the chamber to detect X-rays through a transparent window. Figure 10.28 shows a simple experimental set-up in which single-hit spectroscopy is used to study fluorescence emission from laser interaction with a titanium foil.

Figure 10.29 shows a typical spectrum obtained with the single-hit technique. As one can see from the plot, the technique allows the Kα emission to be easily resolved from the rest of the emission, as expected from the 90-eV resolution that was provided by the single-photon technique in our case.

The same technique shows its highest potential when dealing with multiple emission wavelengths as in the case of irradiation of multilayer targets. An experimental campaign was carried out at the JETI laser system at the IOQ in which the target

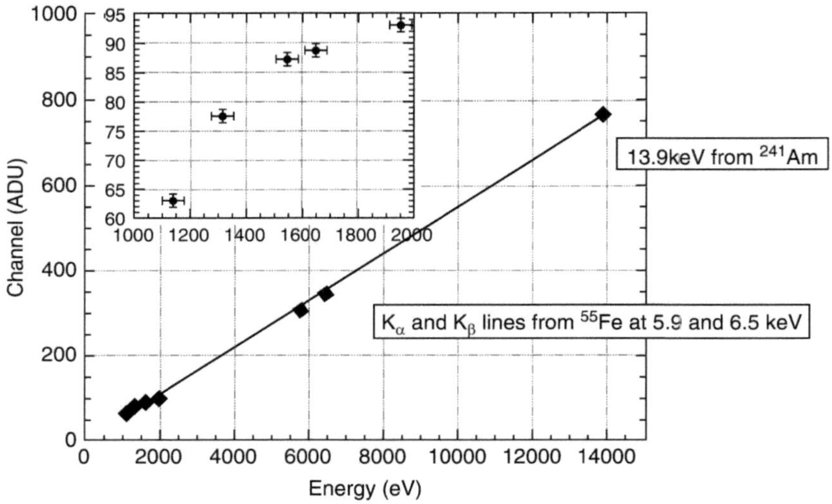

Fig. 10.27 Calibration of a CCD in the 1–14 keV energy range using a laser plasma X-ray source. The result of a linear fit of the points is also shown. In the inset a detail of the calibration points taken using the laser–plasma X-ray source is visible (after [50])

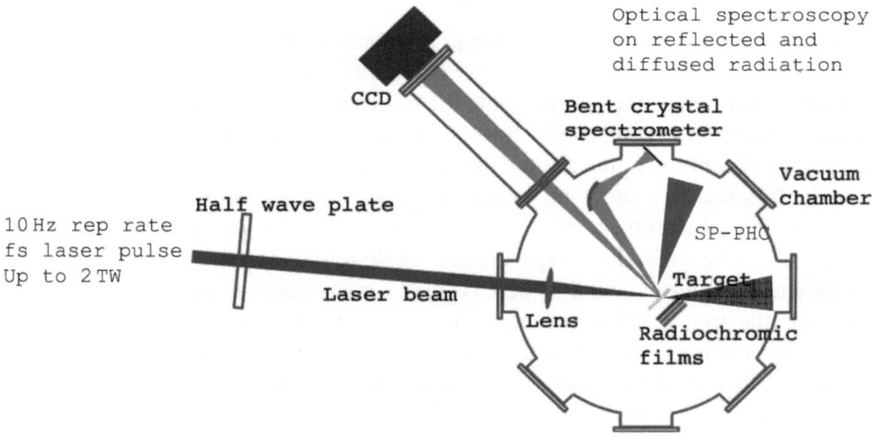

Fig. 10.28 A schematic experimental setup for the generation of fluorescence emission from interaction of intense femtosecond pulse of the ILIL FIXER laser system with a solid Ti target. The intensity on target was 2×10^{17} W/cm^2 and the pulse duration was 67 fs

consisted of three layers of different metals characterized by different fluorescence emission wavelength and shown in Fig. 10.30. In addition to the spectroscopic measurements we also used a pin-hole camera equipped with a 5-μm diameter pinhole to obtain an image of the source [51]. The image is attenuated using filters, down to a level required by the single-hit detection. A sufficient number of images

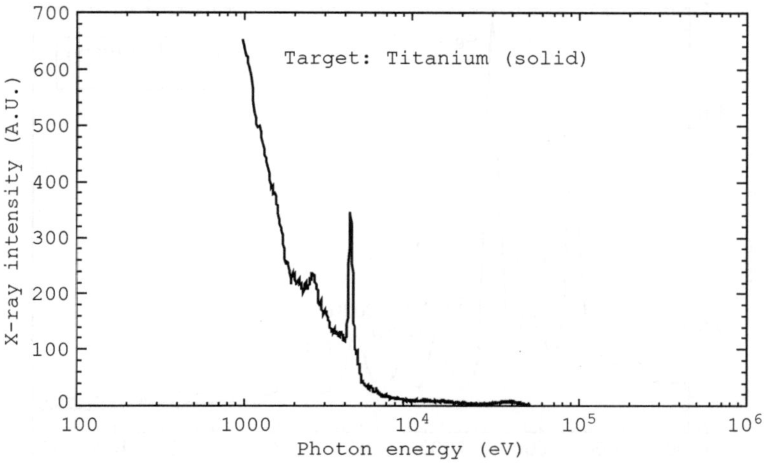

Fig. 10.29 Kα emission from titanium target irradiated with the Ti:Sa FIXER laser system of the ILIL lab, as measured using single hit spectroscopy with a CCD. The intensity of target was 2×10^{17} W/cm^2 and the laser pulse duration was 67 fs

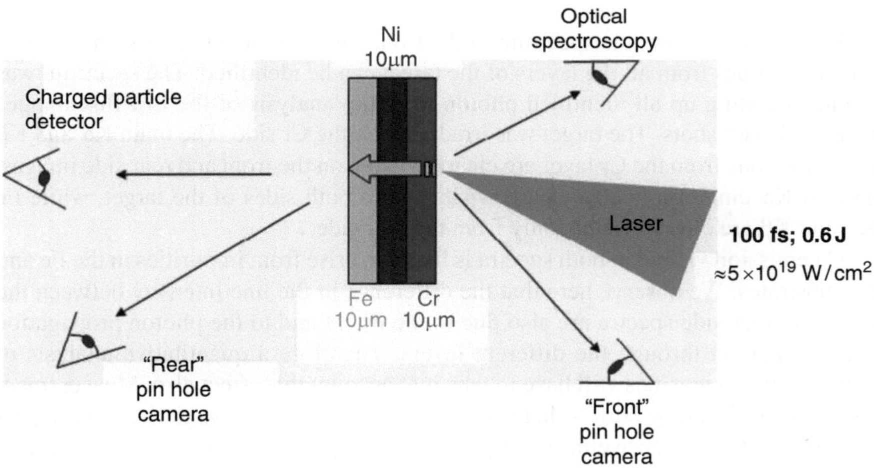

Fig. 10.30 Irradiation and detection setup for the experimental campaign on fast electron transport in solids carried out at using the JETI laser system. Imaging and spectroscopy based on single-hit detection of X-ray photons were implemented to recover full information on the production and transport for fast electrons

are captured resulting from respective laser shots, to obtain the number of photons required to build the full image at a given photon energy. In this configuration, the pin-hole camera acts as an energy-encoded imaging device (EEPHC) that can be used to obtain spectrally resolved images with 5-μm resolution.

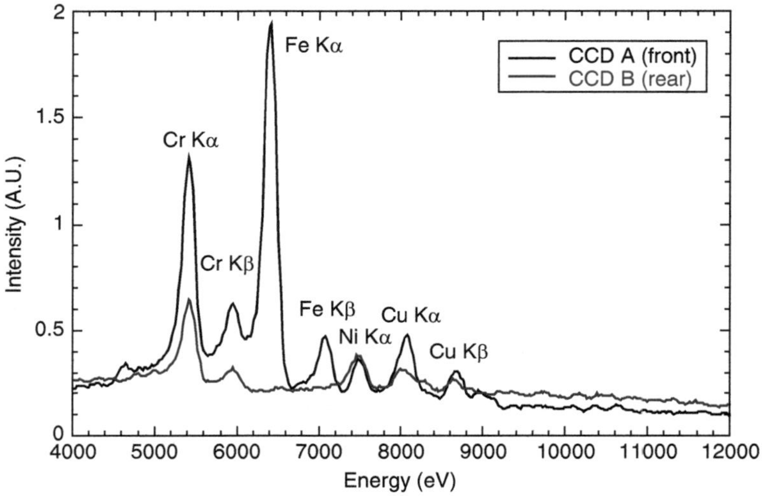

Fig. 10.31 Spectrum of the X-ray radiation detected from the front and the rear side of a three-layer Cr − Ni − Fe target of Fig. 10.30, irradiated at an intensity of $5 \times 10^{19}\,\mathrm{W/cm^2}$. Note that the relative intensities of the two spectra are not to scale

Figure 10.31 shows the spectrum of the front and the rear side emission in which all the Kα lines from all the layers of the target can be identified. The spectrum was obtained adding up all identified photon from the analysis of the low flux images from 350 laser shots. The target was irradiated on the Cr side. The main Kα and Kβ line emissions from the Cr layer are clearly visible on the front and rear side images. The Ni Kα emission is also clearly visible from both sides of the target, while Fe Kα and Kβ are clearly visible only from the rear side.

Cu emission visible in both spectra is likely to arise from impurities in the Fe and Ni substrates. We observe here that the difference in the line intensity between the front and rear side spectra are also due to the filters and to the photon propagation and attenuation through the different layers. Therefore a quantitative analysis of these relative intensities will have to take into account this effect also. Measurement like this immediately shows that the electrons propagate through the three layers because the fluorescence emission peaks originating from all the three (cold) layers are identified.

Finally, energy-resolved images of the source at any given energy can be obtained by selecting the corresponding photons from the spectral distribution, by plotting them according to their landing position on the CCD detector and by taking into account the geometry of the pin-hole camera imaging system, including the magnification. Figure 10.32 shows the energy-resolved full imaging of the X-ray source at the Kα and Kβ line energy of the three materials, as seen by the front and rear side pin-hole cameras. In this case, a total of 1,600 single-shot images were used.

The upper side of Fig. 10.32 shows the images obtained selecting X-ray photons in an energy window around each Kα line (±230 eV). The lower side of the figure shows the images obtained selecting X-ray photons in a spectral region where no

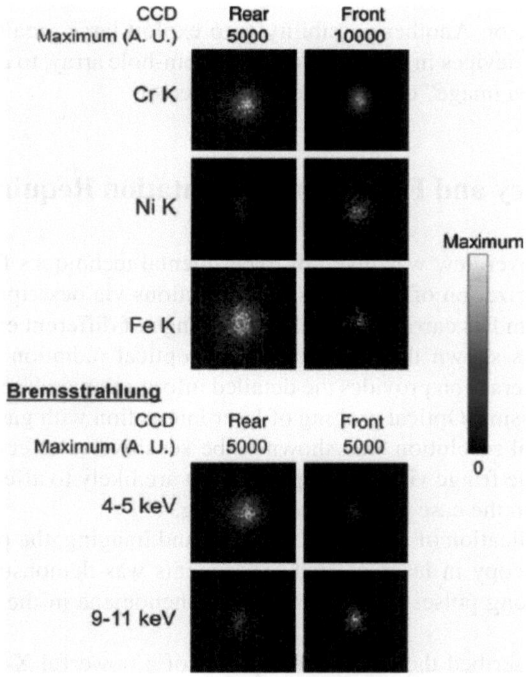

Fig. 10.32 Energy-resolved images of the Kα sources of Cr, Ni and Fe as seen from the rear (*left*) and the front of the target obtained selecting all the photons within a spectral window of ± 230 eV. The lower part of the figure shows the images of the source obtained selecting the photons in two regions where no or weak line emission is expected to occur and where the main contribution is expected to be given by Bremsstrahlung emissions. The scale of the edge of each image window is 200 μm

spectral lines are expected, i.e., below the Cr$-$Kα and above the Ni$-$Kα. Therefore, contribution to these images is expected to come from continuum radiation, possibly arising from Bremsstrahlung emission.

Higher-energy photons are also present in the spectrum; however, for these energetic photons the 2.5-μm thick, pinhole Pt substrate starts becoming transparent and the contrast of the pinhole image consequently decreases.

From the spreading of electrons as visible fro the images of Fig. 10.32, the dynamics of propagation of the electrons can be inferred. This is a complex task that also requires a comparison with a detailed numerical modeling and additional information on the analysis of these data can be found elsewhere [51]. Here the potential of this diagnostic technique is outlined and how the experimental results can be used to clarify the fast electron propagation dynamics.

The technique "as it is" has currently a "multishot" basis due to the need for a statistically significant number of photons to be detected. The number of shots can be reduced by using a higher magnification, which would enable to collect a greater number of photons per shot, while satisfying the single-photon condition on the CCD detector. Another possibility is to exploit large area CCDs or similar position sensitive devices in combination with a pin-hole array, to collect more than one "single-photon image" on the same CCD detector.

10.5 Summary and Future Instrumentation Requirements

In summary, an overview was given of experimental techniques for the investigation and characterization of laser–plasma interactions via description of a number of experimental studies carried out in the wide range of different experimental configurations. It was shown that spectroscopy of optical radiation scattered during the laser–solid interaction provides the detailed information on laser–plasma energy coupling mechanisms. Optical probing of laser interaction with gases with the femtosecond temporal resolution was shown to be sensitive to effects like the probe transit time and the fringe visibility depletion that are likely to affect measurements much more than in the case of long-pulse probing.

As for the application of X-ray spectroscopy and imaging, the potential of time-resolved spectroscopy in laser–plasma experiments was demonstrated even when using relatively long pulses to detect transient phenomena in the evolution of the plasma.

Finally, we described the latest development of a powerful X-ray investigation technique based on single X-ray photon measurements that promises to become a complete tool for imaging and spectroscopy in the X-ray range, ideal for FI relevant studies.

The experiments discussed here were also obtained through a number of collaboration experiments carried out at the Intense Laser Irradiation Laboratory at Istituto Nazionale di Ottica – CNR (Italy) and at several high-power laser facilities worldwide including the Central Laser Facility at STFC (UK), the JETI laser facility at IOQ, University of Jena (Germany), the SLIC facility at CEA (Saclay), the LOA at Ecole Polytechnique (Palaiseau), the TITAN laser at LLNL (USA). The main results of those experiments are the subject of several publications on international journals as pointed out throughout the text. The reader interested in expanding the topics presented here is encouraged to refer to those publications and references therein.

References

1. M. Tabak, J. Hammer, M.E. Glinsky, W.L. Kruer, S.C. Wilks, J. Woodworth, E.M. Campbell, M.D. Perry, R.J. Mason, Phys. Plasmas **1**, 1626 (1994)
2. S. Atzeni, J. Meyer-ter-Vehn, *The Physics of Inertial Fusion* (Oxford University Press, Oxford, 2004)
3. D. Giulietti, M. Galimberti, A. Giulietti, L.A. Gizzi, F. Balcou, A. Rousse, J.Ph. Rousseau, M.Borghesi, Phys. Rev. E **64**, 015402(R) (2001)
4. T. Ceccotti et al., Phys. Rev. Lett. **99**, 185002 (2007)

5. A. Gizzi, D. Giulietti, A. Giulietti, T. Afshar-Rad, V. Biancalana, P. Chessa, C. Danson, E. Schifano, S.M. Viana, O. Willi, Phys. Rev. E, **49**, 5628 (1994)
6. R.A. London, M. Rosen, Phys. Fluids **29**, 3813 (1986)
7. J.P. Christiansen, D.E.F.T. Ashby, K.V. Roberts, Comput. Phys. Commun. **7**, 271 (1974)
8. G.J. Pert, J. Comput. Phys. **43**, 111 (1981)
9. D. Giulietti, L.A. Gizzi, A. Giulietti, A. Macchi, D. Teychenné, P. Chessa, A. Rousse, G. Cheriaux, J.P. Chambaret, G. Darpentigny, Phys. Rev. Lett. **79**, 3194 (1997)
10. F. Brunel, Physical Review Letters **59**, 52 (1987)
11. L.A. Gizzi, D. Giulietti, A. Giulietti, P. Audebert, S. Bastiani, J.P. Geindre, A. Mysyrovicz, Phys. Rev. Lett. **76**, 2278 (1996)
12. D. Giulietti, M. Galimberti, A. Giulietti, L.A. Gizzi, P. Tomassini, M. Borghesi, V. Malka, S. Fritzler, M. Pittman, K. Taphouc, Letter on Phys. Plasmas **9**, 3655 (2002)
13. A. Rousse et al., Phys. Rev. E **50**, 2200 (1994)
14. W.L. Kruer, *The Physics of Laser-Plasma Interactions* (Addison-Wesley, New York, 1988)
15. L.A. Gizzi, D. Batani, V. Biancalana, A. Giulietti, D. Giulietti, Laser Part. Beams **10**, 65 (1992)
16. P. Köster, K. Akli, D. Batani, S. Baton, R.G. Evans, A. Giulietti, D. Giulietti, L.A. Gizzi, J.S. Green, M. Koenig, L. Labate, A. Morace, P. Norreys, F. Perez, J. Waugh, N. Woolsey, K.L. Lancaster, Plasma Phys. Control. Fusion **51**, 014007 (2009)
17. R. Neutze et al., Nature **406**, 752 (2000)
18. A. Rousse et al., Nature **410**, 65 (2001)
19. C. Rose-Petruck et al., Nature **398**, 310 (1999)
20. C. Rischel et al., Nature **390**, 490 (1997)
21. M.F. DeCamp et al., Nature **413**, 825 (2001)
22. P.E. Young et al., Phys. Rev. Lett. **61**, 2766 (1998)
23. R.P. Drake, R.E. Turner, B.F. Lasinski, E.A. Williams, W.L. Kent Estabrook, Kruer, E.M. Campbell, T.W. Johnston, Phys. Rev. A **40**, 3219 (1989)
24. D. Giulietti, D. Batani, V. Biancalana, A. Giulietti, L.A. Gizzi, $3\omega/2$, Il Nuovo Cimento **13D**, 845 (1991)
25. L.A. Gizzi, M. Galimberti, A. Giulietti, D. Giulietti, P. Koester, L. Labate, P. Tomassini, AIP Conf. Proc. **827**, 505–511 (2006)
26. L.A. Gizzi, A. Giulietti, D. Giulietti, P. Koester, L. Labate, T. Levato, F. Zamponi, T. Kämpfer, I. Uschmann, E. Förster, R. Sauerbre, SPIE Proc. **6634** (2007) (editor: S.B. Dabagov)
27. L.A. Gizzi, F. Anelli, C. Benedetti, C.A. Cecchetti, A. Clozza, G. Di Pirro, N. Drenska, R. Faccini, D. Giulietti, D. Filippetto, S. Fioravanti, A. Gamucci, L. Labate, T. Levato, V. Lollo, P. Londrillo, E. Pace, G. Turchetti, C. Vaccarezza, P. Valente, C. Vicario, Il Nuovo Cimento C, **32**, 433 (2009)
28. W.P. Leemans et al., Nat. Phys. **2**, 696 (2006)
29. V. Malka, J. Faure, Y.A. Gauduel, E. Lefebvre, A. Rousse, Kim Ta Phuoc Nat. Phys. **4**, 447–453 (2008)
30. L.A. Gizzi, A. Bacci, S. Betti, C.A. Cecchetti, M. Ferrario, A. Gamucci, A. Giulietti, D. Giulietti, P. Koester, L. Labate, T. Levato, V. Petrillo, L. Serafini, P. Tomassini, C. Vaccarezza, Eur. Phys. J. Spec Top **175**, 3–10 (2009)
31. P. Tomassini, A. Bacci, J. Cary, M. Ferrario, A. Giulietti, D. Giulietti, L.A. Gizzi, L. Labate, L. Serafini, V. Petrillo, C. Vaccarezza, IEEE Trans. Plasma Sci. **36**, 1782 (2008)
32. A. Bacci, F. Broggi, C. DeMartinis, D. Giove, C. Maroli, V. Petrillo, A.R. Rossi, L. Serafini, P. Tomassini, L. Cultrera, G. Di Pirro, M. Ferrario, D. Filippetto, G. Gatti, E. Pace, C. Vaccarezza, C. Vicario, F. Bosi, D. Giulietti, L.A. Gizzi, P. Oliva, Nucl. Instrum. Meth. Phys. Res, Sect. A **608**, S90 (2009)
33. M. Galimberti, J. Opt. Soc. Am. A **24**, 304–310 (2007)
34. L.A. Gizzi, M. Galimberti, A. Giulietti, D. Giulietti, P. Köster, L. Labate, P. Tomassini, Ph. Martin, T. Ceccotti, P. D'Oliveira, P. Monot, Phys. Rev. E **74**, 036403 (2006)
35. A. Giulietti, D. Giulietti, D. Batani, V. Biancalana, L. Gizzi, L. Nocera, E. Schifano, Phys. Rev. Lett. **63**, 524–527 (1989)
36. M. Galimberti et al., SHEEBA: a Spatial High Energy Electron Beam Analyzer. Rev. Sci. Instrum. **76**, 053303 (2005)

37. P. Tomassini, A. Giulietti, L.A. Gizzi, M. Galimberti, D. Giulietti, M. Borghesi, O. Willi, Appl. Opt. **40**, 6561 (2001)
38. P. Tomassini, A. Giulietti, Optics Comm. **199**, 143 (2001)
39. A. Giulietti, P. Tomassini, M. Galimberti, D. Giulietti, L.A. Gizzi, P. Köster, L. Labate, T. Ceccotti, P. D'Oliveira, T. Auguste, P. Monot, Ph. Martin, Phys. Plasmas **13**, 093103 (2006)
40. M. Takeda, H. Ia, S. Kobayashi, J. Opt. Soc. Am. **72**, 156 (1988)
41. D. Giulietti, L.A. Gizzi, La Rivista del Nuovo Cimento **21**, 1 (1998)
42. Lee et al., J. Quant. Spectrosc. Radiat. Transf. **32**, 91 (1984)
43. L.A. Gizzi, C.A. Cecchetti, M. Galimberti, A. Giulietti, D. Giulietti, L. Labate, S. Laville, P. Tomassini, Lett. Phys. Plasmas **10**, 4601 (2003)
44. L. Labate, C.A. Cecchetti, M. Galimberti, A. Giulietti, D. Giulietti, L.A. Gizzi, Phys. Plasmas **12**, 083101 (2005)
45. L.A. Gizzi, A. Giulietti, O. Willi, D. Riley, Phys. Rev. E **62**, 2721 (2000)
46. L.A. Gizzi, A.J. Mackinnon, D. Riley, S.M. Viana, O. Willi, Laser Part. Beams **13**, 511 (1995)
47. K. Lancaster et al., Phys. Rev. Lett. **98**, 125002 (2007)
48. F. Ewald, H. Schwoerer, R. Sauerbrey, Europhys. Lett. **60**, 710 (2002)
49. L.A. Gizzi, A. Giulietti, D. Giulietti, P. Köster, L. Labate, T. Levato, F. Zamponi, A. Lübcke, T. Kämpfer, I. Uschmann, E. Förster, A. Antonicci, D. Batani, Plasma Phys. Control. Fusion **49**, B211–B221 (2007)
50. L. Labate, M. Galimberti, A. Giulietti, D. Giulietti, L.A. Gizzi, P. Tomassini, G. Di Cocco, Nucl. Instr. Meth. A **495**, 148 (2002)
51. L. Labate, A. Giulietti, D. Giulietti, P. Koester, T. Levato, L.A. Gizzi, F. Zamponi, A. Luebcke, T. Kaempfer, I. Uschmann, E. Foerster, Rev. Sci. Instrum. **78**, 103506 (2007)
52. D. Strickland, G. Mourou, Opt. Commun. **56**, 212 (1986)

Index

Ablation, 219
Ablation rate, 298, 300
Above-threshold ionization (ATI), 4, 8, 59, 179
Absorption edges, 204, 213
Acoustic-optic (AO) crystal, 153
Acousto-optic programmable dispersive filters (AOPDF), 153, 161
AC Stark effect, 166
Adaptive control, 165
Adaptive shaping, 157
Adiabatic alignment, 22
Adiabatic decoupling, 45
Ammosov, Delone, and Krainov (ADK), 67, 246
Amplified spontaneous emission (ASE), 278
AOPDF, *see* Acousto-optic programmable dispersive filters
ASE contrast, 279
Atomic, molecular, and optical (AMO) phsics, 45
Attochirp, 100, 104
Attophysics, 98
Attosecond pulses, 256
Auger recombination, 233, 234
Autocorrelation, 179
Autocorrelator, 177
Avalanche ionization, 246
Avoided crossings, 131–133

Background reservoir, 249, 251
Band gap, 245
Binding energy, 42
Bio-agents, 266
Bound-state intensity parameter, 43
Bragg-type diffraction, 294
Breakdown, 252, 254, 255
Bremsstrahlung, 287
Brunel effect, 281

Canonical momentum, 55
Carbon nanotubes, 210
C_2-elimination, 121, 127, 129
CES amplitude, 36
CH_4, 264, 266
Channel closings, 14, 56
Charge coupled device (CCD), 300
Chemical and biological agents, 264
Chirped pulse amplification (CPA), 277
Clean fluorescence, 252, 254, 255, 264
CO_2 laser, 246
Collisional absorption, 282
Collisional ionization, 246
Conduction band, 245
Continuum electronic states, 119
Conventional accelerators, 288
Correlated energy sharing (CES), 29
Correlation effect, 20
Cost function, 140
Coulomb explosions, 167, 170, 179, 196
Coulomb gauge, 55
Critical density for the plasma, 244
Critical power, 242
Cross-correlation, 279, 280
Cross-phased modulation, 253
Cs, 96
Curved crystals, 300
Cylindrical coordinates, 63
Cylindrical symmetry, 290

Damage threshold, 281
De Broglie, 107
De-excitation, 46
Delayed detection, 269
Difference Frequency Generation, 86
Dipole approximation, 43
Dipole phase, 180
Dirac equation, 47

K. Yamanouchi (ed.), *Lectures on Ultrafast Intense Laser Science 1*, Springer Series in Chemical Physics 94, DOI 10.1007/978-3-540-95944-1,

Dissociative ionization, 197
Doorway, 24
Double ionization, 187
Dual transformation method, 113
Dynamic high harmonic generation, 22
Dynamic probe, 204
Dynamic X-ray diffraction theory, 235

Egg white, 266
Einsteinian photoelectric effect, 5
Electric multipoles, 45
Electromagnetic potentials, 44
Electron acceleration, 281
Electron energy distribution, 228
Electron rest energy, 42
$(3/2)\omega$ Emission, 283
Energy-encoded imaging, 303
Energy reservoir, 249
Enhanced ionization, 113, 119
Extended XAFS (EXAFS), 205, 206, 216

Fast electrons, 278, 281, 282
Fast ignition (FI), 278
Femtosecond breakdown, 266
Femtosecond interaction X-ray emitter (FIXER) laser system, 279, 280
Femtosecond laser-induced breakdown spectroscopy (FIBS), 266
Fermi golden rule, 6
Feynman rules, 79
FFT analysis, 292
Field-induced potential distortion, 130
Figure-8 motion, 77
Filament, 247
Filamentation, 241
Filamentation nonlinear optics, 260
Final state Volkov dressing, 35
Finger print fluorescence, 264, 269
Floquet method, 18
Fluorescent protein, 170
Fourier plane, 136
Fourier synthesize, 138
Four-wave mixing, 260
Free-electron intensity parameter, 43
Frequency–intensity illustration, 44
Frequency-resolved optical grating (FROG), 158, 163
Fringe visibility, 291
FROG trace, 159
Fundamental mode, 258

Gain-narrowing effect, 163
Gas-jets, 287, 288
Gauge-invariance, 56
Gauge transformation, 46
Generalized Bessel functions, 11, 59
Genetic algorithm, 140
Gerchberg–Saxton algorithm, 139
Gouy phase, 182
Green's function, 25
Grid point method, 112, 113
Group delay dispersion, 100
Group velocity dispersion (GVD), 243

Halocarbons, 264
Harmonic cutoff, 98
Hartree–Fock wave functions, 65
$h_g(1)$, 127–129
$h_g(1)$ vibrational mode, 127
HHG signal, 17
High Compton-harmonics, 4
Higher-order perturbation theory, 42
Highest occupied molecular orbital (HOMO), 106
High-frequency approximation (HFA), 51
High harmonic generation (HHG), 9
High harmonics, 98
High-order harmonics, 256
HOMO, *see* Highest occupied molecular orbital
Hydrodynamics code, 296
Hydrogenic states, 65

Image processing, 301
Impulsive Raman excitation, 128, 129
IMST wave function, 26
Inertial confinement fusion (ICF), 278
I^n law, 6
Instantaneous electronic Hamiltonian, 115, 116
Intense-field many-body S-matrix theory (IMST), 1
Intensity clamping, 245, 252
Intensity parameters, 43
Interacting states, 49
Interaction Hamiltonian, 49
Interferometric autocorrelation, 199
Intramolecular electron transfer, 112
Inverse Bremsstrahlung, 246, 255
Ionization blue shift, 281
Ionization rate, 94

Jones matrix, 142

Kα radiation, 283
K-B microscope, 219
Keldysh–Faisal–Reiss (KFR), 42
Keldysh parameter, 19, 43, 85, 87, 176
Keldysh scaling, 90
KFR model, 1, 9
KFR theory, 246
Kinetic shift, 123, 125
Kirkpatrick-Baez (K-B) mirrors, 221
Klein–Gordon equation, 79
KTA, *see* Potassium Titanyl Arsenate, $KTiOAsO_4$

Laboratory gauge, 52
Laser heating process, 233
Laser-induced breakdown, 254
Laser–matter interactions, 276, 277
Laser-produced plasma, 207
Laser-produced plasma x-rays, 203
Laser pump and X-ray probe techniques, 223
Laser–solid interactions, 277
Lattice expansion and compression, 232
Length gauge, 42
Lewenstein's theory, 106
LIDAR, 265, 266
Light dressed state, 166
Light–quantum hypothesis, 3
Light shift, 17
Lippmann–Schwinger equation, 25
Longitudinal field, 54
Lorentz force, 77
Low energy structure, 97
Lowest order mode, 258

Magnetic multipoles, 45
Many-fold ionization, 22
Mass spectrometry, 123, 129
Mid-infrared, 85
Mid-infrared sources, 86
Momentum-space wave function, 59
Multiphoton absorption, 245
Multiphoton domain, 42
Multiphoton ionization, 167, 244
Multiphoton processes, 3
Multiple-ionization, 36

Nanoholes, 210
Negative energy states, 79
Neutral explosion, 271
Neutral fragmentation, 271
Nonadiabatic electronic dynamics, 112
Nonadiabatic electronic transitions, 130
Nonadiabatic transitions, 117, 119, 131–133, 170
Nondipole effects, 43
Noninteracting states, 48
Nonlinear sampling, 260
Nonlinear Thomson–Compton scattering, 4
Nonperturbative field effect, 20
Nonperturbative intensity parameter, 43
Nonperturbative methods, 42
Nonsequential double ionization (NSDI), 22, 179
Nuclear packets, 166
Nuclear wave packets, 114, 117, 131, 132
Null mask, 146

Odd harmonics, 16
Optimization control, 155
Orbitals, 106

Parallel momentum, 64
Perelomov, Popov, and Terent'ev (PPT), 67
Perpendicular momentum, 64
Perturbation theory, 41
Phase matching, 106, 181
Phase-only mask, 139
Phase-shift, 289
Phase-space, 59
Phonon propagation, 235
Photoelectron, 90
Photo-excited carriers, 233
Photon, 42
Photon number, 226
Pin-hole array, 306
Pixel gap, 153
Plasma mirror, 279
Plasma probing, 287
Plume of the expanding electrons, 229
Polarization, 282
Polarization shape, 142
Ponderomotive energy, 11, 42, 175
Ponderomotive potential, 225
Positive energy states, 79
Post series, 26
Potassium Titanyl Arsenate, $KTiOAsO_4$ (KTA), 86
PPT, *see* Perelomov, Popov, and Terent'ev
Preplasma, 209
Prepulse, 207, 208
Principal quantum number, 65
Prior series, 26
Pulse width of X-rays, 226

Pump-and-probe, 292
Pump-probe-type experiment, 204

Quantum emission, 223, 224
Quantum remnants, 19
Quantum transition amplitude, 42
Quiver radius, 11

Radiation cooling, 297
Radiation gauge, 55
Radiation pressure, 82
Radiochromic films, 289
Raman active vibrational modes, 127
Raman instability, 283
Rb, 96
Reaction vessel, 39
Recombination, 46
Refocusing, 254
Regenerative amplifier, 157
Relativistic domain, 43
Relativistic effects, 43
Remote sensing, 264, 269
Replica pulse, 149
Rescattering, 29, 42
Rescattering model, 92
Rescattering plateau, 94
Resonance absorption, 281
Rice–Ramsperger–Kassel–Marcus (RRKM) theory, 123, 125

Scalar field, 54
Scalar potential, 49
Scattering operator, 23
Second harmonic emission, 283
Second quantization, 79
Self-compression, 256
Self-defocusing, 244, 245
Self-focusing, 241, 242
Self-focusing distance, 247
Self-group phase locking, 253, 258
Self-guiding, 181
Self-pulse compression, 253, 256
Self-remote projections, 252
Self-spatial filtering, 253, 257
Self-stabilization, 263
SFA, *see* Strong-field approximation
Shake-off (SO) mechanism, 27
Shaping mask, 139
Shock compression, 237
Shock wave, 237, 289
Shock wave propagation, 239
Short and long trajectories, 102
Simpleman's theory, 91
Simulated annealing algorithm, 140
Sinc function, 139, 149
Single-hit spectroscopy, 301
Single mode, 257
Skeletal bond types, 130
Slice-by-slice self-focusing, 247, 251
S-matrix, 46
S-matrix theory, 90
Space-resolved spectrum, 284
Spatial light modulator (SLM), 136, 142, 153
Spatial modes, 257, 258
Spatiotemporal coupling, 144, 145
Spatiotemporal shaping, 144
Spectral interference, 160
Spectrum, 60
Spherical coordinates, 59
Stabilization, 51
Statistical fragmentation, 129, 130
Strong field approximation (SFA), 42, 180
Strong-field physics, 41
Supercontinuum, 253
Super-excited states, 271
Superfluorescence, 86
Suppressed ionization, 22
Synchrotron radiation, 271

TADPOLE, 160, 163
Temporal resolution, 296, 299
Teramobile, 253
The first-order configuration interaction method, 131
Thin foil, 281
Third harmonic generation, 258
Time-dependent adiabatic states, 115, 117–121, 124–127, 131, 132
Time-dependent Schrödinger equation, 111, 179
Time evolution operator, 45
Time-of-flight (TOF), 123, 129, 130
Time-resolved electron radiography, 237
Time-resolved images, 229
Time-resolved spectra, 297, 299
Time-resolved X-ray diffraction, 231
Time-reversed transition amplitude, 49
Transform limited pulse, 172
Transient lattice deformation, 233
Transition states, 24
Transverse field, 54
Tunneling, 87, 88
Tunneling domain, 42
Tunneling time, 71

Tunnel ionization, 175, 179, 244–246
Two-dimensional spatial light modulator, 144
Two-photon above threshold ionization process, 193
Two-photon double ionization, 188
Two-photon excited fluorescence, 170
Two-plasmon decay, 283
Two-slit interference phenomena, 22
$0-\pi$ type phase mask, 145, 146, 148
$4f$-type pulse shaper, 136

Unitary transformation, 46

Valence band, 245
Vector potential, 49
Velocity gauge, 46
Velocity of sound, 234
Vibronic ionization, 22
Virtual fragment, 24
Virtual state, 5
Volkov solution, 54
Volkov spectrum, 12
Volkov state, 13

Water-window, 211
Wavelength scaling, 92
White light chirped laser pulse, 250
White light laser pulse, 250
White light short laser pulse, 253

X-ray absorption fine structure (XAFS), 205
X-ray absorption near-edge structure (XANES), 206
X-ray imaging, 298, 300

Yeast, 266
Yeast powders, 266
Yield, 105

MIX
Papier aus verantwortungsvollen Quellen
Paper from responsible sources
FSC® C105338

If you have any concerns about our products,
you can contact us on
ProductSafety@springernature.com

In case Publisher is established outside the EU,
the EU authorized representative is:
Springer Nature Customer Service Center GmbH
Europaplatz 3, 69115 Heidelberg, Germany

Printed by Libri Plureos GmbH
in Hamburg, Germany